Mathematics in Industry

Volume 34

Mathematics in Industry focuses on the research and educational aspects of mathematics used in industry and other business enterprises. Books for *Mathematics in Industry* are in the following categories: research monographs, problem-oriented multi-author collections, textbooks with a problem-oriented approach, conference proceedings. Relevance to the actual practical use of mathematics in industry is the distinguishing feature of the books in the *Mathematics in Industry* series.

More information about this series at http://www.springer.com/series/4650

Simone Göttlich · Michael Herty ·
Anja Milde

Editors

Mathematical Modeling, Simulation and Optimization for Power Engineering and Management

Springer

Editors
Simone Göttlich
Department of Mathematics, Scientific
Computing Research Group
University of Mannheim
Mannheim, Germany

Michael Herty
Department of Mathematics, Group
Continuous Optimization
RWTH Aachen University
Aachen, Nordrhein-Westfalen, Germany

Anja Milde
Computing Centre (URZ), Unit Technology
Transfer and Cooperation
Heidelberg University
Heidelberg, Germany

ISSN 1612-3956 ISSN 2198-3283 (electronic)
Mathematics in Industry
ISBN 978-3-030-62734-8 ISBN 978-3-030-62732-4 (eBook)
https://doi.org/10.1007/978-3-030-62732-4

This Springer imprint is published by the registered company Springer Nature Switzerland AG
The registered company address is: Gewerbestrasse 11, 6330 Cham, Switzerland

Preface

The transformation of energy systems is the central part of Germany's High-Tech Strategy 2025. In the near future, a successful transformation guarantees a secure, economical, and environmentally compatible energy supply. A key aspect in the transition of energy systems is to enable green technologies and expand the range of renewable energies at the expense of fossil fuels. New storage concepts and intelligent energy networks are central hardware components also required in a transformed energy system. Alongside the expansion of renewable energy sources, a significant increase in energy efficiency as well as stability and resilience are further challenges.

Research and innovation are drivers of progress and in particular applied mathematics has been used in the past as methods-based, cross-sectional science that may enable novel results for sustainable and climate-friendly solutions also in the realm of energy systems. Aim of this book is, therefore, to show the relevance of mathematics in innovation by means of introducing a successful cooperation with industrial partners of the energy sector in order to show the wide range of applications. In many of the discussed problems, the use of modern mathematics in modeling, simulation, and optimization is shown to be a crucial factor for success. Besides novel approaches, the present book also shows the variety of mathematical techniques and disciplines involved in these activities. The selected scientific presentations highlight cooperation projects between mathematics and industry as a two-way transfer of technology and knowledge, providing the industry with applicable solutions and providing mathematics with novel research topics and inspiring new methodologies.

The starting point for this Springer publication was the KoMSO e.V. workshop on "Mathematical Modeling of Energy Systems" in March 2019 at the University of Mannheim. Partners from industry, universities, and research institutes discussed expected challenges and presented their new concepts for reliable energy predictions. We also acknowledged support by the Federal Ministry of Education and Research (BMBF) within the program "Mathematics for Innovations" (https://math4innovation.de).

The diversity of collaborations between mathematics and industry for different energy-related and economic applications is portrayed in the broad scope of topics discussed in this book, ranging from multi-energy systems to energy market models. These facets are presented in three chapters which are called **Economic Aspects**, **Technical Applications**, and **Energy Networks**. All contributions address the advancement of novel mathematical models, modern efficient numerical methods for simulation and optimization arising in energy systems and management, as well as the demands for future development of science and technology in the field of energy research.

Mannheim, Germany   Simone Göttlich
Aachen, Germany   Michael Herty

About KoMSO e.V.

KoMSO is the Committee for Mathematical Modeling, Simulation, and Optimization. It acts as a Germany-wide network of scientists from the field of mathematical modeling, simulation, and optimization (MSO) and potential users of this technology (https://www.komso.org).

KoMSO aims to promote scientific research and cooperation in the field of MSO between science, economy, society, and politics. In addition, the relevance of MSO for society and the economy is to be demonstrated through targeted public relations work.

KoMSO was set up in May 2011 as a result of the Mathematics 2020 Strategy Day initiated by the Federal Ministry of Education and Research (BMBF) as part of the German government's High-Tech Strategy 2020. Since March 2018 KoMSO e.V. is a non-profit registered association.

A central task of KoMSO is to initiate innovative scientific projects in the field of MSO—by bringing together representatives from science and industry at events such as the KoMSO Challenge Workshops. Over the years, numerous successful collaborations have been established, which are described in several KoMSO Success Stories.

KoMSO has access to a broad national and international network of partners from universities, research institutions, and industry, and is expanding this network through ongoing activities. The fields of expertise include aerospace, automotive industry, health care, pharmaceutical industry, industrial robotics, and many more.

On an international level, KoMSO acts as the German network node in the European Service Network of Mathematics for Industry and Innovation (EU-MATHS-IN), which consists of various national or multinational industrial mathematics networks. The declared goal of EU-MATHS-IN is to "leverage the impact of mathematics on innovations in key technologies [through] enhanced communication and information exchanges".

Heidelberg, Germany   Anja Milde

Contents

Contributors

Tim Aschenbruck Automatic Control and System Dynamics Laboratory, Technische Universität Chemnitz, Chemnitz, Germany

Jan Backhaus Institute of Propulsion Technology, German Aerospace Center (DLR), Köln, Germany

Anton Baldin SCAI: Fraunhofer SCAI, Sankt Augustin, Germany

Manuel Baumann Max Planck Institute for Dynamics of Complex Technical Systems, Magdeburg, Germany

Peter Benner Max Planck Institute for Dynamics of Complex Technical Systems, Magdeburg, Germany;
Otto von Guericke University Magdeburg, Magdeburg, Germany

Matthias Bolten Bergische Universität Wuppertal, Fakultät für Mathematik und Naturwissenschaften, IMACM, Wuppertal, Germany

Tanja Clees SCAI: Fraunhofer SCAI, Sankt Augustin, Germany;
Bonn-Rhine-Sieg University of Applied Sciences, Sankt Augustin, Germany

Sema Coskun Department of Mathematics, University of Kaiserslautern, Kaiserslautern, Germany

Jörg Dickert ENSO NETZ GmbH, Dresden, Germany

Matthias Ehrhardt Bergische Universität Wuppertal, Fakultät für Mathematik und Naturwissenschaften, IMACM, Wuppertal, Germany

Matthias Eimer Fraunhofer ITWM, Kaiserslautern, Germany

Benedikt Engel University of Nottingham, Gasturbine and Transmission Research Center (G2TRC), Nottingham, UK

Willem Esterhuizen Automatic Control and System Dynamics Laboratory, Technische Universität Chemnitz, Chemnitz, Germany

Timm Faulwasser Institute for Energy Systems, Energy Efficiency and Energy Economics, TU Dortmund University, Dortmund, Germany

Bartosz Filipecki Faculty of Mathematics, Technische Universität Chemnitz, Chemnitz, Germany

Eike Fokken Department of Mathematics, University of Mannheim, Mannheim, Germany

Christian Frey Institute of Propulsion Technology, German Aerospace Center (DLR), Köln, Germany

Dominik Garmatter Technische Universität Chemnitz, Chemnitz, Germany

Philipp Gerstner Heidelberg University, Heidelberg, Germany

Simone Göttlich Department of Mathematics, University of Mannheim, Mannheim, Germany

Hanno Gottschalk Bergische Universität Wuppertal, Fakultät für Mathematik und Naturwissenschaften, IMACM, Wuppertal, Germany

Ria Grindel Fraunhofer ITWM, Kaiserslautern, Germany

Sara Grundel Max Planck Institute for Dynamics of Complex Technical Systems, Magdeburg, Germany

Michael Günther Bergische Universität Wuppertal, Fakultät für Mathematik und Naturwissenschaften, IMACM, Wuppertal, Germany

Camilla Hahn Bergische Universität Wuppertal, Fakultät für Mathematik und Naturwissenschaften, IMACM, Wuppertal, Germany

Mirko Hahn Otto von Guericke University Magdeburg, Magdeburg, Germany

Gerd Heilscher Institute of Stochastics, Ulm University, Ulm, Germany

Christoph Helmberg Faculty of Mathematics, Technische Universität Chemnitz, Chemnitz, Germany

Michael Herty Institut für Geometrie und Praktische Mathematik, RWTH Aachen, Aachen, Germany

Vincent Heuveline Heidelberg University, Heidelberg, Germany

Christian Himpe MPI: Max Planck Institute for Dynamics of Complex Technical Systems, Magdeburg, Germany

Wieger Hinderks Fraunhofer ITWM, Kaiserslautern, Germany

Basem Idlbi Institute of Stochastics, Ulm University, Ulm, Germany

Peter Jaksch Siemens AG, Power and Gas, Mülheim an der Ruhr, Germany

Jens Jäschke Bergische Universität Wuppertal, Fakultät für Mathematik und Naturwissenschaften, IMACM, Wuppertal, Germany

Tom Kirstein Institute of Stochastics, Ulm University, Ulm, Germany

Bernhard Klaassen SCAI: Fraunhofer SCAI, Sankt Augustin, Germany

Kathrin Klamroth Bergische Universität Wuppertal, Fakultät für Mathematik und Naturwissenschaften, IMACM, Wuppertal, Germany

Oliver Kolb Department of Mathematics, University of Mannheim, Mannheim, Germany

Ralf Korn Department of Mathematics, University of Kaiserslautern, Kaiserslautern, Germany

Leonid Kuoza PSI Gas & Oil, Essen, Germany

Ferdinand Küsters ITWM: Fraunhofer Institute for Industrial Mathematics ITWM, Kaiserslautern, Germany

Christian Leithäuser Fraunhofer ITWM, Kaiserslautern, Germany

Alexander Liefke Siemens AG, Power and Gas, Mülheim an der Ruhr, Germany

Dominik Linn Fraunhofer ITWM, Kaiserslautern, Germany

Daniel Luft Universität Trier, Fachbereich IV, Research Group on PDE-Constrained Optimization, Trier, Germany

Lucas Mäde Siemens Gas and Power GmbH & Co. KG, Probabilistic Design, GP PGO TI TEC PRD, Berlin, Germany

Andrea Maggi Max Planck Institute for Dynamics of Complex Technical Systems, Magdeburg, Germany

Vincent Marciniak Siemens AG, Power and Gas, Mülheim an der Ruhr, Germany

Nicole Marheineke UTrier: Universität Trier, Trier, Germany

Nico Meyer-Hübner Karlsruhe Institute of Technology, Karlsruhe, Germany

Jan Mohring Fraunhofer ITWM, Kaiserslautern, Germany

Shaimaa Monem Max Planck Institute for Dynamics of Complex Technical Systems, Magdeburg, Germany;
Otto von Guericke University Magdeburg, Magdeburg, Germany

Tillmann Mühlpfordt Institute for Automation and Applied Informatics, Karlsruhe Institute of Technology, Karlsruhe, Germany

Siegfried Müller Institut für Geometrie und Praktische Mathematik, RWTH Aachen, Aachen, Germany

Igor Nikitin SCAI: Fraunhofer SCAI, Sankt Augustin, Germany

Lialia Nikitina SCAI: Fraunhofer SCAI, Sankt Augustin, Germany

Jonas Pade HUB: Humboldt University of Berlin, Berlin, Germany

René Pinnau Department of Mathematics, TU Kaiserslautern, Kaiserslautern, Germany

Marco Reese Bergische Universität Wuppertal, Fakultät für Mathematik und Naturwissenschaften, IMACM, Wuppertal, Germany

Markus Rein Fraunhofer ITWM, Kaiserslautern, Germany

Tobias K. S. Ritschel Max Planck Institute for Dynamics of Complex Technical Systems, Magdeburg, Germany

Holger Ruf Institute of Stochastics, Ulm University, Ulm, Germany

Sebastian Sager Otto von Guericke University Magdeburg, Magdeburg, Germany

Philipp Sauerteig Faculty of Mathematics and Natural Sciences, Technische Universität Ilmenau, Ilmenau, Germany

Volker Schmidt Institute of Stochastics, Ulm University, Ulm, Germany

Sebastian Schmitz Siemens Gas and Power GmbH & Co. KG, Probabilistic Design, GP PGO TI TEC PRD, Berlin, Germany

Johanna Schultes Bergische Universität Wuppertal, Fakultät für Mathematik und Naturwissenschaften, IMACM, Wuppertal, Germany

Volker Schulz Universität Trier, Fachbereich IV, Research Group on PDE-Constrained Optimization, Trier, Germany

Nils Schween Max Planck Institute for Nuclear Physics, Heidelberg, Germany

Norbert Siedow Fraunhofer ITWM, Kaiserslautern, Germany

Aleksey Sikstel Institut für Geometrie und Praktische Mathematik, RWTH Aachen, Aachen, Germany

Nadine Stahl UTrier: Universität Trier, Trier, Germany

Johannes Steiner Siemens Gas and Power GmbH & Co. KG, Probabilistic Design, GP PGO TI TEC PRD, Berlin, Germany

Michael Stiglmayr Bergische Universität Wuppertal, Fakultät für Mathematik und Naturwissenschaften, IMACM, Wuppertal, Germany

Martin Stoll Technische Universität Chemnitz, Chemnitz, Germany

Stefan Streif Automatic Control and System Dynamics Laboratory, Technische Universität Chemnitz, Chemnitz, Germany

Christian Strohm HUB: Humboldt University of Berlin, Berlin, Germany

Kai Sundmacher Max Planck Institute for Dynamics of Complex Technical Systems, Magdeburg, Germany;
Otto von Guericke University Magdeburg, Magdeburg, Germany

Onur Tanil Doganay Bergische Universität Wuppertal, Fakultät für Mathematik und Naturwissenschaften, IMACM, Wuppertal, Germany

Caren Tischendorf HUB: Humboldt University of Berlin, Berlin, Germany

Freimut von Loeper Institute of Stochastics, Ulm University, Ulm, Germany

Andreas Wagner Fraunhofer ITWM, Kaiserslautern, Germany

Marcus Wenzel Max Planck Institute for Dynamics of Complex Technical Systems, Magdeburg, Germany

Andreas Wirsen ITWM: Fraunhofer Institute for Industrial Mathematics ITWM, Kaiserslautern, Germany

Karl Worthmann Faculty of Mathematics and Natural Sciences, Technische Universität Ilmenau, Ilmenau, Germany

Part I
Economic Aspects

Chapter 1
Modeling the Intraday Electricity Demand in Germany

Sema Coskun and Ralf Korn

Abstract Future electricity markets face new challenges such as increasing variation in supply due to the dominance of renewable energy providers or variation in
demand due to the presence of price sensitive customers. In this contribution, we survey the first step to modeling the current demand process for electricity in Germany.
Besides standard affine-linear diffusion processes, we aim to model the intraday
electricity demand via a Jacobi process that has attractive properties for our applications. Further, we demonstrate the usefulness of the new models by conducting a
comprehensive data analysis.

Keywords Electricity demand · Intraday market · Jacobi process · Stochastic
differential equations

1.1 The ENets-Project—Modeling the Microstochastics of Intraday Electricity Demand and Intraday Electricity Prices

The BMBF-funded project ENets[1] has the aim to model the energy markets of the
future and the corresponding supply networks. The main aspect of our part in this
project is the task to model the micro-stochastics of the intraday demand and the
electricity prices at the intraday market. Due to many diverse reasons such as e.g.
the increased uncertainty about the production when renewable energy providers
dominate the market, the occurrence of price sensitive demand caused by smart

---

[1] See https://math4innovation.de/index.php?id=15.

S. Coskun · R. Korn (✉)
Department of Mathematics, University of Kaiserslautern, 67663 Kaiserslautern, Germany
e-mail: korn@mathematik.uni-kl.de

S. Coskun
e-mail: coskun@mathematik.uni-kl.de

© Springer Nature Switzerland AG 2021
S. Göttlich et al. (eds.), *Mathematical Modeling, Simulation and Optimization
for Power Engineering and Management*, Mathematics in Industry 34,
https://doi.org/10.1007/978-3-030-62732-4_1

grids, or the mainly non-storable character of electricity, there is still uncertainty about production and demand. This requires new stochastic models for electricity prices and the demand forecast.

This study lays the foundations of the main aims and introduces stochastic models for the current situation. Beneath standard Ornstein-Uhlenbeck (OU) type models we also consider the suitability of Cox-Ingersoll-Ross (CIR) type approaches (see Sect. 1.5) for the detailed introduction of these two process classes) and—as a new feature—Jacobi processes as stochastic models for the electricity demand.

1.2   Introduction—Demand and Electricity Prices

In economic markets commodity prices are often determined by the equilibrium of supply and demand. To a certain degree, this also holds for electricity markets. The following properties of electricity markets, particularly the German one, hint at the challenges of modeling electricity prices:

(1) The largest share of electricity is traded in auction-type markets. As a result of an auction, the price per unit of electricity (say, 1 GW) is then set to the price of the highest bid that is still needed to satisfy the total electricity demand.
(2) As the renewable energy law in Germany requires that all renewable energy produced has to enter the market first, the price is determined by the residual demand, i.e. the prices offered by non-renewable energy providers.
(3) Due to the uncertainty that comes with the production of wind or solar energy, we have an intrinsically stochastic component on the supply side.

As a consequence of the German energy transition to renewable resources (*Energiewende*[2]) the share of renewable energy production has already increased from 31.6 percent in 2016 to 36.2 percent in 2017 [25]. However, the renewable energy production results in a more volatile market environment due to forecast errors of the timing and amount of production. Together with the non-storable nature of electricity, this stochastic component on the supply side has already led to the introduction of the German intraday market.

Intraday markets allow the participants to react to the latest events such as weather changes or surprising demands. Typically, owners of renewable energy resources tend to trade in the intraday market [2]. In particular, in the German intraday market the trading continues up to 30 min before the delivery. So the market participants have the opportunity of reacting to the forecasted offer of renewable electricity production even closer to real-time, a very attractive feature. As a consequence, the traded volume

---

[2]The term Energiewende refers to the reforms caused by the German Renewable Energy Sources Act which are designed to gradually transform the energy production methods from the conventional fossil fuel methods to sustainable and renewable energy resources as e.g. wind and solar power.

in the German/Austrian[3] spot electricity market has increased from 41 TWh in 2016 to 47 TWh in 2017 [22].

As this development will continue in the future, it is necessary to introduce a model for both the spot price and the intraday electricity demand which closely captures the real dynamics of the German intraday market. To gain the necessary insights, we perform a time series analysis of the actual consumption data. By considering publicly available data we want to extract the stylized features of the electricity demand. A possible model framework for the intraday electricity demand V_t is

$$V_t = \Lambda_t + X_t \tag{1.1}$$

where Λ_t is a seasonality component and X_t a mean-reverting stochastic process that models the fluctuations around the seasonality part.

In the next sections, we introduce these components rigorously and judge their suitability on the basis of a detailed data analysis.

1.3 Basics on the Electricity Markets and Models

In this section we give a brief description of the German spot electricity markets and on our suggested stochastic process models for the electricity demand.

1.3.1 German Spot Electricity Markets

Let us briefly introduce to the spot electricity markets and their mechanism. In Germany, there are two main spot markets for trading electricity, namely the day-ahead and the intraday markets. Electricity trading in the German day-ahead markets is completely held by auctions. On the other hand, trading in the intraday market is mainly continuous. Recently, to enhance the flexibility in the German electricity market which is highly driven by renewables, the intraday auctions for 15-min periods are introduced. The particular reason behind this improvement is to provide a better balancing opportunity for the market participants against the solar ramps [21], i.e. the short-term influence of clouds on the solar energy production. The timing mechanism of electricity spot markets is summarized in Fig. 1.1 which is a modification of the figure given in [17].

As our main concern is the intraday electricity demand, we give further details about the German intraday market (see also [23]). In principle, electricity is traded

---

[3]The German and the Austrian electricity markets are considered as one bidding zone by the EPEX SPOT. Thus, the total traded volume is given as the sum of traded volumes in both countries. However, in autumn 2018, EPEX SPOT implemented the so-called split of the German-Austrian bidding zone, following a request of the regulators of these two countries [24].

Fig. 1.1 Timing of spot
electricity markets

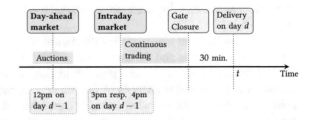

depending on the delivery time which in spot markets usually is the next day (e.g. day d in Fig. 1.1). The delivery time (e.g. time t in Fig. 1.1) can be a full hour, a 15-min period or a block of hours. In the German intraday market, the market participants can trade all of these delivery time alternatives. Moreover, in the German intraday market each hour, 15-min periods or block of hours can be continuously traded until 30 min before the delivery begins. This leads to the possibility of face-to-face-trades comparable to the situation on stock exchanges. This possibility will be used in the final part of the ENets project when we include price sensitive traders and demand sensitive pricing. Trading of full hours of the following day (e.g. day d in Fig. 1.1) starts at 3pm on the current day (e.g. day $d - 1$ in Fig. 1.1). Furthermore, starting at 4pm on the current day all 15-min periods of the following day can be traded in the German intraday market. Moreover, as opposed to the stock markets the trading continues for 7 days a week and 24 hours a day, i.e. electricity is traded also on the weekends.

1.3.2 Structural Models for Electricity Prices

In nearly all markets, prices and the relation between supply and demand are closely connected. Electricity price models which are based on the equilibrium between supply and demand are known as *structural* or *supply/demand* models. Barlow [4] models the electricity demand by an Ornstein-Uhlenbeck (OU) process. In combination with a deterministic supply function it is considered as the only driver of the electricity prices. However, considering the demand as the only driver of electricity prices might be insufficient. To overcome this, many researchers considered additional underlying factors. For instance, in [6] electricity demand and capacity have been presented as the drivers of the spot electricity prices. Also, in [1] demand is combined with the prices of different fuels to introduce a structural spot price model. Coulon and Howison [8] adopt a parametric approach to the bid stack function, i.e. the marginal cost of electricity supply, by allowing multiple fuel prices as underlying driving factors. They further consider the capacity or margin issues such as outages in the electricity supply. Especially, in the German market with its high part of renewable energy providers, a sophisticated modeling and prediction of the electricity demand is indispensable. For that reason, Wagner [27] proposes a comprehensive model for residual demand which is obtained by subtracting the renewable infeed generated by solar and wind power plants from the total demand.

1.3.3   The Jacobi Process as a New Modeling Ingredient

While popular processes for demand modeling such as the OU process are analytically very tractable, they do not consider practical bounds on demand such as its non-negativity or the installed capacity of electricity production. A stochastic process that is tailored to the situation of a bounded state space is the Jacobi process.

The Jacobi diffusion belongs to the class of Pearson diffusions. In its general form, a Pearson diffusion is a stationary solution to the following stochastic differential equation (SDE)

$$dX_t = -\kappa(X_t - \theta)dt + \sqrt{2\kappa(aX_t^2 + bX_t + c)}dW_t \tag{1.2}$$

where $\kappa > 0$ and the coefficients a, b, c ensure that the square root is well defined when X_t is in its state space [12]. The parameters also determine the state space of the diffusion as well as the shape of the invariant distribution. The Jacobi diffusion is defined as the stationary solution to the SDE

$$dJ_t = \kappa(\theta - J_t)dt + \sigma\sqrt{J_t(1 - J_t)}dW_t \tag{1.3}$$

where the parameters κ, θ and σ ensure existence of the stationary distribution which indeed is the Beta distribution. Conditions for existence and uniqueness of the solution to Eq. (1.3) are given in [12], see section "Appendix: The Jacobi Process". Obviously, J_t is only defined for values in $(0, 1)$. If we want to use the process for modeling electricity demand D_t that is known to be in the interval (D_m, D_M), then the transformation

$$D_t = D_m + (D_M - D_m)J_t \tag{1.4}$$

that leads to the SDE

$$dD_t = \kappa(D_\mu - D_t)dt + \sigma\sqrt{(D_t - D_m)(D_M - D_t)}dW_t \tag{1.5}$$

is the appropriate rescaling. Note that now D_μ has the role of the mean demand level, D_m of the minimum and D_M of the maximal demand level, while κ is the mean reversion speed of demand. This transformation is motivated by an application in interest rate modeling by Delbaen and Shirakawa [11]. For modeling (the logarithm of) exchange rates that should be kept in a target zone, a similarly transformed Jacobi diffusion is used by [16, 18]. In a slightly different framework, a recent study by [3] utilizes a Jacobi-type process in order to introduce a probabilistic day-ahead forecasting model for the solar irradiation. More technical details on the properties of the Jacobi process are provided in the appendix.

Table 1.1 Descriptive statistics of the ex-post consumption data in MWh per 15 min on a yearly basis and for 2015–2018. ♯ and NaN denote the number of observations and the number of missing values, respectively

Year	♯(NaN)	Mean	Std	Med	Min	Max
2015	35032 (8)	13,640.34	2,502.70	13,543.37	7,458.50	19,159.00
2016	35128 (8)	13,693.03	2,453.53	13,582.37	7,824.50	18,952.75
2017	35032 (8)	14,078.24	2,572.08	14,022.00	7,363.25	19,870.25
2018	35023 (17)	14,516.28	2,479.28	14,419.25	8,025.75	19,698.25
	140215 (41)	13,981.74	2,526.82	13,906.00	7,363.25	19,870.25

1.4 Data Analysis—Stylized Facts of German Electricity Demand

In this section, we present the results of our detailed analysis of the electricity consumption per quarter hour (i.e. we have a 15 min resolution) for the time span from 01.01.2015 to 31.12.2018. The data set for Germany is provided by Smard.[4] As the data correspond to the actually realized amounts of intraday electricity consumption, they are called ex-post data.

Missing values in Table 1.1 (and duplicate values) in years 2015, 2016 and 2017 are only due to daylight saving practice in Germany, i.e. the change from winter to summer time. As the proportion of missing values in the whole data set is fairly low (<0.03%), we proceed with our analysis by sampling out the missing values.

There is a slight upward movement in the quarter hourly electricity consumption as the mean value has increased almost 900 MWh in four years, see also Fig. 1.2. Next we are going to present the main characteristics of the electricity demand in Germany, its so-called *stylized facts*.

Weekend Effect For German day-ahead electricity prices, a weekend effect (i.e. prices are lower on the weekend) is shown in [15]. Given the mechanism for the German electricity prices, the natural reason for this is a smaller demand on the weekend. We have discovered such a weekend effect for the demand (i.e. lower demand on weekends) and have illustrated it in Fig. 1.3.

The reasons for this are mainly stopped industry and business processes. Outlier values on weekends (i.e. surprisingly high 15-min demands) may be due to some nationwide specific events such as football matches, etc. The assertions on the differing demands for weekends and weekdays are also confirmed in Table 1.2.

Seasonality It is well documented in the literature that electricity price series exhibit strong seasonality which varies depending on the electricity consumption with different time scales. For a general overview of the seasonality of the electricity prices and different functions used to model this behavior we refer to [15, 29]. For the electricity demand, we first focus on the yearly seasonality. Geman and Roncoroni

---

[4]This platform is operated by the Federal Network Agency (Bundesnetzagentur) and the data is obtained directly from the European Transmission System Operators Association (ENTSO-E).

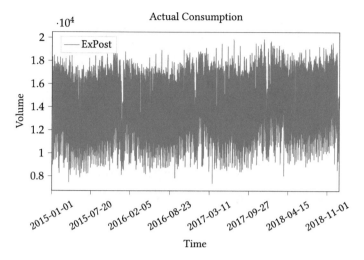

Fig. 1.2 Electricity consumption data in Germany for 15-min slots (2015–2018)

Fig. 1.3 Summary of the data set with respect to days of a week

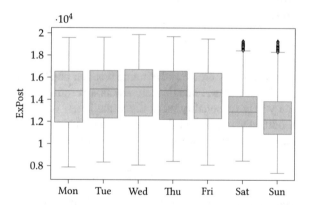

[13] argue that electricity prices typically follow a periodic path having two maxima per year of possibly different magnitude, i.e. a 12-month and a 6-month periodicity which roughly account for winter and summer peaks in demand. Furthermore, they also integrate a linear trend function into the seasonality component.

Figure 1.4 illustrates the monthly fluctuations in the electricity demand separately for the years in the data set. Yearly seasonality in electricity consumption occurs

Table 1.2 Comparison of the 15-min electricity demand in Germany on weekdays and weekends

	Mean	Std	Min	Med	Max	Skewness	Kurtosis
Weekdays	14,446.53	2,501.21	7,878.75	14,876.75	19,870.25	−0.2598	1.9339
Weekends	12,821.09	2,195.46	7,363.25	12,557.37	19,299.25	0.4437	2.6516

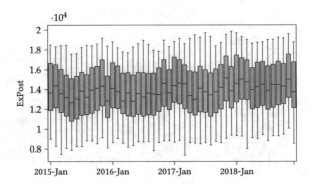

Fig. 1.4 Monthly deviation of the German 15-min electricity consumption

Table 1.3 Yearly seasonality coefficients determined via the ordinary least square (OLS) regression fit leading to an adjusted $R^2 = 0.072$ for 2015–2018

	a	b	c_1	c_2
Coefficients	13,950	0.0116	798.0500	153.7067
Std error	15.325	0.000	10.816	10.816

mainly due to climate and extreme weather conditions. Keeping this in mind, we see in Fig. 1.4 that although there exists a certain sinusoidal behavior in the data set, the yearly seasonality has a moderate level. We employ the following function—a modified version of the function given in [13] and similar to the one given in [20]—to capture the yearly seasonality in our data set

$$g(t) = a + bt + c_1 \cos(2\pi t) + c_2 \cos(4\pi t) . \tag{1.6}$$

Here, the linear trend is also contained in the function.

The results given in Table 1.3 confirm the observation of [20] regarding the poor contribution of the yearly seasonality to the overall variability of the electricity prices. Although we consider the electricity consumption data, this observation is still valid and in particular confirmed by the very small value of the adjusted R^2 that indicates nearly no (linear) predictability of demand by yearly seasonality.

Figure 1.5 contains the fitted yearly seasonality function with coefficients from Table 1.3. Besides this yearly periodic behavior, electricity demand also exhibits a weekly seasonal pattern, see for example Fig. 1.3. As we dropped the weekends in our analysis, we eliminated the weekly seasonality.

The most interesting seasonality behavior for our analysis is the intraday cyclic behavior of electricity consumption. It is usually associated with the working hours during a day. This pattern is also realized in the German intraday market by distinguishing the peak hours which cover the time range from 9 to 20 o'clock and the remaining off-peak hours. Meanwhile, the base load covers the hours from 1 to 24 o'clock in the German intraday market. To capture the hourly effect, we use dummy

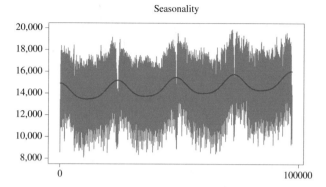

Fig. 1.5 Yearly seasonality (red line) of the German 15-min electricity consumption in 2015–2018 with coefficients from Table 1.3

Table 1.4 Intraday seasonality coefficients determined via OLS regression fit leading to an adjusted $R^2 = 0.697$ for 2015–2018

	Coefficients		Coefficients		Coefficients		Coefficients
D_0	−2722.19	D_6	−633.02	D_{12}	2195.77	D_{18}	1430.58
D_1	−3183.48	D_7	743.53	D_{13}	1961.50	D_{19}	1282.63
D_2	−3356.97	D_8	1448.52	D_{14}	1639.27	D_{20}	617.16
D_3	−3413.75	D_9	1707.18	D_{15}	1407.19	D_{21}	−87.52
D_4	−3069.97	D_{10}	2034.13	D_{16}	1217.14	D_{22}	−775.24
D_5	−2335.91	D_{11}	2321.63	D_{17}	1372.58	D_{23}	−1800.40

variables that assign an indicator function for each of the different time points as follows:

$$h(t) = \sum_{i=0}^{23} \alpha_i D_i(t) \tag{1.7}$$

Here, $D_i(t)$ is the indicator function for each hour and α_i, $i = 0, \ldots, 23$ are the coefficients which have to be estimated. Using OLS regression, we fit the dummy variables to the data, where we have eliminated the yearly seasonality with the sinusoidal function. The results are given in Table 1.4. From Table 1.4, it can be seen that starting from 7 am the electricity consumption gets higher which refers to the peak hours classification. Moreover, after 6 pm there seems to be a slight increase in the electricity consumption.

It can also be seen from Fig. 1.6 that there are usually two peak times in the intraday electricity demand, one is reached around noon and the second one occurs mostly after 6 pm. This might be related to the demand of the households after the working hours on a day. As a result, we conclude that our seasonality function consists of two parts, $g(t)$, $h(t)$ given by

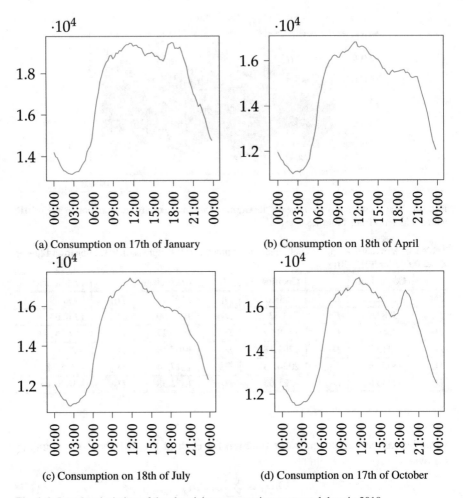

Fig. 1.6 Intraday deviation of the electricity consumption on several days in 2018

$$\Lambda_t = g(t) + h(t) = a + bt + c_1 \cos(2\pi t) + c_2 \cos(4\pi t) + \sum_{i=0}^{23} \alpha_i D_i(t) . \quad (1.8)$$

To analyze the remaining part of the series after removing the trend and seasonality of time series, the histogram of the residuals is presented in Fig. 1.7. Obviously, the residuals are not normally distributed. When we empirically fitted several distributions to the residuals, the best empirical fit is achieved by a beta distribution.

Fig. 1.7 Histogram and fitted empirical kernel density estimation for residuals

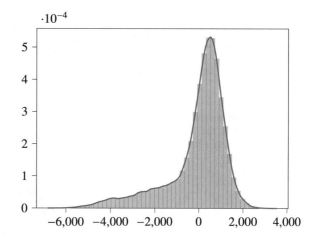

The method of moments [5] resulted in a beta distribution on $[-6515, 3592]$ with parameters $(p, q) = (6.955, 3.592)$.

Mean Reversion The mean reverting behavior of a stochastic process implies that it tends to revert to a certain constant or time varying mean level. In [30] a Hurst analysis shows that the return of electricity prices exhibits a mean-reverting property. To check the validity of the mean reversion assumption, we tested the stationarity of the time series in our data set [5]. For this, we used the Augmented Dickey-Fuller (ADF) test for unit roots.[6] The result of the ADF test indicates that the deseasonalized consumption data set exhibits stationary behavior. We can reject the null hypothesis at the 5% significance levels (Test statistic $= -22.5427$, $p = 0.0$, critical value for 5% is -2.8615). We also conducted the ADF test for the non-deseasonalized time series and can again confidently reject the null hypothesis. Hence, our time series can be assumed to be stationary which addresses the mean reverting behavior.

Spikes While electricity price spikes (i.e. sudden upward or downward jumps of a large magnitude) have received a high attention in electricity price modeling (see e.g. [20]), they play no huge role in the electricity demand. To show this, we checked the existence of consumption values which are higher/lower than three times the inter quartile range. Although, there exist values larger than the 1.5 inter quartiles, they still remain in the range of three times the inter quartiles for the complete data set. As these values occur specifically on public holidays, we removed the holidays from our data set.

Distribution and Tail Behavior The descriptive statistics given in Table 1.5 regarding the skewness and kurtosis of the electricity consumption data set imply that the distribution is not Gaussian. Although the skewness of our data set is fairly close to

---

[5]I.e. we choose the parameters p and q of the beta distribution with support [a, b] $= [-6515, 3592]$ such that $E(X) = (b - a)p/(p + q) + a$, $V(X) = (b - a)^2 pq/((p + q + 1)(p + q)^2)$ is satisfied where the mean and the variance are estimated by their empirical counter parts.

[6]The null hypothesis of the Augmented Dickey-Fuller is that there is a unit root (i.e. the time series is non-stationary), with the alternative that there is no unit root (i.e. the time series is stationary).

Table 1.5 Skewness and Kurtosis of the German 15-min electricity consumption

Year	2015	2016	2017	2018	All
Skewness	−0.0298	−0.0271	−0.0273	−0.0596	−0.0314
Kurtosis	1.8453	1.8399	1.9172	1.8454	1.9014

Fig. 1.8 Histogram and fitted empirical kernel density estimation for the 15-min electricity consumption

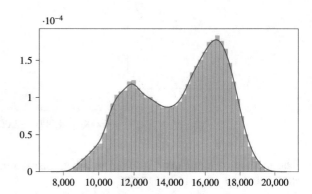

0, the kurtosis statistics is clearly negative. This implies that our data set might have a light tailed distribution.

We further checked the histogram and the corresponding kernel density estimation of the German 15-min electricity consumption. Figure 1.8 shows that its distribution exhibits two peaks.

Autocorrelation In the analysis of electricity price time series, the exponential decay coefficient of the autocorrelation function provides an estimation of the mean reversion speed of the stochastic process. Meyer-Brandis and Tankov [20] argue that the following sum of two exponentials describes quite precisely the observed structure of electricity price series:

$$\rho(t) = w_1 e^{-t/\lambda_1} + w_2 e^{-t/\lambda_2}.$$

In the light of this approach, we fit the following exponential function to our electricity consumption data

$$h(t) = w_1 e^{-t/\kappa} \tag{1.9}$$

with κ being the mean reversion speed of the mean reverting process. The result of the exponential fit implies that the mean reversion speed κ is approximately 82 and the multiplier w_1 is equal to 0.95. We illustrate the exponential fit by Fig. 1.9.

Fig. 1.9 Autocorrelation function of the consumption data and exponential fit

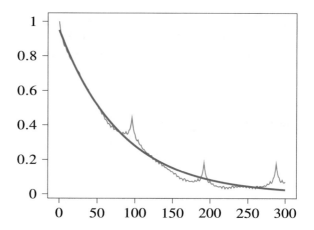

1.5 Case Study: Modeling the Intraday Electricity Demand

In the preceding section, we have seen that the residuals of the electricity consumption became stationary after the introduction of suitable dummy variables. Also, the so modified data exhibited a mean-reverting behaviour.

Consequently, we now consider several mean-reverting polynomial processes as suitable stochastic models for the intraday electricity demand modified by the dummy variables. Subsequently, we calibrate these models to the intraday electricity consumption data as being the indicator of the demand.

Ornstein-Uhlenbeck Process. We start with the simple, but still widely used Ornstein-Uhlenbeck (OU) process. It is a mean reverting stochastic process driven by the SDE

$$dV_t = \kappa(\theta - V_t)dt + \sigma dW_t \tag{1.10}$$

where κ, θ and σ are constant parameters that have to be estimated and further, W_t is a one-dimensional Brownian motion. Here—and also in the processes below—the parameters κ, θ and σ refer to the mean reversion speed, the long term mean reversion level and the volatility of the process, respectively. The explicit form of V_t is given as

$$V_t = \theta\left(1 - e^{-\kappa(t-s)}\right) + V_s e^{-\kappa(t-s)} + \sigma e^{-\kappa t}\int_s^t e^{\kappa u}dW_u \tag{1.11}$$

between any two time instants s and t. Its discrete time correspondence is the autoregressive AR(1) model which reads as

$$V(t_i) = c + bV(t_{i-1}) + \delta\epsilon(t_i) \tag{1.12}$$

with $\Delta t = t_i - t_{i-1}$ an equidistant, constant time step and $\epsilon(t)$ a Gaussian white noise (i.e. $\epsilon \sim \mathcal{N}(0, 1)$). By matching the Eqs. (1.11) and (1.12) and further using the Itô isometry for the volatility parameter, one obtains the following equations

$$c = \theta \left(1 - e^{-\kappa \Delta t}\right), \quad b = e^{-\kappa \Delta t}, \quad \delta = \sigma \sqrt{\left(1 - e^{-2\kappa \Delta t}\right)/2\kappa}.$$

For a time series $V(t_i)$, we calibrate the parameters given above using the ordinary least squares method (OLS). From the parameters c, b and δ, we obtain the parameters for the OU process κ, θ and σ as:

$$\kappa = \frac{-\ln(b)}{\Delta t}, \quad \theta = \frac{c}{(1-b)}, \quad \sigma = \frac{\delta}{\sqrt{\frac{(b^2-1)\Delta t}{2\ln(b)}}}. \tag{1.13}$$

For our data set, the calibration results of the AR(1) estimation yield

$$\kappa = 83.37, \quad \theta = 14515.29, \quad \sigma = 3266.76$$

Furthermore, we estimated the parameters by using the maximum likelihood estimation (MLE) which is given in [5] by the following estimators

$$\hat{b} = \frac{N \sum_{i=1}^{N} V_i V_{i-1} - \sum_{i=1}^{N} V_i \sum_{i=1}^{N} V_{i-1}}{N \sum_{i=1}^{N} V_{i-1}^2 \left(\sum_{i=1}^{N} V_{i-1}\right)^2}$$

$$\hat{\theta} = \frac{\sum_{i=1}^{N} \left[V_i - \hat{b} V_{i-1}\right]}{N(1 - \hat{b})}$$

$$\hat{\delta} = \frac{1}{N} \sum_{i=1}^{N} \left[V_i - \hat{b} V_{i-1} - \hat{\theta}(1 - \hat{b})\right]^2.$$

Here, N is number of observations in the data set, V_i denotes the observation $V(t_i)$ at time t_i. We obtain the required parameters κ and σ using the equations in (1.13). The result of the MLE application for the OU model calibration is very close to that from the AR(1) calibration and reads as

$$\kappa = 83.35, \quad \theta = 14515.14, \quad \sigma = 3266.76$$

CIR Process. The CIR process, which is initially used to model the short rate in [9], has its dynamics driven by the following SDE

$$dV_t = \kappa(\theta - V_t)dt + \sigma \sqrt{V_t} dW_t \tag{1.14}$$

with the constant parameters κ, θ and σ and W_t a one-dimensional Brownian motion. Under the Feller condition, i.e. $2\kappa\theta > \sigma^2$, the CIR process remains strictly positive, otherwise it is only non-negative. Further, the conditional distribution of the discretized CIR process $V(t_i)$ is known explicitly via

$$p_{CIR}(V(t_i) \mid V(t_{i-1})) = ce^{-u-v}\left(\frac{u}{v}\right)^{q/2} I_q(2\sqrt{uv}) \qquad (1.15)$$

with

$$c = \frac{2\kappa}{\sigma^2(1 - e^{-\kappa\Delta t})}, \quad u = cV(t_i)e^{-\kappa\Delta t}, \quad v = cV(t_{i-1}), \quad q = \frac{2\kappa\theta}{\sigma^2} - 1.$$

Here, I_q is the modified Bessel function of the first kind of order q and $\Delta t = t_i - t_{i-1}$. This explicit formula allows us to implement the MLE procedure to calibrate the model for any given data set. By following the approach given by [5], we obtain the initial parameters using the AR(1) fitting given in Eq. (1.12)

$$\kappa_0 = \frac{-\log(b)}{\Delta t} = 83.37, \quad \theta_0 = \mathbb{E}(V_t) = 14514.84, \quad \sigma_0 = \sqrt{\frac{2\kappa_0 \mathbb{V}ar(V_t)}{\theta_0}} = 27.12.$$

For this, we also use the steady state moments of the CIR process

$$\mathbb{E}(V_t) = \theta, \quad \mathbb{V}ar(V_t) = \frac{\theta\sigma^2}{2\kappa}.$$

The MLE for our data set are then obtained as

$$\kappa = 80.98, \quad \theta = 14515.31, \quad \sigma = 27.80.$$

Jacobi Process. As briefly introduced in Sect. 1.3.3, the Jacobi process is suitable to model the stochastic behavior of a process that remains within a bounded region. For simplicity, we consider the Jacobi process remaining in the interval [0, 1], i.e.

$$dV_t = \kappa(\theta - V_t)dt + \sigma\sqrt{V_t(1 - V_t)}dW_t. \qquad (1.16)$$

with constant parameters κ, θ and σ and W_t a one-dimensional Brownian motion. By restricting ourselves to the case where the Jacobi process stays in [0, 1], we omit the estimation of upper or lower bounds of the process. For this, we have mapped the data set to the interval [0, 1] via the transformation given in Eq. (1.4) and subsequently only have to estimate the remaining parameters κ, θ and σ.

For the Jacobi process, the application of MLE is not feasible as its transition probability is not given in an explicit form. However, in [26] an extensive comparison among various estimation methods for a Jacobi process is presented. The suggested method is the quasi-MLE (QMLE) due to its simplicity and its bias and variance properties. Therefore, we implemented the QMLE method proposed in Chap. 5 of [26] (for details, see [26]) to calibrate the Jacobi model to our data set. The only remaining issue related to the QMLE implementation is to identify the initial parameters. For this purpose, we adopt the approach of [5] already explained for the CIR model calibration. Hence, we use the AR(1) estimation for κ_0 and the steady state moments of the Jacobi process for θ_0 and σ_0.

The initial parameters for the QMLE procedure are then given by

$$\kappa_0 = \frac{-\log(b)}{\Delta t}, \quad \theta_0 = \mathbb{E}(V_t), \quad \sigma_0 = \sqrt{\frac{2\kappa_0 \mathbb{V}ar(V_t)}{\theta_0 - \theta_0^2 - \mathbb{V}ar(V_t)}}.$$

The conditional mean and variance, which are derived from the first and second conditional moments of the Jacobi process given in [11], are as follows

$$\mathbb{E}(V_t \mid V_s = V) = \theta + (V - \theta)e^{-\kappa(t-s)},$$

$$\mathbb{V}ar(V_t \mid V_s = V) = \frac{(2\kappa\theta + \sigma^2)\theta}{2\kappa + \sigma^2} - \theta^2 + \left[\frac{(2\kappa\theta + \sigma^2)\theta}{2\kappa + \sigma^2} - 2\theta\right](V - \theta)e^{-\kappa(t-s)}$$

$$+ \left[V^2 - \frac{(2\kappa\theta + \sigma^2)\theta}{2\kappa + \sigma^2}V + \frac{\kappa\theta(2\kappa\theta + \sigma^2)}{(2\kappa + \sigma^2)(\kappa + \sigma^2)}\right]e^{-(2\kappa + \sigma^2)(t-s)} - (V - \theta)^2 e^{-2\kappa(t-s)}.$$

By taking the limit $t \to \infty$ of the conditional moments, we obtain the initial parameters. Unfortunately, due to memory problems,[7] we can here only present a result for a daily example, 17 January 2018, where we have $\theta = 17521.73$ and $\sigma = 0.2535$. Further, we have set κ to its initial value $\kappa_0 = 83.37$. The upper and lower limits for the demand are set manually (see the comment below in the comparison).

Comparison of the three model approaches: All three models have been simulated for 17 January 2018. The corresponding paths together with the realized demand are displayed in Fig. 1.10. All models show a good behaviour with the OU model slightly outperforming the CIR model. The Jacobi model performs best. For the paths denoted by *Jacobi-I*, we have set the upper bound to 20000 and the lower bound to 13000 which leads to better results than the choices for *Jacobi-II* with upper bound of 22000 and lower bound of 11000.

It is a particular attractive feature of the Jacobi process that we have the possibility to set bounds on the demand locally. While in Fig. 1.10 they have been set on the daily level, it is even possible to set them on the hourly level. However, this also poses a challenge on forecasting and on estimating the Jacobi parameters under this additional feature.

Figure 1.11 contains the residuals of the four model variants. It is not surprising that the Jacobi models perform best as the stationary distribution of the Jacobi models is the beta distribution. For this, remember that in the foregoing section we have seen that the residuals of the consumption data are best fitted by the beta distribution.

---

[7]The calibration of the Jacobi model parameters require a lot of nested function evaluations so that we were running out of memory when using the full data set.

Fig. 1.10 Realized and simulated demand for the 15-min electricity consumption on January 17, 2018 (time units are 15 min)

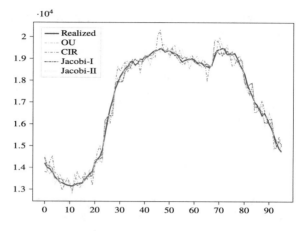

Fig. 1.11 Residuals of the simulated processes on January 17, 2018 (time units are 15 min)

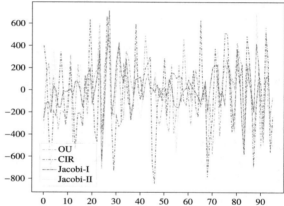

1.6   Challenges and Future Work Packages

So far, we have considered an electricity market where *demand can always be satisfied* and *demand is not price sensitive*, i.e. the customers agree to pay the offered electricity prices. This, however, will not be the case in future electricity markets due to the uncertainty in renewable energy production and the possibility of the customers to organize in groups (*smart grids*) that are able to partially postpone consumption for a certain time span. Those customers have the possiblity to react on weather forecasts and the possible prediction of energy production and prices. We will take this into account and develop models that include

- price sensitive customers, i.e. customers that have a view on the prices they are willing to pay and that are able to partially postpone consumption. Their information about the future electricity production will particularly be based on (local) weather forecasts and their implication on the production.

- demand sensitive energy providers, i.e. providers can to a certain extent control their produced amount of electricity or can decide to convert it into a storable form e.g. by using batteries or power-to-gas production.

This has a particular impact on the trading at the intraday market as the corresponding trades will typically be individual ones. On the technical modeling side we will introduce *bridge processes* conditioned on their *area under the curve*, i.e. their integral over the time span corresponding to the bridge. In particular we assume to know in advance

- the *total demand* of the day corresponding to the integral of the demand process.
- the *final demand* of the day.

Of course, both these quantities are not perfectly known in advance, but can be better predicted than the actual demand at each time point. Possible predictors might be simple ones obtained from historical demand figures, the traded volumes at the day-ahead market or more sophisticated ones such as predictions via well-trained neural networks.

A possible simulation approach for the prediction of the future demand development over a particular day will then consist of first predicting/simulating the daily demand and the final electricity demand at midnight. It will be followed by the simulation of a conditional bridge process such as the constraint Ornstein-Uhlenbeck bridge given the daily and the final electricity demand. For this process, explicit distributional results are given in [19]. Further, there are related results on stochastic control of the OU processes (see [7]). These results can be used to
Challenge 1: infer the price sensitive demand strategy of the price sensitive customers and to
Challenge 2: find a corresponding price offering strategy such that (local) demand and supply coincide.

Solving the two challenges first on the basis of the results on constrained OU bridges will be our next work package. The following one then consists of the extension of our results to the Jacobi setting case. Of course, for this, we also have to derive the relevant results for constraint Jacobi bridges first.

Appendix: The Jacobi Process

In this section, we collect some properties of the Jacobi process given in Eq. (1.3). For further properties and its applications in financial models see [10–12, 28]. For the sake of consistency, we recall the SDE in its standard form given by

$$dX_t = \kappa(\theta - X_t)dt + \sigma\sqrt{X_t(1 - X_t)}dW_t.$$

As shown in [12], this equation has a unique solution, the Jacobi diffusion, with the differential operator $\mathcal{L}$ defined on the set of twice continuously differentiable functions f as

$$\mathcal{L}f(x) = \kappa(\theta - x)\frac{\partial f(x)}{\partial x} + \frac{1}{2}\sigma^2 x(1-x)\frac{\partial f^2(x)}{\partial x^2}.$$

The corresponding Fokker-Planck or forward Kolmogorov equation for the transition density $p(x, t)$ of the Jacobi process is given as

$$\frac{\partial}{\partial t}p(x, t) = -\frac{\partial}{\partial x}p(x, t)[\kappa(\theta - x)] + \frac{1}{2}\frac{\partial^2}{\partial x^2}p(x, t)[\sigma^2 x(1-x)] \qquad (1.1)$$

with the initial condition

$$p(x, 0) = \delta_{x_0}(x)$$

where δ_{x_0} is the Dirac delta function. Further, the transition probability has the following representation, (see Theorem 2.2 in [11]),

$$p(x, s; y, t) = \sum_{n=0}^{\infty} k_n \psi_n(x)\psi_n(y)\omega(y)e^{-\lambda_n(t-s)} \qquad (1.2)$$

where

$$a = \frac{2\kappa\theta}{\sigma^2} > 0, \quad b = \frac{2\kappa(1-\theta)}{\sigma^2} > 0, \quad \lambda_n = \kappa n + \frac{\sigma^2}{2}n(n-1),$$

$$k_n = \frac{(a+b+2n-1)\Gamma(a+n)\Gamma(a+b+n-1)}{n!\Gamma^2(a)\Gamma(b+n)},$$

$$\psi_n(x) = \sum_{k=0}^{n}(-1)^k\binom{n}{k}\frac{(a+b+n-1)_k}{(a)_k}x^k, \quad \omega(x) = x^{a-1}(1-x)^{b-1}.$$

In [14], the following canonical decomposition of the transition density given in Eq. (1.2) is presented and, also, in [26] it is used for the QMLE of the parameters of the Jacobi process

$$p(x, s; y, t) = \tilde{p}(y, t)\left\{1 + \sum_{n=1}^{N}k_n\psi_n(x)\psi_n(y)e^{-\lambda_n(t-s)}\right\}$$

with $\tilde{p}(y, t)$ being the density of the Beta distribution (see [26] for the exact parameterization). It is further recommended to set $N = 4$, for the QMLE method.

References

1. Aïd, R., Campi, L., Huu, A.N., Touzi, N.: A structural risk-neutral model of electricity prices. Int. J. Theor. Appl. Financ. **12**(7), 925–947 (2009)
2. Aïd, R., Gruet, P., Pham, H.: An optimal trading problem in intraday electricity markets. Math. Financ. Econ. **10**(1), 49–85 (2016)
3. Badosa, J., Gobet, E., Grangereau, M., Kim, D.: Day-ahead probabilistic forecast of solar irradiance: a stochastic differential equation approach. In: Drobinski, P., Mougeot, M., Picard, D., Plougonven, R., Tankov, P. (eds.) Renewable Energy: Forecasting and Risk Management, pp. 73–93. Proceedings in Mathematics & Statistics, vol. 254. Springer, Cham (2018)
4. Barlow, M.T.: A diffusion model for electricity prices. Math. Financ. **12**(4), 287–298 (2002)
5. Brigo, D., Dalessandro, A., Neugebauer, M., Triki, F.: A stochastic processes toolkit for risk management: mean reverting processes and jumps. J. Risk Manage. Financ. Inst. **3**(1), 65–83 (2009)
6. Cartea, A., Villaplana, P.: Spot price modeling and the valuation of electricity forward contracts: the role of demand and capacity. J. Bank. Financ. **32**(12), 2502–2519 (2008)
7. Christopeit, N., Helmes, K.: Controlled Ornstein-Uhlenbeck process with chance constraints. J. Optim. Theory Appl. **28**(1), 89–101 (1979)
8. Coulon, M., Howison, S.: Stochastic behaviour of the electricity bid stack: from fundamental drivers to power price. J. Energy Markets **2**(1), 29–69 (2009)
9. Cox, J.C., Ingersoll, J.E., Ross, S.A.: A theory of the term structure of interest rates. Econometrica **53**(2), 385–408 (1985)
10. Cuchiero, C., Keller-Ressel, M., Teichmann, J.: Polynomial processes and their applications to mathematical finance. Financ. Stochast. **16**(4), 711–740 (2012)
11. Delbaen, F., Shirakawa, H.: An interest rate model with upper and lower bounds. Asia-Pac. Financ. Markets **9**, 191–209 (2002)
12. Forman, J.L., Sørensen, M.: The Pearson diffusions: a class of statistically tractable diffusion processes. Scand. J. Stat. **35**(3), 438–465 (2008)
13. Geman, H., Roncoroni, A.: Understanding the fine structure of electricity prices. J. Bus. **79**(3), 1225–1261 (2006)
14. Gouriéroux, C., Jasiak, J.: Multivariate Jacobi process with application to smooth transitions. J. Econom. **131**(1–2), 475–505 (2006)
15. Hinderks, W.J., Wagner, A.: Factor models in the German electricity market: stylized facts, seasonality, and calibration. Energy Econ. (2019)
16. De Jong, F., Drost, F.C., Werker, B.J.M.: A jump-diffusion model for exchange rates in a target zone. Statistica Neerlandica **55**(3), 270–300 (2001)
17. Kiesel, R., Paraschiv, F.: Econometric analysis of 15-minute intraday electricity prices. Energy Econ. **64**, 77–90 (2017)
18. Larsen, K.S., Sørensen, M.: A diffusion model for exchange rates in a target zone. Math. Financ. **17**(2), 285–306 (2007)
19. Mazzolo, A.: Constraint Ornstein-Uhlenbeck bridges. J. Math. Phys. **58**(9) (2017)
20. Meyer-Brandis, T., Tankov, P.: Multi-factor jump-diffusion models of electricity prices. Int. J. Theor. Appl. Financ. **11**(5), 503–528 (2008)
21. EPEX SPOT. 15-minute intraday call auction (2011). https://www.epexspot.com/document/29113/15-Minute_Intraday_Call_Auction. Accessed: 23.07.2019
22. EPEX SPOT. Annual reports (2017). http://static.epexspot.com/document/39308/Annual%20Report%20-%202017
23. EPEX SPOT. Information sheet (2017). http://www.epexspot.com/en/product-info/intradaycontinuous/germany/. Accessed: 23.07.2019
24. EPEX SPOT. Annual reports (2018). http://static.epexspot.com/document/40879/%20Report%20-%202018. Accessed: 10.07.2019
25. Umweltbundesamt. Erneuerbare Energien in Deutschland - Daten zur Entwicklung im Jahr 2017. Technical report, Geschäftsstelle der Arbeitsgruppe Erneuerbare Energien-Statistik (AGEE-Stat) am Umweltbundesamt (2017)

26. Valéry, P.: Simulation-Based Inference and Nonlinear Canonical Analysis in Financial Econometrics. Ph.D. thesis, Université de Montréal (2005)
27. Wagner, A.: Residual demand modeling and application to electricity pricing. Energy J. **35**(2), 45–73 (2014)
28. Ware, T.: Polynomial processes for power prices. Appl. Math. Financ. **26**(5), 453–474 (2019)
29. Weron, R.: Electricity price forecasting: a review of the state-of-the-art with a look into the future. Int. J. Forecast. **30**(4), 1030–1081 (2014)
30. Weron, R., Przybylowicz, B.: Hurst analysis of electricity price dynamics. Physica A **283**(3–4), 462–468 (2000)

Chapter 2
Application of Continuous Stochastic Processes in Energy Market Models

Ria Grindel, Wieger Hinderks, and Andreas Wagner

Abstract Examples for the use of continuous stochastic processes in the modelling of energy markets are discussed. Two practical use-cases are chosen to illustrate applications. For the stochastic modelling of the economics behind energy markets, models for temperature, demand, and renewable electricity generation are discussed. The modelling of prices on energy markets themselves is debated as a second application. Possible open source data sets for an application of the models are listed.

2.1 Introduction

One goal of this book is to illustrate the usage of mathematical tools in energy related modelling applications. This chapter is devoted to the use of the mathematical tool *continuous stochastic process* for modelling energy markets in literature and practice. Energy markets are exposed to different sources of uncertainty from various areas. The analysis of electricity systems as well as risk management require these uncertainties to be modelled and quantified based on observable data. Continuous stochastic processes are one of the main tools used for this purpose. The choice of the exact model often depends on the availability of data. Therefore, we also share possible sources of data and practical experience with the models and the data.

In the first application, we focus on fundamental modelling of energy markets with stochastic processes in the literature. This involves in particular models for the factors driving the markets, like temperature, demand, and renewable generation. The second application shows the use of stochastic processes for modelling energy market prices for use in risk management.

R. Grindel (✉) · W. Hinderks · A. Wagner
Fraunhofer ITWM, Kaiserslautern, Germany
e-mail: ria.grindel@itwm.fraunhofer.de

W. Hinderks
e-mail: hinderks.itwm@gmail.de

A. Wagner
e-mail: andreas.wagner@itwm.fraunhofer.de

© Springer Nature Switzerland AG 2021
S. Göttlich et al. (eds.), *Mathematical Modeling, Simulation and Optimization for Power Engineering and Management*, Mathematics in Industry 34,
https://doi.org/10.1007/978-3-030-62732-4_2

2.2   Application I: Economics Behind Energy Markets

A prominent source of uncertainty in energy markets is consumption, i.e. **energy demand**. Unfortunately and to the discontent of modellers, data availability regarding true demand of single market participants is sparse and not available as a detailed time series. For the European electricity markets, however, there is at least data on **total load** available, defined as "[...] a load equal to generation and any imports deducting any exports and power used for energy storage", see [21]. This value misses some information due to only considering electricity flowing through electricity grids with medium as well as high voltage and leaving low voltage grids left out. Nonetheless, it is still a good indicator for true electricity demand due to non-storability of electricity. Therefore, it will be considered equivalent to true demand in the following. Its development over the last years in Germany is shown in Fig. 2.1.

The fundamental driver underlying demand uncertainty is **temperature uncertainty**. Temperature strongly influences energy consumption: Warm and sunny spring days lead to less heating and therefore demand for gas and electricity than a cold and dark winter day. Another driver for electricity consumption is daylight, which is deterministic, though. Temperature variations over the years are shown in Fig. 2.2.

Another very dominant source of uncertainty in the electricity market nowadays is generation from renewable energy sources. An inherent trait of **wind and solar energy** is their random and not influenceable nature of production. It is not possible to control when the sun is shining or wind is blowing, and consequently there is

Fig. 2.1 The figure shows hourly total load for Germany during the years 2017 to 2019 in MWh. The inherent yearly and weekly seasonality is clearly visible. *Source* entsoe [20]

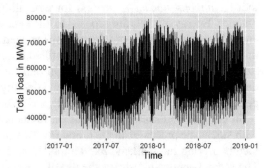

Fig. 2.2 Daily temperature for Germany in °C. The yearly seasonality is clearly visible in the graph. *Source* Deutscher Wetterdienst [17]

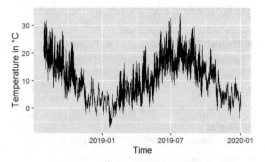

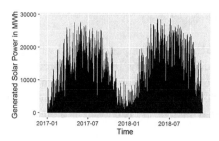
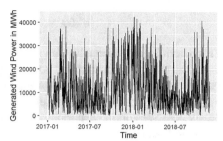

Fig. 2.3 Left: Electricity generated in Germany by solar energy in MWh over the years 2017 to 2019. A strong yearly seasonality of its strength is visible. Right: Electricity generated in Germany by wind energy in MWh, again over the years 2017 to 2019. Seasonality is also present, but much less distinct than for solar. *Source* EEX Transparency [1]

few control on the amount of energy produced by them. Therefore, it is necessary to imitate those inherent patterns when modelling these sources. Electricity infeed from wind and solar power for the years 2017 to 2019 is shown in Fig. 2.3.

In the following subsections, we take a closer look at existing stochastic models for these uncertainty sources. We describe these stochastic models from the literature beginning with temperature and following closely with wind and solar modelling. This section is then continued with an overview on consumption models. Furthermore, each of these subsection is closed by a short overview over possible open data sources containing the data needed for a calibration of the mentioned models. Finally, a short modelling routine with data example is given which describes how to get from a chosen stochastic process to a finished model fitting the underlying data.

2.2.1 Temperature

Temperature has been playing an important role on financial markets at least since 1997, when the first weather option was traded [14]. Temperature observations show, among other patterns, yearly as well as daily seasonalities regarding mean and variance distribution. In this subsection, we discuss and compare the models found in [3, 8, 10, 34]. They all use different stochastic approaches for temperature modelling, which try to capture the inherent temperature dynamics. Furthermore, a collection of possible data sources for model calibration is given.

In the context of adequately pricing weather derivatives, [3] were among the first to model temperature. They observed the afore-mentioned strong yearly seasonal behaviour of temperature as well as a positive trend in mean temperature. A possible reason for the trend, which appears stronger than global warming, is concluded to be the urban heating effect. Based on their observations, the following underlying deterministic model for mean temperature $\eta^T(t)$ at time t is implemented:

$$\eta^T(t) = a + bt + c\sin(\omega t + \varphi). \tag{2.1}$$

This deterministic seasonality function includes a mean temperature level with drift $a + bt$ as well as a sinusoid part covering yearly seasonality. In order to incorporate stochastic deviations from $\eta^T(t)$, an analysis of the variation around the deterministic mean is made. Based on the quadratic variation $(\sigma^T(t))^2 \in \mathbb{R}_+$ of temperature being nearly constant within each month in the used data, $\sigma^T(t)$ is chosen as a piecewise constant function

$$\sigma^T(t) = \begin{cases} \sigma_1, & \text{if } t \text{ is a day in January,} \\ \sigma_2, & \text{if } t \text{ is a day in February,} \\ \vdots & \vdots \\ \sigma_{12}, & \text{if } t \text{ is a day in December,} \end{cases} \tag{2.2}$$

with $\{\sigma_i\}_{i=1}^{12} > 0$. In combination with a Brownian motion (BM) as stochastic driver and the mean-reverting property of the random deviations, the model results in an Ornstein-Uhlenbeck process (OU process) with time-dependent mean as stochastic component of the model:

$$dT(t) = \theta^T(\eta^T(t) - T(t))dt + \sigma^T(t)dB(t). \tag{2.3}$$

For a more detailed description of the OU process the reader is referred to Sect. 2.2.5. Alaton et al. [3] state that modelling (2.3) with a mean reversion speed of θ^T does not necessarily lead to a temperature process with a long run mean equal to $\eta^T(t)$, see [15]. For compensation purposes, the term

$$\frac{d\eta^T(t)}{dt} = b + \omega c\cos(\omega t + \varphi) \tag{2.4}$$

is added. This leads to the following model for temperature $T(t)$:

$$dT(t) = \left(\frac{d\eta^T(t)}{dt} + \theta^T(\eta^T(t) - T(t))\right)dt + \sigma^T(t)dB(t), \quad T(0) = x_0, \tag{2.5}$$

with solution

$$T(t) = (x_0 - \eta^T(0))e^{-\theta^T t} + \eta^T(t) + \int_0^t e^{-\theta^T(t-s)}\sigma^T(s)\,dB(s). \tag{2.6}$$

Alaton et al. [3] then validate their model with data from Sweden and use it as foundation for a weather derivatives pricing model.

A similar approach is employed in [34], the only difference lying in the definition of $\sigma^T(t)$. Here, the volatility is implemented as a stochastic process itself. As the volatility is supposed to display a mean-reverting property as well, the authors decide to again use an OU process for its representation. Discretizing leads them to a model where $\sigma^T(t)$ follows an AR(1) process, but is only simulated once a month. The method is tested with data from Casablanca and performs to the authors' content.

Brody et al. [10] concentrate on covering the long-range temporal correlations present in temperature. Detrended and deseasonalized temperature data from England supports the presence of such long-range dependencies. In order to model them adequately, [10] decide for an OU process driven by a fractional BM:

$$dT(t) = \theta^T(t)(\eta^T(t) - T(t))dt + \sigma^T(t)dB^H(t), \qquad T_0 = x. \qquad (2.7)$$

The fractional BM $B^H(t)$ is characterized by its Hurst exponent $H \in (0, 1)$. This Hurst parameter determines whether the increments of the BM are independent $(H = 0.5)$ or show positive or negative correlation. For a definition of it and the stochastic calculus behind, the interested reader is referred to [10] or [32]. The functions $\theta^T(t)$, $\eta^T(t)$ and $\sigma^T(t)$ are assumed to be bounded, and their choice of $\sigma^T(t)$ and $\eta^T(t)$ is

$$\sigma^T(t) = a_1 + b_1 \sin(\omega t + c_1),$$
$$\eta^T(t) = a_2 + b_2 \sin(\omega t + c_2). \qquad (2.8)$$

This method captures the deterministic seasonality and standard deviation from it on the test data, but tends to underestimate the degree of short-time persistence appearing in the raw data.

Benth and Šaltytė-Benth [8] propose a generalization by introducing a Lévy process as driver instead of a BM. Therefore, temperature is assumed to follow

$$dT(t) = d\eta^T(t) - \theta^T(\eta^T(t) - T(t))dt + \sigma^T(t)dL(t),$$

where $L(t)$ represents Lévy noise with marginal distributions in the class of generalized hyperbolic distributions. They further discuss the related discrete-time version of the model

$$T(t) - \eta^T(t) = -(1 + \theta^T)(\eta^T(t - 1) - T(t - 1)) + \sigma^T(t)\varepsilon_t, \qquad t = 1, 2, \ldots$$

and use it to analyse Norwegian temperature data. They state that a model using normally distributed residuals with constant variance underestimates the variations in the data heavily. Therefore, they combine a seasonally varying variance, based on a truncated Fourier series, with the Lévy distributed residuals. The residuals are modelled using a multiplicative time series model with $\tilde{\varepsilon}_t = \sigma^T(t)\varepsilon_t$ with $\sigma^T(t)$ being deterministic and ε_t having zero mean and variance of one. Furthermore, [8] contemplate over the fact that significant lags are present even for two days, which are not captured by their model. They then seize on the idea of [10] as discussed above, see (2.7). Different from them, though, [8] show for Norwegian temperature time series that the deseasonalized temperature data itself is fractioned instead of fractioning the

Table 2.1 Data sources for the calibration of temperature models

Data source	Website	Area	Resolution
Deutscher Wetterdienst (DWD)	[17]	Mainly Germany (616 stations)	Mixture from yearly to 10 min
European Climate Assessment & Dataset (ECA&D)	[18]	Weather stations over most of Europe (63 countries, 18958 stations)	Daily
National Centers for Environmental Information (NOAA)	[35]	US (224 stations) and world	Down to hourly for US, daily for world

residuals obtained from the regression of temperature changes. Correlation between measure stations is shortly discussed. Ultimately, inherent time lags could not be captured by the model, but it still adds to the understanding of inherent marginal behaviour.

In order to adapt these models to a specific country or city, data from the chosen region are necessary for calibration purposes. To support this, we specify possible data sources and their main temperature data content in the following Table 2.1.

2.2.2 Solar Infeed

Electricity production from solar radiation is based on the photovoltaic (PV) effect which converts energy from light, more specific from solar irradiance, to electricity. For that, photovoltaic cells installed on solar panels are used, that convert solar energy with a certain efficiency to electricity. Solar cell designs for the currently highest possible efficiency with values of around and above 50% are presented by [23].

A strong yearly seasonality is present in the production profile of solar power, see Fig. 2.3. Another dominant feature of solar infeed is its daily seasonality with no infeed at night and a peak at mid-day, see Fig. 2.4. The production profile of solar supports peak times in electricity demand.

Many models rely on the results of Numerical Weather Prediction for estimated values of global horizontal or direct normal irradiance, as these values have been found to influence solar power output the most. Prominent research in this area concentrates on addressing error reduction of these forecasts instead of focussing on modelling power output from solar radiation itself. The interest in the latter has been growing in recent years though, wherefore a short collection of these approaches is presented. We discuss the approaches by [7, 38, 43].

One of the more recent publications in the area of PV power production is the work by [7]. They analyse the German photovoltaic energy $S(t)$ produced in three of

Fig. 2.4 Electricity generated from solar energy in Germany in June 2018 (hourly data in MWh). The daily seasonality is obvious. *Source* EEX Transparency [1]

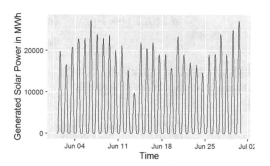

the four TSO areas. Their approach relies on first removing the underlying seasonal pattern $\eta^S(t)$, which is proposed to follow

$$\eta^S(t) := a + bt + c\ln(I(t)),\qquad(2.9)$$

with $I(t)$ representing sun intensity (solar radiation). After removing this deterministic pattern, the stochastic movements are represented by an autoregressive process (AR), which is the discretized version of an OU process in case of AR(1). More precisely, $S(t)$ is noted down as

$$S(t) = \beta_1 S(t-1) + \beta_2 S(t-2) + \beta_3 S(t-3) + \varepsilon_t,\qquad(2.10)$$

where ε_t are the residuals modelled by a skewed normal distribution. Finally, the model is calibrated with the aforementioned data and found to mirror their behaviour quite well.

Accounting for the decentralisation of the German energy system and therefore also accounting for spatial relations, [38] propose a stochastic solar irradiance model which incorporates the correlation between different power generating areas. They rely on the usage of a Gaussian copula C to model the random deviations of the deterministic seasonality of solar irradiance at n places with an appropriate correlation structure, i.e.

$$C(\varepsilon_1, \varepsilon_2, \ldots, \varepsilon_n) = \Phi_n(\Phi^{-1}(\varepsilon_1), \Phi^{-1}(\varepsilon_2), \ldots, \Phi^{-1}(\varepsilon_n)).\qquad(2.11)$$

Here, ε_i, $i = 1, \ldots, n$, are random deviations and Φ and Φ_n denote the n-dimensional multivariate Gaussian distribution function. This is combined with the deterministic seasonality, and finally, the full solar irradiance is translated to solar power at each area. Based on their results, they conclude that not integrating correlation can heavily underestimate the total grid load induced through solar power.

The model used in [43] is also based on stochastic processes, more specifically on OU processes. It intends to model not actual power output, but the realised percentage of installed photovoltaic capacity. This model for the solar load factor first specifies a day counting function with t representing time on a yearly basis:

$$d : [0, T] \mapsto \mathbb{N}_0, \qquad (d_t = n) \equiv (t \text{ is on n-th day}), \tag{2.12}$$

$$d_0 = 0. \tag{2.13}$$

With this, the daily maximum of solar infeed is modelled. $d(t)$ enumerates the days and can therefore serve as input for

$$\tilde{M}^s(d(t)) = \text{logit} \left(\max_{(t:d(t)=i)} (E^s(t)) \right), \quad i = 0, 1, \ldots, d_T, \tag{2.14}$$

where $E^s(t)$ defines the used percentage of installed solar capacity at time t:

$$E^s(t) := \frac{AI^S(t)}{IC^S(t)}, \tag{2.15}$$

with $AI^S(t)$ representing absolute infeed through solar at time t and $IC^S(t)$ installed capacity for solar, respectively. The logit-function is a transformation $(0, 1) \rightarrow \mathbb{R}$ defined as

$$\text{logit}(x) = \ln \frac{x}{1-x}. \tag{2.16}$$

Installed capacity itself is assumed to be constant in [43]. By modelling $\tilde{M}^s(d(t))$ adequately, it is possible to model $E^s(t)$, as the inverse of the logit -function exists. Therefore, the maximum process is first split in its deterministic and stochastic parts:

$$\tilde{M}^s(d(t)) = \eta^s(d(t)) + \bar{M}^s(d(t)), \tag{2.17}$$

where $\eta^s(d(t))$ represents a deterministic seasonal and trend component. Then, $\bar{M}^s(d(t))$ is modelled as an OU-process with zero mean. In order to transform the such gained daily maximum to a daily infeed pattern, a function mapping the maximum and introducing an appropriate decline around it needs to be introduced. Possible candidates for this function δ_i are step functions or a Gaussian function:

$$\delta_i : \{t : d(t) = i\} \times (0, 1) \mapsto [0, 1]. \tag{2.18}$$

Concluding, the solar load factor $E^s(t)$ is modelled by

$$E^s(t) = \delta_{d(t)}(t, \text{logit}^{-1}(\eta^s(d(t)) + \bar{M}^s(d(t)))). \tag{2.19}$$

This model is then used in combination with other models to mirror residual demand dynamics and to deduce electricity prices from that. It is shown that the model suits the German electricity market well.

Again, we present possible data sources for the calibration of the aforementioned models. As for solar modelling, we distinguished between modelling based on solar

Table 2.2 Data sources for the calibration of solar energy models

Data source	Website	Area	Resolution	Data kind
Deutscher Wetterdienst (DWD)	[17]	Mainly Germany (satellite)	Mixture from yearly to 10 min	Solar radiation & cloud coverage
European Climate Assessment & Dataset (ECA&D)	[18]	Weather stations over most of Europe (63 countries, 18958 stations)	Daily	Solar radiation & cloud coverage
National Centers for Environmental Information (NOAA)	[35]	US (224 stations) and world	Down to hourly for US, daily for world	Solar radiation & cloud coverage
entsoe Transparency Platform	[20]	Europe	Quarter hourly	Solar power generation
SMARD	[40]	Germany, Austria, Luxembourg	Quarter hourly	Solar power generation

irradiation data and cloud coverage and modelling based on actual solar generation. Therefore, this distinction is also made in the following Table 2.2.

2.2.3 Wind Infeed

Different from solar energy, wind energy relies on the transition of mechanical energy, i.e. a rotor moving, to electricity. Nonetheless, wind infeed shows a yearly seasonality like solar infeed, even though it is notably narrowed in comparison, see Fig. 2.3. An adequate wind power model should therefore consider this seasonal behaviour. Another prominent feature of wind time series is the existence of autocorrelation, which is also often taken into account.

Models for wind power infeed are manifold and rely on a diverse mathematical background. There are two general approaches to modelling wind infeed: Either by modelling wind speeds and directions and then transforming this through a power curve into power output, or by modelling wind power directly. Both approaches have their advantages and disadvantages and are therefore discussed in this subsection. Some of the modelling methods will be recognized by the attentive reader from the preceding modelling approaches. We discuss the models of [5, 36, 43].

A first approach to wind power modelling for a single wind turbine is introduced by [36] and is based on the Markov Chain Monte Carlo (MCMC) technique. In order to avoid having to model wind speeds themselves, they concentrate on modelling power output directly. Calibration is accomplished by either using wind power data from the outset or by transforming wind speed to wind power data. To achieve the latter, a wind turbine power curve P as a function of wind speed v is introduced, returning wind power for a given speed:

$$P(v) = \begin{cases} 0, & v \leq v_{ci} \text{ or } v > v_{co} \text{ (area A)}, \\ P_N, & v_N \leq v \leq v_{co} \text{ (area B)}, \\ f(v), & v_{ci} \leq v < v_N \text{ (area C)}, \end{cases} \qquad (2.20)$$

where $v_{ci}, v_{co}, v_N > 0$. $P(v)$ is based on the functionality of turbines. The areas in (2.20) are based on the following reasoning: A turbine has a cut-in wind speed v_{ci}, such that it produces energy for wind speeds $v > v_{ci}$—below v_{ci}, it is not profitable to run the turbine. Furthermore, every turbine model has a nominal wind speed v_N, from which onwards the turbine has reached its optimal output and due to mechanical reasons doesn't produce more energy even at higher wind speeds. Finally, a cut-off wind speed v_{co} exists: In case of the wind speed growing over this threshold, the wind turbine has to stop producing in order to spare its material.

By splitting up the reachable power range into these three areas (2.20), a mapping between wind speed and power is only needed in area C for wind speed data transformation. Based on this representation, the different states of the Markov Chain are defined with equal spacing between the discrete states $P(v) \equiv 0$ and $P(v) \equiv P_N$. The Markov Chain of a discrete physical process X can be defined as first or higher order chain with transition probabilities

$$\mathbb{P}(X(t) = j \mid X(t-1) = i) = p_{ij}, \qquad (2.21)$$

$$\mathbb{P}(X(t) = j \mid X(t-1) = i_1, \ldots, X(t-n) = i_n) = p_{i_n, i_n-1, \ldots i_1, j} \qquad (2.22)$$

respectively. Dependence on the last n observations then can be covered by an nth-order Markov Chain. These probabilities can be estimated using maximum likelihood estimation (MLE). Nonetheless, it has to be considered that in the latter case, many of the possible transitions may not have appeared at all in the underlying data and therefore have no proper estimate in the MLE method. The mentioned method can be used to simulate wind speeds as well as wind power, but the latter's sample space is much smaller as very high and very low wind speeds are summarized in the states 0 and P_N. Results in [36] show that using power values produces better fits for the probability density function as well as for the auto-correlation function than first simulating wind speeds and translating afterwards. Furthermore, second or higher order models perform slightly better than first order Markov chain models.

Another paper addressing this approach for a single turbine is [5], who embed the model of [36] in the evaluation of generating capacity assessment adequacy. A difference is the usage of several transition matrices in order to satisfy the changing behaviour of wind production throughout the year. Therefore, 12 matrices, i.e. one for each month, are generated and used. Furthermore, a slightly modified power curve is used to transform wind speed to wind power, i.e.

$$P(v) = \begin{cases} 0, & v \leq v_{ci} \text{ or } v > v_{co} \text{ (area A)}, \\ P_N, & v_N \leq v \leq v_{co} \text{ (area C)}, \\ (a_1(v) + a_2(v)v + a_3(v)v^2)P_N, & v_{ci} \leq v < v_N \text{ (area B)}, \end{cases} \quad (2.23)$$

with $a_i, i = 1, 2, 3$, specified in [5].

Modelling generated wind power for the whole of Germany, [43] chooses a different approach, namely modelling wind power based on OU processes. He models a wind power load factor $E^W(t)$ instead of direct power output. The model is designed to mirror infeed in a whole market, so a collection of many wind turbines in different areas. It is not suited to model a single wind turbine or wind park.

$$E^W(t) := \frac{AI^W(t)}{IC^W(t)}, \quad (2.24)$$

where $AI^W(t)$ denotes absolute infeed at time t and $IC^W(t)$ total installed capacity. It is assumed that $IC^W(t)$ is a deterministic, known function of time. This assumption mirrors reality quite well, as the capacities installed as well as to be installed are known for some time in advance. For further periods, scenarios exist that give a picture of future installations. Stochastic fluctuations are therefore only contained in $AI^W(t)$. The assumption of a normally distributed time series holds approximately true for $E^W(t)$ after a logit transformation. Subsequently, the data is deseasonalized, which results in a weakly stationary stochastic process for the residuals $\bar{E}^W(t)$. The complete model is presented here:

$$\text{logit}(E^W(t)) =: \tilde{E}^W(t) = \eta^W(t) + \bar{E}^W(t), \quad \forall t \in [0, T], \quad (2.25)$$

$$\eta^W(t) = \eta_{t-1}^W = a\cos(2\pi t + b) + c, \quad \forall t \geq 1, \quad (2.26)$$

$$d\bar{E}^W(t) = -\theta^W \bar{E}^W(t)dt + \sigma^W dB(t), \quad \bar{E}_0^W = e_0. \quad (2.27)$$

The transformed load factor is divided in a seasonal component $\eta^W(t)$ and a diffusion component $\bar{E}^W(t)$, where the latter is following an OU process. In combination with (2.19), this model is used to model stochastic renewable infeed in Germany.

This subsection also ends with a short collection of possible data sources for calibration purposes. Here, the data kind distinguishes between collections of wind direction, wind speed and wind power generation. These different sorts of data represent what the mentioned models use as basis. The data sources are found in Table 2.3.

2.2.4 Total Energy Demand

Predicting total demand is a very relevant topic for researchers, see e.g. [4]. The reason for that is a very intuitive economic one: An electricity grid cannot store electricity. Therefore, inflow and outflow (= assumed to be equal to demand) must be the same.

Table 2.3 Data sources for the calibration of wind energy models

Data source	Website	Area	Resolution	Data kind
Deutscher Wetterdienst (DWD)	[17]	Mainly Germany (satellite)	Mixture from yearly to 10 min	Wind speed & direction
European Climate Assessment & Dataset (ECA&D)	[18]	Weather stations over most of Europe (63 countries, 18958 stations)	Daily	Wind speed & direction
National Centers for Environmental Information (NOAA)	[35]	US (224 stations) and world	Down to hourly for US, daily for world	Wind speed & direction
entsoe Transparency Platform	[20]	Europe	Quarter hourly	Wind power generation & installed capacity
SMARD	[40]	Germany, Austria, Luxembourg	Quarter hourly	Wind power generation

Otherwise, e.g. in case of higher inflow than outflow, too much electricity is in the grid and is converted to heat, which eventually destroys the wires. As a consequence, it is of high interest to match inflow and outflow, i.e. generation of and demand for electricity.

We present different approaches to modelling demand. Energy demand for electricity or gas, as intuitively follows from its dependence on temperature, shows a seasonal pattern, which must be incorporated in an appropriate model.

A first, quite simplified modelling approach using stochastic processes is presented in [6]. He denotes demand, as seen before, as an OU process

$$dD(t) := \kappa^D(\mu^D - D(t))dt + \sigma^D dB(t) \tag{2.28}$$

with $\kappa^D, \sigma^D > 0$ and $\mu^D \in \mathbb{R}$ constants. The simplification in this approach is the implicit assumption of no dependence of the demand on the price. Nonetheless, this does picture reality to a certain degree, as most energy consumers do not orientate their consumption on energy prices.

The model specified in [43] for demand is close to the model of [6]. It also assumes a stochastic OU driver

$$dD(t) := -\theta^D D(t)dt + \sigma^D dB(t) \tag{2.29}$$

for daily deviations of the mean. Seasonality is captured by a deterministic function, more specifically

$$\tilde{\eta}^D(t) := a\cos(2\pi t + b) + c \tag{2.30}$$

defining the demand on business days, and a proportion of it is assumed to appear on partial and normal holidays or non-business days:

$$\eta^D(t) := (1 - w_1 \mathbb{1}_{\{t \in B_1\}} - w_2 \mathbb{1}_{\{t \in B_2\}})\tilde{\eta}^D(t), \tag{2.31}$$

where B_1 defines the set of partial holidays and B_2 the set of full holidays and non-business days. This model therefore doesn't assume a different deterministic demand structure for days in these sets, but instead a reduced demand equal in its form to regular business days.

The seasonality $\eta^D(t)$ models daily total demand. Therefore, a function similar to (2.18) is used to model an intra-day load curve

$$L(t) := \delta_{d_t}^L(t, \eta^D(t) + D(t)). \tag{2.32}$$

Here, δ is a step-function

$$\delta_{d_t}(t, x) := x \sum_{k=1}^{24} c_k(d_t) \mathbb{1}_{\{t \in (k\text{-th hour of the day})\}} \tag{2.33}$$

with $c_k(d_t) > 0$ constant for each month.

Almeshaiei and Soltan [4] make a general guide to load prediction available to researchers. They introduce an algorithm showing how to proceed with analysing, modelling and predicting electric load for individual sites. The approach is based on a decomposition of the available time series in different factors, i.e.

$$D(t) = D_n(t) + D_w(t) + D_s(t) + D_r(t) \quad \text{or}$$
$$D(t) = D_n(t) \, F_w(t) \, F_s(t) \, F_r(t).$$

Here, $D_n(t)$ represents the trend component, $D_w(t)$ a weather dependence, $D_s(t)$ a special event component depicting strong deviations from the usual trend and $D_r(t)$ a noise term. The F's are denoting positive correction factors with the same background, respectively. A load forecast for a region is then proposed to be based on the factors of region similarity, contour and proposed related points. An analysis of the method is conducted with data from Kuwait and its results show the practical applicability of the method.

Data sources for energy demand are listed in Table 2.4. In case of demand, data are not as openly accessible as for the other afore-mentioned uncertainty factors.

Table 2.4 Data sources for the calibration of energy demand models

Data source	Website	Area	Resolution	Data kind
entsoe Transparency Platform	[17]	Europe (countries)	Quarter hourly	Vertical grid load
SMARD	[40]	Germany, Austria, Luxembourg	Quarter hourly	Consumption
eurostat	[22]	Europe (countries)	Yearly to monthly	Consumption
U.S. Energy Information Administration (EIA)	[19]	U.S.	Monthly	Consumption

2.2.5 Modelling Routine

As specified beforehand, the models in this section are based on stochastic processes. This subsection is meant to guide the reader through how to get from a chosen stochastic process to its discretization, calibration and model evaluation in order to get a feeling for the modelling peculiarities. This procedure is explained using the example of the **Ornstein-Uhlenbeck process** (OU process) with Brownian motion (BM) as stochastic driver, which is often applied in applications in energy markets. The OU process with BM is defined as the solution of the following stochastic differential equation (SDE)

$$dX_t = \theta(\mu - X_t)dt + \tilde{\sigma}dW_t,$$
$$X_0 = x_0, \tag{2.34}$$

where $\mu \in \mathbb{R}, \theta > 0$ and $\tilde{\sigma} > 0$ are constants, the stochastic driver $\{W_t\}_t$, $t \in [0, T]$, is a BM and x_0 is the initial value. Here, μ represents the mean of the process, θ the speed at which the process returns to this mean after a deviation and $\tilde{\sigma}$ the strength of the diffusion.

The OU process is relevant due to its mean reverting property: the term $\theta(\mu - X_t)dt$ pushes X_t upwards if it is lower than its mean μ, and pushes X_t downwards if it is bigger. Furthermore, the process admits a stationary probability distribution.

In order to simulate the process, the time interval $[0, T]$ is divided in N sub-intervals, which all have a length of $\delta t = T/N$. For BM, it holds true that

$$W_{t_n} - W_{t_{n-1}} = \tilde{\varepsilon}_n, \quad \tilde{\varepsilon}_n \sim \mathcal{N}(0, \delta) \quad \forall n \in \{1, \dots, N\}. \tag{2.35}$$

Therefore, it is possible to simulate exact BM paths at selected points only. In order to discretize the continuous stochastic process, different methods can be applied. An overview of discretization methods is e.g. found in [28]. The basic scheme to solve SDEs numerically is called Euler-Maruyama scheme. Applying it to (2.34), we find

$$X_{t_n} = \theta(\mu - X_{t_{n-1}})\delta + \tilde{\sigma}\tilde{\varepsilon}_n \tag{2.36}$$

with δ and $\tilde{\varepsilon}_n$ as above. This is equivalent to an **autoregressive process of order 1** (AR[1] process), which is in general defined as follows:

$$Y_{t_n} = c + a_1 Y_{t_{n-1}} + \varepsilon_n. \tag{2.37}$$

$c, a_1 \in \mathbb{R}$ and ε_t is a noise term with $\mathbb{E}(\varepsilon_t) = 0\,\forall t$ and $\mathrm{Var}(\varepsilon_t) = \sigma^2 < \infty$. With $\sigma^2 = \tilde{\sigma}^2\delta$ the discretized OU-process in (2.36) is equivalent to the AR[1] time-series model.[1]

Now, with an appropriate discretization scheme at hand, the next step is to calibrate parameters of the process for a set of observations $O = \{(t_0, x_0), \ldots, (t_N, x_N)\}$. We assume equidistant time increments $t_n - t_{n-1} = \delta\,\forall n \in \{1, \ldots, N\}$. Prominent calibration procedures are maximum likelihood estimation (MLE) or least squares regression (LSR). For maximum likelihood estimation we use the conditional distribution function of the continuous process X_t and maximize its log-likelihood based on the observation set O. Therefore, the set of calibrated values is chosen such that it has the highest probability for the observed set O. LSR on the other hand is based on the discretized version of the process and makes use of the linear relationship between $Y_{t_{n+1}}$ and Y_{t_n}. The goal of LSR is to minimize squared error distances between fit and real observations.

For observations $(t_0, y_0), \ldots, (t_N, y_N)$, we follow [41] and define $S_y = \sum_{i=1}^{N} y_i$, $S_x = \sum_{i=1}^{N} y_{i-1}, S_{yy} = \sum_{i=1}^{N} y_i^2, S_{xx} = \sum_{i=1}^{N} y_{i-1}^2$ and $S_{xy} = \sum_{i=1}^{N} y_i y_{i-1}$. The optimal parameters for both cases are given in the following.

- MLE: Optimizing the log-likelihood leads us to

$$\theta = -\frac{1}{\delta}\ln\left(\frac{S_{xy} - \mu(S_x + S_y) + N\mu^2}{S_{xx} - 2\mu S_x + N\mu^2}\right),$$

$$\mu = \frac{S_y S_{xx} - S_x S_{xy}}{N(S_{xx} - S_{xy}) - (S_x^2 - S_x S_y)},$$

$$\tilde{\sigma}^2 = \frac{2\theta}{N(1 - e^{-2\theta\delta})}(S_{yy} - 2e^{-\theta\delta}S_{xy} + e^{-2\theta\delta}S_{xx} + N\mu^2(1 - e^{-\theta\delta})^2 - 2\theta(1 - e^{-\theta\delta})(S_y - e^{-\theta\delta}S_x)).$$

- LRS: The least squares fit leads us to

---

[1] $c := \theta\mu\delta$ and $a_1 := -\theta\delta$.

$$c = \frac{S_y - a_1 S_x}{N},$$

$$a_1 = \frac{N S_{xy} - S_x S_y}{N S_{xx} - S_x^2},$$

$$\sigma^2 = \frac{N S_{yy} - S_y^2 - a_1(N S_{xy} - S_x S_y)}{N(N - 2)}.$$

From these values, the original values of the discretized process are derived through

$$\theta = \frac{1 - a_1}{\delta},$$

$$\mu = \frac{c}{\theta \delta},$$

$$\tilde{\sigma} = \frac{\sigma}{\sqrt{\delta}}.$$

For testing reasons, it is often sensible to split the observation set in a training set $\{(t_0, y_0), \ldots, (t_{n-1}, y_{n-1})\}$ and a test set $\{(t_n, y_n), \ldots, (t_N, y_N)\}$ with $0 < n < N$. The calibration procedure is then only performed on the observations from the training set. The test set is used for model validation. A popular approach is to compare simulated and realized paths and quantiles of the real time series on the test set with quantiles of the calibrated model using e.g. qqplots. Comparing the load duration curves of simulated paths with those from the test set also helps to assess realistic behaviour of the simulations.

Using data on the entsoe Transparency Platform for solar infeed in Germany, we practically apply the mentioned routine to the solar power model from [43], which is also discussed in Sect. 2.2.2. The data considered contains 15-minute intervals from 2017 through 2020. For analysis, they are aggregated to hourly values. Therefore, the observation set consists of 26, 280 observations. All observations are plotted in Fig. 2.5.

In [43] and as described in Sect. 2.2.2, the data which is modelled as an OU process represents daily maxima of solar infeed, which reduces the cardinality of the modelled observation set to 1, 095. These maxima are scaled by the total installed capacity in Germany, with the resulting percentage being called efficiency (often also called *load factor*). The efficiency necessarily is in $[0, 1]$. Assuming the probability of the interval boundaries to be 0, we can assume the efficiency to be in $(0, 1)$. This allows the usage of a logit transformation for variance stabilisation. The result of these transformations is visible in the left figure in Fig. 2.5, and these observations are denoted by $O_{1095} = \{(t_0, y_0), \ldots, (t_{1095}, y_{1095})\}$. Obviously, this transformed time series neither admits the assumption of a mean 0 nor the assumption of stationarity. Therefore, a deseasonalization is performed to remove any deterministic seasonality before modelling this data with an OU process. This separation of deterministic and stochastic contributions to the time series is expressed mathematically in equation (2.17). The chosen deterministic seasonality $\eta^s(t)$ is of the form

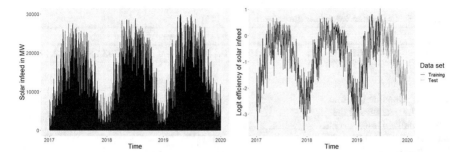

Fig. 2.5 Left: Hourly solar infeed data in Germany in MWh for the years 2017 to 2020. Right: This data is transformed by first taking only the maxima of each day, dividing them through the installed solar capacity in Germany during the mentioned years and then applying a logit transform. Furthermore, they are split into training set $\{(t_0, y_0), \ldots, (t_{900}, y_{900})\}$ and test set $\{(t_{901}, y_{901}), \ldots, (t_{1095}, y_{1095})\}$

Fig. 2.6 Deseasonalized logit transformed efficiency maxima of solar infeed in Germany

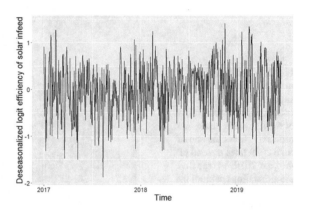

$$\eta^s(t) = b_0 + b_1 \sin\left(\frac{2\pi t}{365}\right) + b_2 \cos\left(\frac{2\pi t}{365}\right) + b_3 \sin\left(\frac{4\pi t}{365}\right) + b_4 \cos\left(\frac{4\pi t}{365}\right),$$
$$(2.38)$$

where t represents a day-count for the whole year. This formulation makes use of truncated Fourier series to capture yearly and half yearly seasonalities. We first calibrate the values $b_0, \ldots, b_4$ with normal least squares regression based on the values from the training set, i.e. O_{900}. This deterministic seasonality is calculated for all $t_0, \ldots, t_{900}$ and deducted from the corresponding $y_0, \ldots, y_{900}$, leading to $\tilde{O}_{900} = \{(t_0, \tilde{y}_0), \ldots, (t_{900}, \tilde{y}_{900})\}$. $\tilde{O}_{900}$ is visible in Fig. 2.6. These residuals are then assumed to follow an OU process defined as in (2.34), and are calibrated with MLE. The results of the calibration are presented in Table 2.5.

In order to assess the model, a quantile-quantile plot is used. It compares the theoretical quantiles with the empirical quantiles on the deseasonalized test set. The result is given in Fig. 2.7, and it is noticeable that the quantiles match well. Therefore, the chosen model should not be rejected based on the quantile analysis.

Table 2.5 Calibrated values for the solar model in [43]

	b_0	b_1	b_2	b_3	b_4	θ	$\tilde{\sigma}$
Solar	−0.8461	0.1971	−1.1192	0.1168	−0.3781	243.2590	12.1721

Fig. 2.7 Quantile-quantile plot of the quantiles of the deseasonalized test set versus the theoretical quantiles of the calibrated OU process

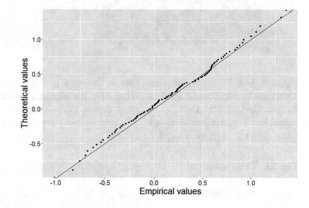

Fig. 2.8 Ten daily maximum solar efficiency simulations over time in colour, with the true values of the test set in black. The time axis shows the end of June 2019 until the end of the year 2019. We see that the simulations capture the pattern of the observed maxima quite nicely

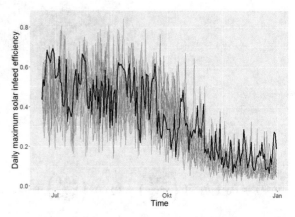

Furthermore simulated paths are compared to realized paths. Example paths, which are compared to the test set, are shown in Fig. 2.8. As neither Fig. 2.7 nor Fig. 2.8 show a mismatch of chosen model and test set, the model is deemed well fit to model the underlying data.

2.3 Application II: Risk Management on the Electricity Market

Following the worldwide deregulation of energy markets in the 1990s, electricity prices moved from being a governmentally regulated product to being agreed upon on the basis of supply and demand. As one of the natural consequences, the uncertainty regarding electricity prices increased. With that, the need to quantify electricity price uncertainty and to mathematically model electricity prices arose. This topic is especially relevant for risk managers of companies engaging in buying or selling big amounts of electricity. Therefore, this application is concerned with modelling electricity prices in a risk management context, again on the basis of continuous stochastic processes.

For electricity price models, the earliest works date back to the late 1990s [13, 16, 37, 39]. The first methods focussed on incorporating prominent patterns seen in real electricity price data into the models. These patterns are called stylized facts hereinafter. Examples for such stylized facts are short-living and extreme spikes in the prices as well as mean-reverting behaviour. The class of models which are based on stochastic processes is today called the class of **factor models**, see e.g. [9] and the references therein. Still based on stochastic processes, but reaching out with a different approach are **structural models**. Both classes are explained in more detail in the first subsection in combination with a short analysis of shortcomings and benefits of each, whereas the second subsection accompanies this by elaborating on the usage of one specific model in an industrial application.

Electricity is usually traded on three different markets: the futures market, the day-ahead market and the intraday market. They can be distinguished by the time between trading and the actual delivery of the electricity. On the futures market, electricity can be traded up to six years ahead. A **future** is a contract specifying a delivery period (e.g. 1/1/2021 to 31/12/2021), a specified power (e.g. 5 MW) and a price per MWh at which the electricity will be delivered (e.g. 35 EUR/MWh). These contracts are traded on the exchange. Closer to the delivery date, the day-ahead market comes into play: As the name already suggests, electricity with a delivery at the next day is traded. In most markets, all 24 h of the next day are traded separately. At the European markets, the day-ahead market is in a single auction at 12:00 am the day before. The resulting prices from the auction are called **spot prices**. Finally, closest to the delivery time, the intraday market opens and offers continuous trading of hours or even quarters up to short before actual delivery. Results from the day-ahead market are considered the reference price. In recent years, also due to the increasing share of renewables, the intraday market is gaining popularity.

As all traded futures have delivery periods (e.g. years, months, weeks), they do not provide information on the price for electricity in e.g. a single hour. However, for many applications, future price expectations are needed with a finer granularity like days or hours. Therefore, market participants use the traded futures contracts and break them down on hourly prices using historical hourly prices from the day-ahead market. The resulting price curve is called **(hourly) price forward curve (PFC)**.

In the section on price modelling in the following, we always model the spot price, i.e. prices on the day-ahead market.

2.3.1 Price Models

The class of factor models is based on a direct modelling of the day-ahead spot price or its logarithm. It assumes that all information necessary to characterize typical price movements is included in the price data itself and does not need any additional information. The modelling is based on the usage of a sum of several stochastic processes. Those processes are also called factors and are usually modelled by OU processes, see Eq. (2.34). These OU processes are allowed to have different types of driving processes, e.g. standard BMs or compound Poisson processes. More elaborated stochastic drivers such as Lévy processes are possible, but are not commonly used by practitioners.

Using the OU processes, we model the day-ahead spot price $S(t)$ by an **arithmetic** n-factor model through

$$S(t) := \Lambda(t) + \sum_{i=1}^{n} X_i(t)$$

or by a **geometric** n-factor model through

$$\log S(t) := \Lambda(t) + \sum_{i=1}^{n} X_i(t).$$

Arithmetic models can model prices on the whole real line, whereas geometric models are limited to $(0, \infty)$. In equity markets, geometric models obviously are in favour, as a stock price may not become negative. Electricity, however, in particular in markets with a high share of renewables, is regularly traded at negative prices. Therefore the e.g. German market is usually modelled using an arithmetic approach.

The seasonality function $\Lambda(t)$ captures yearly and weekly seasonalities of the day-ahead spot prices. For a quality comparison of different seasonality functions and factor models based on German day-ahead spot price data, we refer the interested reader to [25].

Even though factor models are capable of reproducing patterns observed in electricity price paths, it is intuitive to incorporate knowledge about other structures into the pricing model on which the price depends. Based on this idea, the work of [6] in the beginning of the 2000s sparked a new stochastic modelling approach: the structural models. This approach is based on the inelasticity of the electricity market, i.e. it relies on the fact that demand is almost completely independent from the market price, but the price is very much dependent on the demand. Therefore, one can equivalently model the demand process and then use a non-linear transformation to compute the day-ahead spot price from the modelled demand. Often, the demand process $D(t)$ is again modelled by a factor model of some kind. As demand follows steadier patterns than prices themselves, they can be replicated with less effort. Therefore, the used factor models often only have BMs as stochastic drivers. We then define the day-ahead spot price by

$$S(t) := \Lambda(t) + g(D(t)),$$

where g is the function mapping the supply-demand-curve or the merit-order-curve. This function tends to be similarly shaped to $f(x) = x^3$ due to the shape of the merit-order-curve.[2] $\Lambda(t)$ again denotes the deterministic seasonality function. Throughout the years, several extended modelling approaches of this basic idea have been proposed, cf. [2, 31, 43]. The methods presented in Sect. 2.2 are used as groundwork for pricing models in this category by enabling a more sophisticated demand simulation.

Both modelling approaches have their benefits and shortcomings. Factor models often allow an analytical computation of derivatives (futures, options) prices and follow the lines of classical financial models from equity markets or interest rates. However, to build factor models with realistic day-ahead spot price behaviour, i.e. to capture all stylized facts, one needs to use complex OU processes. The consequence is that the calibration procedure is often not obvious or might actually be impossible. In comparison, structural models are unique to the electricity market and often do not offer closed-form solutions, which makes an analytical derivatives computation impossible. Nonetheless, structural models are usually based on simpler stochastic drivers, which are easier to calibrate, but still are able to generate realistic prices through the usage of a non-linear transformation. Furthermore, the calibration procedure makes use of both demand and spot price data. This has the advantage of a bigger data pool for model calibration.

In practice, we find that almost all evaluations are based on Monte-Carlo simulations of the price model, so the availability of closed-form solutions is apparently not the most important issue. Therefore, we suggest to let the model choice depend on the availability of data. If reliable supply and demand data is available, we suggest to use structural models for electricity markets. In contrast, if there is only price information and some derivatives prices available, factor models are to be preferred.

Recent attempts to combine day-ahead spot price modelling with the other markets such as the intraday and futures market have been made in the form of **Heath-Jarrow-Morton (HJM) approaches**, see [11, 12, 27, 29, 30]. The work of [24] fits both factor models and structural models within the HJM framework.

---

[2]The merit-order curve describes the marginal cost of electricity production from different technologies like nuclear, coal, gas, etc.

2.3.2 Industrial Application

This subsection introduces a model which is used in a commercial energy risk management software for electricity, gas, and emissions, named Commodity Risk Manager (CRM). More information about the CRM is found on [42].

The price model has to fulfil the following requirements: explainable to professional risk managers (often non-mathematicians), applicable to different commodities and market areas, realistic short- and long-term price dynamics, and displaying consistency with a given (hourly) price forward curve. Moreover, the calibration on market prices must be reliable, fast, and without human interaction. The risk management software is using the model not only for electricity price modelling, but to quantify different types of risks as well. As they are not in the scope of this chapter, we refer the interested reader for more details on the different types of risk to [26].

To meet the criteria described above, we use a 3-factor model for the spot prices. It is capable of modelling the dynamics of electricity, gas, and the European emissions market EUA. Calibration is done individually for each commodity, whereas different market areas are distinguished by their price forward curve.

The 3-factor spot price model is defined as follows:

$$S(t) = X_1(t) \, \Lambda(t) + X_2(t) + X_3(t), \tag{2.39}$$

where Λ is a price forward curve incorporating the seasonality. This function accounts for the seasonality within the year such as the more expensive electricity prices in winter than in summer, and also replicates the effects of Christmas holidays[3] or weekends.[4] The seasonality function also accounts for normal holidays as well as partial holidays[5] and even the bridge days, i.e. the days between a holiday and a weekend, which appear to have a very similar electricity price pattern to partial holidays. In conclusion, this seasonality function exhibits fairly realistic behaviour and manages to replicate historical price movements very well. In practice we use an individual price forward curve for each commodity and market area for the seasonality function.

As described above, the model (2.39) has three stochastic factors, each of which is responsible for modelling a different aspect of the price dynamics:

- X_1 models the long-term price development, with

$$dX_1(t) = \sigma_1 X_1(t) \, dW_1(t), \quad X_1(0) = 1$$

being a geometric BM without drift,

---

[3]The Christmas holidays usually have lower prices than the rest of December or January.

[4]The weekend effect is the phenomenon that the prices on Saturdays and Sundays are in general lower than the prices during the week, which usually reach their maximum on Wednesday.

[5]The holidays that are holidays in some but not all German federal states.

Fig. 2.9 Simulated
day-ahead spot prices from
the 3-factor model

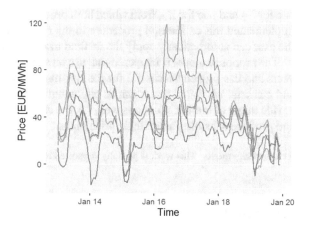

- X_2 models the regular price dynamics caused by supply and demand balancing, with

$$dX_2(t) = -\alpha_2 X_2(t)\,dt + \sigma_2\,dW_2(t), \quad X_2(0) = 0$$

being a Gaussian OU process and W_2 independent of W_1,
- and X_3 causes spikes in the price from occasional short-term imbalances, with

$$X_3(t) = \sum_{n=1}^{N(t)} D_n$$

being a jump OU process, where the driving process is modelled by a compound Poisson process with normally distributed jumps. Here, $D_1, D_2, \ldots$ is a sequence of i.i.d. $\mathcal{N}(\mu_3, \sigma_3^2)$-distributed random variables, and $N(t)$ is a homogenous Poisson process of rate $\lambda_3 > 0$.

In particular, this approach allows the jumps to be positive as well as negative. This is important, since in recent years e.g. the German electricity price exhibits (mainly negative) price spikes due to the increasing renewable energy infeed. Moreover, as the chosen model is a factor model, futures prices can be computed and expressed in the seasonality function $\Lambda(t)$ and the three factors. This leads to consistency between futures and spot prices, which is important for risk management on spot and futures positions. Spot price simulations of the model are shown in Fig. 2.9.

The hurdle which has to be overcome is the calibration of this model: Only few available literature discusses the calibration of more than two factors, or non-Gaussian factors for that matter. As a detailed description is not in the scope of this chapter, we only outline the chosen procedure: we based our method on the work of [33] and found a calibration procedure that separates our Gaussian OU process X_2 from the jump OU process X_3 based on a maximum likelihood estimation. Starting with the single (deseasonalized) price time series, this method generates two paths, one for X_2 and one for X_3, from which both processes can be estimated properly. We implemented this calibration procedure in the risk management software, such that the user can automatically apply the method to recalibrate her model parameters.

The proposed model is in operational use on a day-to-day basis for more than four years and has proven successful for the risk management of different commodities and markets. We found that it can be calibrated to different markets and can handle a wide range of market situations well, i.e. the calibrated parameters are meaningful and produce sensible and realistic price paths.

Acknowledgements This work is partially supported by BMBF grant 05M18AMC.

References

1. EEX Transparency Data. https://www.eex-transparency.com/power/. Accessed: 2020-01-08
2. Aïd, R., Campi, L., Huu, A.N., Touzi, N.: A structural risk-neutral model of electricity prices. Int. J. Theor. Appl. Financ. **12**(07), 925–947 (2009). https://doi.org/10.1142/s021902490900552x
3. Alaton, P., Djehiche, B., Stillberger, D.: On modelling and pricing weather derivatives. Appl. Math. Financ. **9**(1), 1–20 (2002). https://doi.org/10.1080/13504860210132897
4. Almeshaiei, E., Soltan, H.: A methodology for electric power load forecasting. Alex. Eng. J. **50**(2), 137–144 (2011). https://doi.org/10.1016/j.aej.2011.01.015
5. Almutairi, A., Ahmed, M.H., Salama, M.M.A.: Use of MCMC to incorporate a wind power model for the evaluation of generating capacity adequacy. Electr. Power Syst. Res. **133**, 63–70 (2016). https://doi.org/10.1016/j.epsr.2015.12.015
6. Barlow, M.T.: A diffusion model for electricity prices. Math. Financ. **12**(4), 287–298 (2002). https://doi.org/10.1111/j.1467-9965.2002.tb00125.x
7. Benth, F.E., Ibrahim, N.A.: Stochastic modeling of photovoltaic power generation and electricity prices. J. Energy Markets **10**(3), 1–33 (2017). https://doi.org/10.21314/JEM.2017.164
8. Benth, F.E., Šaltytė-Benth, J.: Stochastic modelling of temperature variations with a view towards weather derivatives. Appl. Math. Financ. **12**(1), 53–85 (2005). https://doi.org/10.1080/1350486042000271638
9. Benth, F.E., Šaltytė-Benth, J., Koekebakker, S.: Stochastic Modelling of Electricity and Related Markets. World Scientific (2008). https://doi.org/10.1142/6811
10. Brody, D.C., Syroka, J., Zervos, M.: Dynamical pricing of weather derivatives. Quant. Financ. **2**(3), 189–198 (2002). https://doi.org/10.1088/1469-7688/2/3/302
11. Broszkiewicz-Suwaj, E., Weron, A.: Calibration of the multi-factor HJM model for energy market. Acta Physica Polonica **37**(5), 1455–1466 (2006). http://www.actaphys.uj.edu.pl/fulltext?series=Reg&vol=37&page=1455
12. Caldana, R., Fusai, G., Roncoroni, A.: Electricity forward curves with thin granularity: theory and empirical evidence in the hourly EPEXspot market. Eur. J. Oper. Res. **261**(2), 715–734 (2017). https://doi.org/10.1016/j.ejor.2017.02.016
13. Clewlow, L., Strickland, C.: Valuing energy options in a one factor model fitted to forward prices. SSRN Electron. J. (1999). https://doi.org/10.2139/ssrn.160608
14. Considine, G.: Introduction to Weather Derivatives (2000). https://www.cmegroup.com/trading/weather/files/WEA_intro_to_weather_der.pdf
15. Davis, M.H.A.: Pricing weather derivatives by marginal value. Quant. Financ. **1**(3), 305–308 (2001). https://doi.org/10.1080/713665730
16. Deng, S.-J.: Stochastic models of energy commodity prices and their applications: mean-reversion with jumps and spikes. In: Working Paper PWP-073 of Program on Workable Energy Regulation (POWER) (2000)

17. DWD. Deutscher Wetterdienst climate data center. https://opendata.dwd.de/climate_environment/CDC. Accessed: 2020-01-16
18. ECA. European Climate Assessment & Dataset. https://www.ecad.eu/. Accessed: 2020-01-16
19. EIA. U.S. Energy Information Administration. https://www.eia.gov/. Accessed: 2020-01-16
20. entsoe. Available data. https://transparency.entsoe.eu/. Accessed: 2020-01-16
21. entsoe. entsoe business glossary: total load definition. https://docstore.entsoe.eu/data/data-portal/glossary/Pages/home.aspx. Accessed: 2020-01-08
22. eurostat. eurostat: European Statistics from the European Commission. https://ec.europa.eu/eurostat/web/energy/data/database. Accessed: 2020-01-16
23. Geisz, J.F., Steiner, M.A., Jain, N., Schulte, K.L., France, R.M., McMahon, W.E., Perl, E.E., Friedman, D.J.: Building a six-junction inverted metamorphic concentrator solar cell. IEEE J. Photovolt. **8**(2), 626–632 (2018). https://doi.org/10.1109/JPHOTOV.2017.2778567
24. Hinderks, W.J., Korn, R., Wagner, A.: A structural Heath–Jarrow–Morton framework for consistent intraday spot and futures electricity prices. Quant. Financ. **116**, 1–11 (2019). https://doi.org/10.1080/14697688.2019.1687927
25. Hinderks, W.J., Wagner, A.: Factor models in the German electricity market: stylized facts, seasonality, and calibration. Energy Econ. 104351 (2019). https://doi.org/10.1016/j.eneco.2019.03.024
26. Hinderks, W.J., Wagner, A., Oktoviany, P.: Handbook of Energy Finance, chapter Energy Risk Management in Practice. World Scientific (2020). https://doi.org/10.1142/11213
27. Kiesel, R., Schindlmayr, G., Börger, R.H.: A two-factor model for the electricity forward market. Quant. Financ. **9**(3), 279–287 (2009). https://doi.org/10.1080/14697680802126530
28. Kloeden, P.E., Platen, E.: Numerical solution of stochastic differential equations. In: Stochastic Modelling and Applied Probability, 1st edn., vol. 23, pp. XXXVI, 636. Springer, Berlin, Heidelberg (1992). https://doi.org/10.1007/978-3-662-12616-5
29. Koekebakker, S., Ollmar, F.: Forward curve dynamics in the Nordic electricity market. Manag. Financ. **31**(6), 73–94 (2005). https://doi.org/10.1108/03074350510769703
30. Latini, L., Piccirilli, M., Vargiolu, T.: Mean-reverting no-arbitrage additive models for forward curves in energy markets. Energy Econ. **79**, 157–170 (2019). https://doi.org/10.1016/j.eneco.2018.03.001
31. Lyle, M.R., Elliott, R.J.: A 'simple' hybrid model for power derivatives. Energy Econ. **31**(5), 757–767 (2009). https://doi.org/10.1016/j.eneco.2009.05.007
32. Mandelbrot, B.B., Van Ness, J.W.: Fractional brownian motions, fractional noises and applications. SIAM Rev. **10**(4), 422–437 (1968). http://www.jstor.org/stable/2027184
33. Meyer-Brandis, T., Tankov, P.: Multi-factor jump-diffusion models of electricity prices. Int. J. Theor. Appl. Financ. **11**(05), 503–528 (2008). https://doi.org/10.1142/s0219024908004907
34. Mraoua, M., Bari, D.: Temperature stochastic modeling and weather derivatives pricing: empirical study with Moroccan data. Afr. Stat. **2**(1), 22–43 (2007). https://doi.org/10.4314/afst.v2i1.46865
35. NOAA. National Centers for Environmental Information. https://www.ncdc.noaa.gov/. Accessed: 2020-01-16
36. Papaefthymiou, G., Klöckl, B.: MCMC for wind power simulation. IEEE Trans. Energy Convers. **23**(1), 234–240 (2008). https://doi.org/10.1109/TEC.2007.914174
37. Pilipovic, D.: Energy Risk: Valuing and Managing Energy Derivatives. McGraw-Hill, New York (1997)
38. Schermeyer, H., Schwarz, H., Bertsch, V., Fichtner, W.: Proceedings of the Smarter Europe Conference, pp. 2–16, Duisburg-Essen (2015). DuEPublico. https://doi.org/10.5445/IR/1000062641
39. Schwartz, E., Smith, J.E.: Short-term variations and long-term dynamics in commodity prices. Manag. Sci. **46**(7), 893–911 (2000). https://doi.org/10.1287/mnsc.46.7.893.12034
40. SMARD. Bundesnetzagentur: SMARD Strommarktdaten. https://smard.de/. Accessed: 2020-05-11
41. van den Berg, T.: Calibrating the Ornstein-Uhlenbeck (Vasicek) model (2011). https://www.statisticshowto.com/wp-content/uploads/2016/01/Calibrating-the-Ornstein.pdf. Accessed: 2020-06-02

42. Wagner, A.: Commodity Risk Manager. https://www.itwm.fraunhofer.de/en/departments/fm/
    mathematics-power-industry/risk-controlling.html. Accessed: 2020-05-25
43. Wagner, A.: Residual demand modeling and application to electricity pricing. Energy J. **35**(2)
    (2014). https://EconPapers.repec.org/RePEc:aen:journl:ej35-2-03

Chapter 3
Probabilistic Analysis of Solar Power Supply Using D-Vine Copulas Based on Meteorological Variables

Freimut von Loeper, Tom Kirstein, Basem Idlbi, Holger Ruf, Gerd Heilscher, and Volker Schmidt

Abstract Solar power generation at solar plants is a strongly fluctuating non-deterministic variable depending on many influencing factors. In general, it is not clear which and how certain variables influence solar power supply at feed-in points in a distribution network. Therefore, analyzing the dependence structure of measured solar power supply and other variables is very informative and can be helpful in designing probabilistic prediction models. In this paper multivariate D-vine copulas are fitted to investigate the relationship between solar power supply and certain meteorological variables in the current time period of one hour length as well as solar power supply in previous time periods. The meteorological variables considered in this analysis are global horizontal irradiation, temperature, wind speed, humidity, precipitation and pressure. By applying parametric D-vine copulas useful insight is gained into the dependence structure of solar power supply and the considered meteorological variables. The main goal lies in determining suitable explanatory variables for the design of probabilistic prediction models for solar power supply at single feed-in points and analyzing their impact on the validation of conditional level-crossing probabilities.

Keywords D-vine copula · Dependence structure · Solar power supply · Meteorological variable · Conditional level-crossing probability

3.1 Introduction

In recent years global warming was acknowledged as a serious problem becoming a topic of public concern. To reduce carbon dioxide emissions and limit further global warming, alternative energy sources such as renewable energy are required. For that reason, the renewable energy sector has the support of the governments in many

F. von Loeper (✉) · T. Kirstein · B. Idlbi · H. Ruf · G. Heilscher · V. Schmidt
Institute of Stochastics, Ulm University, Helmholtzstraße 18, 89069 Ulm, Germany
e-mail: freimut.von-loeper@uni-ulm.de

© Springer Nature Switzerland AG 2021
S. Göttlich et al. (eds.), *Mathematical Modeling, Simulation and Optimization for Power Engineering and Management*, Mathematics in Industry 34, https://doi.org/10.1007/978-3-030-62732-4_3

countries and is growing rapidly [4]. Especially the solar energy sector has been reporting record growth for many years [24].

However, higher solar power penetration might lead to new problems for distribution network operators. Since solar power generation strongly depends on weather conditions, the prospective solar power supply of the distribution network cannot be easily taken into account. Thus, it might be difficult to predict excessive power flow in the grid and to prevent voltage violations and overloading of power lines and transformers which destabilize the electricity supply and cause economic damages [14]. To regulate fluctuations in power and load of distribution networks, automated and more economic applications for smart technologies are needed [5]. Smart grid management can use solar power forecasting with hourly forecasting horizons for power system operation such as economic dispatch and unit commitment [26].

To compute probabilistic predictions regarding the generation of solar power supply at certain feed-in points, several high-dimensional stochastic models are considered in the literature. In [2] a non-parametric quantile regression forest is used to predict solar power supply based on several meteorological variables such as temperature, wind speed, humidity, sea level pressure and cloud cover at different levels. The prediction model proposed in [12] takes solar radiation, temperature, cloud ice water content and wind speed as input parameters to compute prediction intervals based on k-nearest neighbor regression. In [3] wind and solar power are computed by applying a combination of the gradient boosting tree algorithm and feature engineering techniques, where the authors of [3] concluded that information about the forecast grid further improves the prediction results. In [28] Gaussian conditional random fields are used to model the spatial-temporal dependence at neighboring feed-in points. However, the papers mentioned above give little insight into the dependence structure of solar power supply and meteorological variables. In particular, it is not clear which explanatory variables are suitable for probabilistic prediction of solar power supply.

An alternative approach can be given by multivariate copulas which are applied to compute the joint distribution of interdependent random variables [20]. In the literature of renewable energy modeling copulas are mostly used to account for spatial, temporal or spatio-temporal dependence at neighboring wind parks or feed-in points of solar power. In [18] and [21] Gaussian copulas and R-vine copulas are applied to model the spatial dependence of the wind power supply generated at many different wind parks. In [6] and [10] Gaussian and D-vine copulas are utilized to model the temporal and spatio-temporal dependence of these characteristics. Furthermore, in [8] and [9] Gaussian and R-vine copulas are applied to model the spatial dependence of solar power supply at neighboring feed-in points. However, multivariate copulas are not only helpful to analyze and model the spatial or temporal dependence of wind or solar power supply at neighboring feed-in points, but also the relationship between power supply and other influencing factors, such as meteorological variables.

In the present paper, we determine suitable explanatory variables for the design of probabilistic prediction models for solar power supply at single feed-in points, extending the methodology which has recently been used in [17]. For that purpose, multivariate D-vine copulas are applied to analyze the dependence structure of the considered meteorological variables and the solar power supply in the current time period of one hour length as well as in previous hourly periods. The stepwise fitting of D-vine copulas helps us in analyzing the interdependence of the considered variables in detail. Moreover, conditional level-crossing probabilities are computed and validated with prediction scores to determine suitable explanatory variables for the design of prediction models which compute probabilities for critical amounts of solar power supply. Last but not least the effect of the considered month and time of day on the probabilistic prediction of solar power supply is investigated.

The rest of the paper is organized as follows. In Sect. 3.2 the data is described and analyzed empirically. Section 3.3 introduces the mathematical methodology applied in this paper. In Sect. 3.4 the results are presented and discussed. Section 3.5 concludes.

3.2 Data

The measurements of solar power supply and reverse power utilized in this paper were collected in cooperation with the local distribution network operator Stadtwerke Ulm/Neu-Ulm Netze GmbH (SWUN). Additional meteorological information concerning global horizontal irradiation (GHI) was gathered by satellites where temperature, wind speed, humidity, precipitation and pressure were computed by reanalysis and published as an open source, see [7, 19] respectively. The datasets of both sources are described in Sect. 3.2.1 and empirically analyzed in Sect. 3.2.2.

3.2.1 Data Description

Solar power supply of a community near the city of Ulm, called Hittistetten, is measured for the years 2016–2018. The location is a test site defined on the website of SWUN [25]. The power measurements are taken by smart meters at 14 PV systems in Hittistetten. Afterwards they are summed over all PV systems in the community to get the representative solar power supply of the community which is considered in the following. Furthermore, power measurements at the local area transformers are evaluated and daily accumulated reverse energy is estimated. Note that there exist some gaps in the collected data, because of some possible errors in the smart meter infrastructure.

The hourly measurements of solar power supply, the hourly meteorological data and the daily accumulated reverse energy measurements are rescaled to the unit interval using a linear transformation. All time stamps are converted to Central European Time (CET). The min-max-rescaled measurements of hourly aggregated solar power supply and data concerning meteorological variables are interpreted as realizations of random variables, see [17] for details. In Table 3.1, the mathematical symbols P_t and M_i are introduced for the random variables considered in the present paper.

Table 3.1 Considered variables and their physical meaning

Variable	Physical meaning
P_t	Solar power supply in the period t hours before the current time period of one hour length
M_1	GHI at ground level in the current period
M_2	Temperature at 2 m above ground in the current period
M_3	Precipitation at ground level in the current period
M_4	Relative humidity at 2 m above ground in the current period
M_5	Wind speed at 10 m above ground in the current period
M_6	Pressure at ground level in the current period

Fig. 3.1 Daily accumulated reverse energy for each month

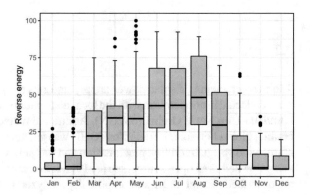

Furthermore, rescaled power exceeding 70 percent of the highest recorded amount of solar power supply is seen as a critical amount of solar power supply. Therefore, in the following the value of 0.7 is considered as an exemplary high threshold of rescaled power supply.

3.2.2 Empirical Data Analysis

In Fig. 3.1, the daily accumulated reverse energy is visualized. The box plots indicate that during the winter months of January, February, November and December the amounts of reverse power are very low in comparison to the remaining months. In particular, there are no critical overloading events for the distribution network happening within these months. Therefore, since these months are not interesting for the distribution network operators, they are excluded in the present study. Furthermore, in Fig. 3.2 the means of hourly aggregated solar power supply are plotted for each time of day. For all months considered in Fig. 3.2 the means of the hourly aggregated solar power supply are always below the threshold of 0.7 for the time periods of 6–9

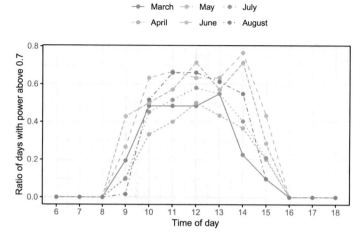

Fig. 3.2 Ratio of considered days with solar power supply above the threshold of 0.7 for each time of day

CET and 16–18 CET. Hence, our analysis is limited to the months from March to August and the time of day from 9–16 CET.

For the solar power supply P_0 in the current time period and the meteorological variables $M_1, \ldots, M_6$, the panels in Fig. 3.3 show pairwise scatter plots, Kendall's rank correlation coefficients and histograms of these variables. In particular, in the panels below the diagonal the pairwise scatter plots provide detailed information about the kind of pairwise dependence between the considered variables, which is further quantified by pairwise Kendall's rank correlation coefficients in the panels above the diagonal. Note that the Kendall's rank correlation coefficient measures the rank adjusted correlation between two random variables X and Y and can be estimated by

$$\widehat{\tau} = \frac{2}{d(d-1)} \sum_{i<j} \mathrm{sgn}(x_i - x_j)\, \mathrm{sgn}(y_i - y_j) \tag{3.1}$$

for given realizations $(x_1, \ldots, x_d)$ and $(y_1, \ldots, y_d)$ of X and Y, respectively.

Figure 3.3 shows that the meteorological variables temperature (M_2), precipitation (M_3), and humidity (M_4) are stronger correlated with solar power supply than wind speed (M_5) and pressure (M_6). Furthermore, the histograms in the panels on the diagonal give a rough idea about the shape of the density of the corresponding variables. Solar power supply (P_0), GHI (M_1) and temperature (M_2) might have a bimodal distribution, whereas the distributions of precipitation (M_3), humidity (M_4), wind speed (M_5) and pressure (M_6) appear to be unimodal.

In Fig. 3.4, the Kendall's rank correlation coefficients of the measurements of solar power supply in the current period of one hour length and in certain previous

Fig. 3.3 Pairwise
scatterplots, histograms and
Kendall's rank correlation
coefficients for solar power
supply and meteorological
variables in the current time
period

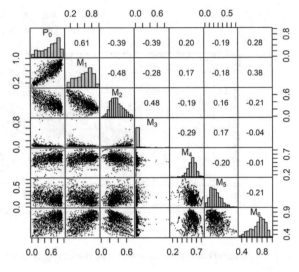

Fig. 3.4 Kendall's rank
correlation coefficients for
solar power supply in the
current time period of one
hour length and solar power
supply in previous hourly
periods

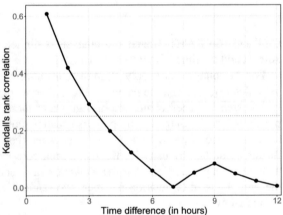

hourly periods are plotted. As expected the correlation gets weaker the larger the
time difference is between both measurements.

In the following analysis our focus is put on variables which have a rank correlation
coefficient larger than 0.25 to solar power supply in the current period. The remaining
variables are too weakly correlated to have sufficiently much influence on solar power
supply. Based on the correlation plots of Figs. 3.3 and 3.4 only the variables M_1, M_2,
M_3, M_4, P_1, P_2, and P_3 fulfill the requirement mentioned above.

3.3 Methodology

If several explanatory variables are considered, high pairwise correlation does not necessarily mean that all of them are good explanatory variables. In the worst case it might happen that all helpful information within an explanatory variable is part of other explanatory variables. To analyze the dependence structure of interdependent random variables in-depth, our modeling approach is based on D-vine copulas which will be explained in this section.

3.3.1 Modeling Approach

The main goal of the present paper is to determine those components of the random vector $E = (M_1, M_2, M_3, M_4, P_1, P_2, P_3)$ which are useful explanatory variables for the design of probabilistic prediction models for solar power supply. In particular, their impact on the probabilistic prediction of critical amounts of solar power supply is investigated. That impact is quantified by computing conditional level-crossing probabilities of solar power supply P_0 given that $E = e$ for some realization $e \in \mathbb{R}^7$ of the random vector E. In particular, for the exemplary threshold 0.7 we get that

$$P(P_0 \geq 0.7 \mid E = e) = \int_{0.7}^{1} f_{P_0|E=e}(x)\, dx = \int_{0.7}^{1} \frac{f_{P_0,E}(x, e)}{f_E(e)}\, dx$$

$$= \int_{0.7}^{1} \frac{f_{P_0,E}(x, e)}{\int_0^1 f_{P_0,E}(y, e)\, dy}\, dx = \frac{\int_{0.7}^{1} f_{P_0,E}(x, e)\, dx}{\int_0^1 f_{P_0,E}(y, e)\, dy}.$$

To determine the joint probability density $f_{P_0,E}$ of the random vector (P_0, E) the marginal densities of $P_0, M_1, M_2, M_3, M_4, P_1, P_2$ and P_3 are estimated in a first step and D-vine copulas are applied in a second step. Note that our approach based on D-vine copulas gives us in-depth insight into the dependence structure of solar power supply and the considered meteorological variables.

3.3.2 D-Vine Copulas

A function $C : [0, 1]^d \to [0, 1]$ with $d \geq 2$ is called a *d-dimensional copula* if C is the joint cumulative distribution function (CDF) of a d-dimensional random vector with uniformly distributed marginals in $[0, 1]$. Copulas are a powerful tool to parametrically model multivariate CDFs with non-Gaussian marginals. The reason for this is the following fundamental *theorem of Sklar*, see [13, 20].

Let $F : [0, 1]^d \to [0, 1]$ be an arbitrary multivariate CDF of a random vector $(X_1, \ldots, X_d)$. Then, for the marginal CDFs $F_i(x) = P(X_i \leq x)$ with $i \in \{1, .., d\}$

there exists a d-dimensional copula C such that F can be expressed as

$$F(x_1, .., x_d) = C(F_1(x_1), \ldots, F_d(x_d)) \quad \text{for each } (x_1, \ldots, x_d) \in \mathbb{R}^d. \quad (3.2)$$

Thus, the CDF F of an arbitrary random vector $X = (X_1, \ldots, X_d)$ can be represented as superposition of the corresponding marginal CDFs $F_1, \ldots, F_d$ and a certain copula C which models the dependence structure of the components $X_1, \ldots, X_d$. The parameters of marginal distributions and copula can be estimated separately from data which reduces the number of parameters fitted simultaneously, see [13, 20] for further details.

In the present paper a special class of d-dimensional copulas, so-called D-vine copulas, is used which are a popular type of multivariate copulas, see e.g. [6, 10]. We assume that the CDF F has a d-dimensional density $f_{1,\ldots,d}$ and denote the 1-dimensional densities of the corresponding marginal CDFs $F_1, \ldots, F_d$ by $f_1, \ldots, f_d$, respectively. To obtain the D-vine structure the d-dimensional density $f_{1,\ldots,d}$ is decomposed into conditional densities by

$$f_{1,\ldots,d} = f_{d|1:d-1} \cdots f_{2|1:1} f_1, \quad (3.3)$$

where $f_{i|1:i-1}$ is the conditional density given by $f_{i|1:i-1} = f_{1,\ldots,i}/f_{1,\ldots,i-1}$ for $i = 2, \ldots, d$. In the next step, Sklar's theorem is applied to the conditional density $f_{i,j|i+1:j-1}$ of the random vector (X_i, X_j) with $i + 1 < j$ given that $X_{i+1:j-1} = x_{i+1:j-1}$, where $X_{i+1:j-1} = (X_{i+1}, \ldots, X_{j-1})$ and $x_{i+1:j-1} = (x_{i+1}, \ldots, x_{j-1})$. Considering probability densities instead of distribution functions on both sides of Eq. 3.2 we get that

$$f_{i,j|i+1:j-1} = c_{i,j|i+1:j-1}(F_{i|i+1:j-1}, F_{j|i+1:j-1}) f_{i|i+1:j-1} f_{j|i+1:j-1}, \quad (3.4)$$

where $c_{i,j|i+1:j-1}$ is some bivariate copula density, $F_{i|i+1:j-1}$ and $F_{j|i+1:j-1}$ denote the conditional CDFs of X_i and X_j, respectively, given that $X_{i+1:j-1} = x_{i+1:j-1}$, and $f_{i|i+1:j-1}$ and $f_{j|i+1:j-1}$ are the corresponding conditional marginal densities. This results in

$$f_{j|i:j-1} = \frac{f_{i,j|i+1:j-1}}{f_{i|i+1:j-1}} = c_{i,j|i+1:j-1}(F_{i|i+1:j-1}, F_{j|i+1:j-1}) f_{j|i+1:j-1}. \quad (3.5)$$

Note that for $j = i + 1$ Sklar's theorem is used for the (unconditional) bivariate density of the random vector (X_i, X_j) and $i + 1 : j - 1$ is the empty set. Finally, Eq. 3.5 is repeatedly applied to the conditional densities on the right-hand side of Eq. 3.3 which leads to

$$f_{1,\ldots,d} = \prod_{k=1}^{d-1} \prod_{i=1}^{d-k} c_{i,i+k|i+1:i+k-1}(F_{i|i+1:i+k-1}, F_{i+k|i+1:i+k-1}) \prod_{l=1}^{d} f_l. \quad (3.6)$$

To estimate the copula densities $c_{i,i+k|i+1:i+k-1}$ in Eq. 3.6 we assume that they do not depend on specific values of $x_{i+1:i+k-1}$. Further mathematical details regarding copula theory and, in particular, D-vine copulas can be found e.g. in [13].

3.3.3 Fitting Procedure

To select parametric families of (marginal) distributions for the components $X_1, \ldots,$ X_d of the random vector $X = (X_1, \ldots, X_d)$ considered in Sect. 3.3.2, histograms are computed for visual inspection based on all available data $(x_1, \ldots, x_d)$. Thus, the dimension d and the random variables $X_1, \ldots, X_d$ have to be specified, see Table 3.2. Their histograms give us a rough idea of the shapes of the densities to be fitted and, in particular, exhibit distributional properties such as multimodality. This information is helpful in determining distribution types which might give a good fit to the data considered in this paper. Once candidates for the distribution types are selected, we compute the *Bayesian information criterion* (BIC) defined as

$$BIC = k \ln d - 2 \ln L \tag{3.7}$$

for each distribution type with k parameters, where L is the maximized value of the likelihood function. The distribution type with the smallest BIC value is a reasonable choice, see [15]. In the next step the parameters of unimodal distributions are estimated by maximizing the likelihood and the parameters of bimodal distributions are fitted by applying the expectation-maximization algorithm, see [11, 16].

Finally, to fit a multivariate D-vine copula two further steps have to be carried out:

(1) Select parametric families of copula densities for the bivariate copula densities $c_{i,i+k|i+1:i+k-1}$ considered in Eq. 3.6.

(2) Estimate the parameter(s) for each copula density $c_{i,i+k|i+1:i+k-1}$ in the D-vine structure.

According to the fitting procedure described above, we first specify the input variables $X_1, \ldots, X_d$, before selecting the types of marginal distributions and fitting their parameters. We put $d = 8$ and consider P_0 and M_1, see Table 3.1, which are the most important input variables. Secondly, the meteorological input variables M_2, M_3 and M_4 are taken into account. To complete the specification of input variables, the solar power inputs P_1, P_2 and P_3 at previous periods of time are added to the other variables according to their temporal ordering. The chosen specification of all considered variables is summarized in Table 3.2.

Finally, the conditional bivariate copula densities considered in Eq. 3.6 are determined sequentially, see [13] for details. In particular, for each bivariate copula the copula type and its parameter(s) are determined by maximizing the corresponding likelihood. As candidates the Archimedean copula types Joe, Frank, Clayton and Gumbel are considered, which guarantee a large variety of possible tail dependen-

Table 3.2 Specification of input variables in the D-vine copula structure

General notation	X_1	X_2	X_3	X_4	X_5	X_6	X_7	X_8
Specification	P_0	M_1	M_2	M_3	M_4	P_1	P_2	P_3

cies. A detailed description of the fitting procedure of bivariate copulas can be found e.g. in [17]. The results are obtained using the VineCopula package in R, see [22].

3.4 Results and Discussion

In this section we apply the D-vine copula model, which has been described in Sect. 3.3, to a sample of 8-dimensional data vectors, in order to get deeper insight into the dependence structure of solar power supply and the considered meteorological variables.

3.4.1 Model Fitting and Validation

The parametric families of beta, mixed beta, log-normal, Weibull and gamma distributions are chosen as candidates for the marginal distributions of the random vector $(P_0, M_1, M_2, M_3, M_4, P_1, P_2, P_3)$, based on the histograms visualized in Fig. 3.3. To determine the most suitable distribution type, the BIC score given in Eq. 3.7 is computed for each of the 7 considered hours of day separately. In Fig. 3.5, the obtained BIC scores are visualized by means of box-plots. Most of the BIC scores of the mixed beta distribution are clearly smaller than those of other distribution types, in particular, for solar power supply, GHI and precipitation. However, for humidity and temperature most distribution types have similar BIC scores. For all input variables the distribution type with the smallest average BIC is chosen, where averaging is taken over the 7 considered hours of day. As a result, Weibull distributions are fitted for humidity and mixed beta distributions for solar power supply, GHI, precipitation and temperature.

In Fig. 3.6 the fitted marginal densities and the underlying histograms are visualized for solar power supply, GHI, humidity, precipitation and temperature regarding the exemplary hour of day from 14–15 CET. Moreover, to compare the quality of the fits for each input variable the parametric density with the second best average BIC is visualized as well.

Figure 3.7 shows the D-vine copula structure fitted to the input variables specified in Table 3.2. Each panel entitled with $i, j|i+1, \ldots, j-1$ visualizes the fitted conditional copula density $c_{i,j|i+1:j-1}$, see Sect. 3.3.2 for $j = k + i$. The bivariate copulas with symmetric tail-dependence, see Fig. 3.7, are modelled using the Frank

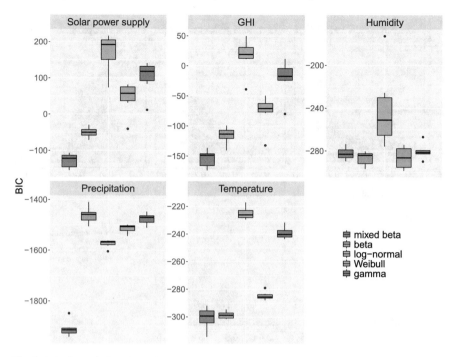

Fig. 3.5 BIC scores for beta, mixed beta, log-normal, Weibull and gamma distributions

copula, see e.g. the copula entitled with 4,5, whereas the bivariate copulas with asymmetric tail-dependence are modelled using Joe, Clayton or Gumbel copulas, see e.g. the copula entitled with 1,2.

In Table 3.3 pairwise Kendall's rank correlation coefficients are presented which have been computed based on original data and 5000 simulated realizations drawn from the D-vine copula for the input variables on the diagonal of the corresponding rows/columns, using the algorithm explained in [1, 13] and its implementation in R, see the VineCopula package in [22]. By comparing the Kendall's rank correlation coefficients it becomes apparent that the coefficients of the original data are very similar to the coefficients based on the simulations of the fitted D-vine copula model. Therefore, the D-vine copula represents the dependence structure of the measurements of solar power supply and meteorological data sufficiently well with respect to Kendall's rank correlation coefficient.

3.4.2 Conditional Means of Solar Power Supply

In this section we further analyze the dependence structure of solar power supply in the current time period and various other input variables. For this purpose, 250000

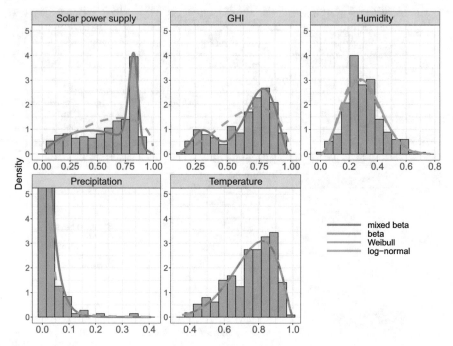

Fig. 3.6 Fitted marginal densities at the exemplary hour of day 14–15 CET. Solid lines correspond to the best average BIC, whereas dashed lines correspond to the second best average BIC

Table 3.3 Pairwise Kendall's rank correlation coefficients of original data (below the diagonal) and of realizations drawn from the fitted D-vine copula model (above the diagonal) computed for the input variables on the diagonal of the corresponding rows/columns

P_0	0.54	0.25	−0.38	−0.38	0.50	0.43	0.43
0.57	M_1	0.34	−0.39	−0.38	0.54	0.47	0.43
0.29	0.36	M_2	−0.11	−0.09	0.26	0.28	0.23
−0.42	−0.38	−0.10	M_3	0.67	−0.46	−0.48	−0.48
−0.41	−0.35	−0.07	0.63	M_4	−0.47	−0.47	−0.44
0.59	0.54	0.31	−0.43	−0.44	P_1	0.64	0.51
0.52	0.48	0.33	−0.47	−0.45	0.64	P_2	0.65
0.49	0.45	0.30	−0.48	−0.44	0.54	0.67	P_3

realizations were drawn from the D-vine copula shown in Fig. 3.7, using the algorithm explained in [1, 13] and its implementation in R, see the VineCopula package in [22]. The conditional empirical means of solar power supply are computed in dependence of GHI, for given lower, middle and upper values of further input variables, see Fig. 3.8. To plot the graphs shown in Fig. 3.8, a partition of the unit interval into 40 parts is used.

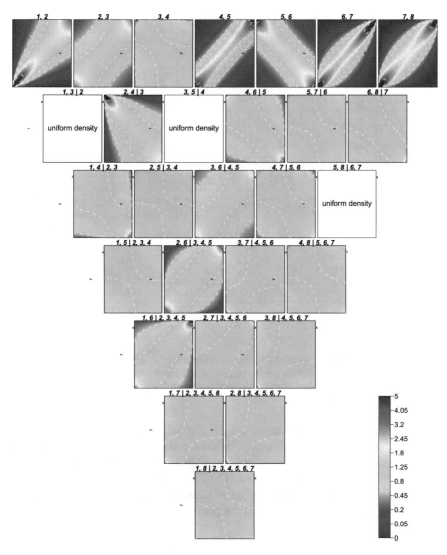

Fig. 3.7 The D-vine structure fitted to the variables specified in Table 3.2 for the exemplary hour of day 14–15 CET

Figure 3.8 visualizes how strongly the conditional means of solar power supply depend on further input variables for given values of GHI in the exemplary time period between 14 and 15 CET. The larger the discrepancy is between the red and blue lines the stronger is the conditional correlation, see [23]. If the red line is above the corresponding blue line in Fig. 3.8 the conditional correlation is positive and otherwise negative. Figure 3.8 indicates that the conditional means of solar power input are independent of temperature, whereas solar power supply in the previous

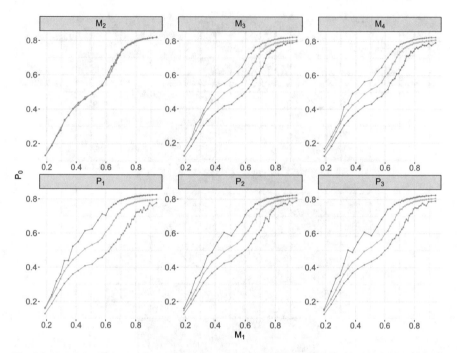

Fig. 3.8 Conditional empirical means of solar power supply in the current period of one hour length (P_0) depending on global horizontal irradiation (M_1) given that the value of a further input variable (specified in the title of the panel) is in the highest 25 percent (red), the middle 50 percent (green), the lowest 25 percent (blue)

hourly period and in the hourly period before the previous hourly period are positively correlated with the conditional means of solar power supply. On the other hand, precipitation and humidity are negatively correlated to the conditional means of solar power supply for given values of GHI. Thus, our analysis shows that temperature does not provide additional information to GHI for the probabilistic prediction of solar power supply in the considered time period.

3.4.3 Validation Scores for Conditional Level-Crossing Probabilities

To quantify the impact of the considered explanatory variables on the probabilistic prediction of solar power supply conditional level-crossing probabilities are computed by means of the fitted D-vine copula model and compared to corresponding quantities obtained from the measurements of solar power supply using various validation scores. For example, the bias, Brier score (BS), Brier skill score (BSS), empirical correlation coefficient (Corr), reliability (Rel), resolution (Res) and uncertainty (Unc) are considered. Clearly, the bias should be near zero, BS and Rel as low as possible, and BSS and Res as high as possible. Note that BS shows the accuracy of the computed level-crossing probabilities and its information can be decomposed into Rel, Res and Unc. Furthermore, Unc does not depend on the selected model but shows the variability of the measurements. In addition, the considered BSS is used to compare the BS of the D-vine copula model with the BS of the climatological mean. For definitions and further details we refer to [17, 27].

3.4.3.1   Analysis of the Goodness-of-Fit for Different Time Frames

In this section the goodness of model fit is investigated for different time frames, which are described in Table 3.4. By comparing different hourly time periods the impact of the solar elevation angle on solar power supply can be determined. In addition, the comparison of the results obtained for different monthly time periods shows the impact of seasonal changes.

The D-vine copula model fitted for time frame No. 2 in Table 3.4 leads to the best average scores for bias, BS, BSS, Corr and Rel, whereas time frame No. 5 leads to a slightly better Res than time frame No. 2. Based on the results shown in Table 3.5, it becomes apparent that fitting the model for each hour of day separately improves the conditional level-crossing probabilities of solar power supply. On the other hand, no further improvement is obtained if each month is considered separately. Thus, the information regarding the solar elevation angle seems to be more important than the information regarding seasonal changes.

Table 3.4 Time frames for model fitting

Time frame	Description
1	Each hour of day and month separately
2	Each hour of day separately, whereas combining all months
3	Each month separately, whereas combining all hours of day
4	Two consecutive hourly time periods and months
5	Three consecutive hourly time periods and months

Table 3.5 Validation scores for conditional level-crossing probabilities of the fitted multivariate D-vine model for different time frames

Time frame	BS	BSS	Corr	Bias	Rel	Res	Unc
1	0.120	0.495	0.712	0.027	0.004	0.122	0.238
2	0.101	0.575	0.760	0.021	0.003	0.140	0.238
3	0.121	0.493	0.719	0.075	0.011	0.128	0.238
4	0.106	0.553	0.746	0.028	0.003	0.135	0.238
5	0.104	0.564	0.759	0.053	0.007	0.142	0.238

3.4.3.2 Analysis of the Goodness-of-Fit for Different Copula Models

In this section, the validation scores bias, BS, Brier BSS, Corr, Rel, Res and Unc of the conditional level-crossing probabilities obtained from the multivariate D-vine copula model fitted in Sect. 3.4.1 are compared with those obtained from a bivariate Frank copula, which merely models the dependence between solar power supply and GHI. Note that a Frank copula was proposed as prediction model for solar power supply in [17]. By comparing the scores computed based on both models the impact of further input variables on the probabilistic prediction of solar power supply can be quantified.

In Tables 3.6 and 3.7 the validation scores for conditional level-crossing probabilities are computed for each hourly time period separately and, on the other hand, for the entire time period from 9–16 CET. Note that the scores for the entire time period are usually not the averages over all scores obtained for the hourly time periods. The validation scores show that the multivariate D-vine copula model leads to a

Table 3.6 Validation scores for conditional level-crossing probabilities of the fitted bivariate Frank copula model

Period of time	BS	BSS	Corr	Bias	Rel	Res	Unc
09–10	0.083	−0.146	0.353	0.072	0.031	0.021	0.072
10–11	0.140	0.436	0.673	0.029	0.038	0.148	0.249
11–12	0.111	0.554	0.748	−0.022	0.012	0.151	0.248
12–13	0.104	0.580	0.762	−0.001	0.015	0.158	0.248
13–14	0.109	0.562	0.755	0.028	0.030	0.171	0.249
14–15	0.135	0.452	0.682	0.057	0.017	0.127	0.247
15–16	0.113	−0.012	0.310	0.036	0.028	0.025	0.112
09–16	0.114	0.523	0.726	0.028	0.006	0.131	0.238

Table 3.7 Validation scores for conditional level-crossing probabilities of the fitted multivariate D-vine model

Period of time	BS	BSS	Corr	Bias	Rel	Res	Unc
09–10	0.075	−0.047	0.401	0.056	0.021	0.019	0.072
10–11	0.108	0.566	0.757	0.041	0.010	0.151	0.249
11–12	0.114	0.540	0.735	0.002	0.009	0.145	0.248
12–13	0.101	0.593	0.773	0.007	0.013	0.160	0.248
13–14	0.103	0.585	0.772	0.016	0.017	0.162	0.249
14–15	0.114	0.539	0.742	0.022	0.013	0.146	0.247
15–16	0.093	0.172	0.421	0.006	0.010	0.027	0.112
09–16	0.101	0.575	0.760	0.021	0.003	0.140	0.238

considerable improvement compared to the bivariate Frank copula model. Thus, the impact of the additionally considered input variables on the probabilistic prediction of solar power supply is important.

3.5 Conclusion

Using the D-vine copula model considered in this paper, we determined suitable explanatory variables for the design of probabilistic prediction models for solar power supply at single feed-in points. This knowledge is very important because most probabilistic prediction models considered in the literature have limitations regarding the selection of input variables which can be modelled in a reasonable way. For that purpose, the dependence structure of meteorological input variables, solar power supply measured in previous hourly periods and solar power supply measured for the exemplary time period between 14 and 15 CET was analyzed in detail.

Based on our analysis it turned out that temperature is indeed correlated with solar power supply, but has no additional impact on the probabilistic prediction of solar power supply if considered in combination with GHI. Humidity and precipitation were determined as conditionally negatively correlated to solar power supply in the exemplary time period between 14 and 15 CET for given values of GHI, whereas solar power supply in previous hourly periods seemed to be conditionally positively correlated for given values of GHI. Thus, humidity, precipitation, and solar power supply in the previous hourly period indicate a clear impact on the probabilistic prediction of solar power supply in the current time period.

Moreover, the goodness of model fit was investigated for different time frames and for different copula models. For that purpose, conditional level-crossing probabilities were computed for the exemplary high threshold of 0.7 where various scores were used to validate them. Our analysis showed that the goodness of model fit improves if hours of day are considered separately, whereas considering months separately gives no further improvement. Furthermore, it turned out that considering other input variables, besides GHI, clearly has a positive effect on the accuracy of the computed conditional level-crossing probabilities.

References

1. Aas, K., Czado, C., Frigessi, A., Bakken, H.: Pair-copula constructions of multiple dependence. Insur. Math. Econ. **44**(2), 182–198 (2009)
2. Almeida, M.P., Perpinan, O., Narvarte, L.: PV power forecast using a nonparametric PV model. Solar Energy **115**, 354–368 (2015)
3. Andrade, J.R., Bessa, R.J.: Improving renewable energy forecasting with a grid of numerical weather predictions. IEEE Trans. Sustain. Energy **8**(4), 1571–1580 (2017)
4. Balasubramanian, T.N., Appadurai, A.N.: Climate policy. In: Venkatramanan, V., Shah, S., Prasad, R. (eds.) Global Climate Change and Environmental Policy, pp. 37–54. Springer (2020)

5. Bayindir, R., Colak, I., Fulli, G., Demirtas, K.: Smart grid technologies and applications. Renew. Sustain. Energy Rev. **66**, 499–516 (2016)
6. Bessa, R.J.: On the quality of the Gaussian copula for multi-temporal decision-making problems. In: 2016 Power Systems Computation Conference (PSCC), pp. 1–7 (2016)
7. Copernicus Atmosphere Monitoring Service: Open source global horizontal irradiation data. http://www.soda-pro.com/web-services/radiation/cams-radiation-service
8. Golestaneh, F., Gooi, H.B.: Multivariate prediction intervals for photovoltaic power generation. In: 2017 IEEE Innovative Smart Grid Technologies-Asia (ISGT-Asia), pp. 1–5. IEEE (2017)
9. Golestaneh, F., Gooi, H.B., Pinson, P.: Generation and evaluation of space-time trajectories of photovoltaic power. Appl. Energy **176**, 80–91 (2016)
10. Haghi, H.V., Lotfifard, S.: Spatiotemporal modeling of wind generation for optimal energy storage sizing. IEEE Trans. Sustain. Energy **6**(1), 113–121 (2014)
11. Hastie, T., Tibshirani, R., Friedman, J.: The Elements of Statistical Learning: Data Mining, Inference and Prediction. Springer (2009)
12. Huang, J., Perry, M.: A semi-empirical approach using gradient boosting and k-nearest neighbors regression for GEFCom2014 probabilistic solar power forecasting. Int. J. Forecast. **32**(3), 1081–1086 (2016)
13. Joe, H.: Dependence Modeling with Copulas. Chapman and Hall/CRC (2014)
14. Karimi, M., Mokhlis, H., Naidu, K., Uddin, S., Bakar, A.: Photovoltaic penetration issues and impacts in distribution network - a review. Renew. Sustain. Energy Rev. **53**, 594–605 (2016)
15. Konishi, S., Kitagawa, G.: Information Criteria and Statistical Modeling. Springer (2008)
16. Leisch, F.: A general framework for finite mixture models and latent glass regression in R. J. Stat. Softw. **11**(8), 1–18 (2004)
17. von Loeper, F., Schaumann, P., de Langlard, M., Hess, R., Bäsmann, R., Schmidt, V.: Probabilistic prediction of solar power supply to distribution networks, using forecasts of global horizontal irradiation. Solar Energy **203**, 145–156 (2020)
18. Lu, Q., Hu, W., Min, Y., Yuan, F., Gao, Z.: Wind power uncertainty modeling considering spatial dependence based on pair-copula theory. In: PES General Meeting| Conference & Exposition, pp. 1–5. IEEE (2014)
19. Modern-Era Retrospective Analysis for Research and Applications Version 2: Open source meteorlogical data. https://gmao.gsfc.nasa.gov/reanalysis/MERRA-2/
20. Nelsen, R.B.: An Introduction to Copulas. Springer (2006)
21. Papaefthymiou, G., Kurowicka, D.: Using copulas for modeling stochastic dependence in power system uncertainty analysis. IEEE Trans. Power Syst. **24**, 40–49 (2009)
22. R Core Team: R: A Language and Environment for Statistical Computing. R Foundation for Statistical Computing, Vienna, Austria (2014). http://www.R-project.org/
23. Rässler, S.: Statistical Matching: A frequentist Theory, Practical Applications, and Alternative Bayesian Approaches. Springer (2012)
24. SolarPower Europe: Global market outlook 2018–2022 (2017). http://www.solarpowereurope.org/wp-content/uploads/2018/09/Global-Market-Outlook-2018-2022.pdf
25. Stadtwerke Ulm/Neu-Ulm Netze GmbH: test areas smart grids. https://www.ulm-netze.de/unternehmen/projekt-smart-grids
26. Wan, C., Zhao, J., Song, Y., Xu, Z., Lin, J., Hu, Z.: Photovoltaic and solar power forecasting for smart grid energy management. CSEE J. Power Energy Syst. **1**(4), 38–46 (2015)
27. Wilks, D.S.: Statistical Methods in the Atmospheric Sciences. Academic Press (2011)
28. Zhang, B., Dehghanian, P., Kezunovic, M.: Spatial-temporal solar power forecast through use of gaussian conditional random fields. In: IEEE Power and Energy Society General Meeting (PESGM), vol. IEEE, pp. 1–5 (2016)

Part II
Technical Applications

Chapter 4
GivEn—Shape Optimization for Gas Turbines in Volatile Energy Networks

Jan Backhaus, Matthias Bolten, Onur Tanil Doganay, Matthias Ehrhardt, Benedikt Engel, Christian Frey, Hanno Gottschalk, Michael Günther, Camilla Hahn, Jens Jäschke, Peter Jaksch, Kathrin Klamroth, Alexander Liefke, Daniel Luft, Lucas Mäde, Vincent Marciniak, Marco Reese, Johanna Schultes, Volker Schulz, Sebastian Schmitz, Johannes Steiner, and Michael Stiglmayr

Abstract This paper describes the project GivEn that develops a novel multiobjective optimization process for gas turbine blades and vanes using modern "adjoint" shape optimization algorithms. Given the many start and shut-down processes of gas power plants in volatile energy grids, besides optimizing gas turbine geometries for efficiency, the durability understood as minimization of the probability of failure is a design objective of increasing importance. We also describe the underlying coupling structure of the multiphysical simulations and use modern, gradient based multiobjective optimization procedures to enhance the exploration of Pareto-optimal solutions.

J. Backhaus · C. Frey
Institute of Propulsion Technology, German Aerospace Center (DLR), 51147 Köln, Germany
e-mail: jan.backhaus@dlr.de

C. Frey
e-mail: christian.frey@dlr.de

M. Bolten · O. Tanil Doganay · M. Ehrhardt · H. Gottschalk (✉) · M. Günther · C. Hahn ·
J. Jäschke · K. Klamroth · M. Reese · J. Schultes · M. Stiglmayr
Bergische Universität Wuppertal, Fakultät für Mathematik und Naturwissenschaften, IMACM,
Gaußstrasse 20, 42119 Wuppertal, Germany
e-mail: hgotsch@uni-wuppertal.de

M. Bolten
e-mail: bolten@uni-wuppertal.de

O. Tanil Doganay
e-mail: doganay@uni-wuppertal.de

M. Ehrhardt
e-mail: ehrhardt@uni-wuppertal.de

M. Günther
e-mail: guenther@uni-wuppertal.de

C. Hahn
e-mail: chahn@uni-wuppertal.de

S. Göttlich et al. (eds.), *Mathematical Modeling, Simulation and Optimization for Power Engineering and Management*, Mathematics in Industry 34,
https://doi.org/10.1007/978-3-030-62732-4_4

4.1 Introduction

According to the Federal Government's plans for the energy turnaround, the share of renewable energy in the German electricity grid is to rise to over 60% in 2050. Since wind, photovoltaics and solar thermal energy supply fluctuating energy yields, this also means an increase in volatile energy supply to up to 60% of total output. Flexible reserve power plants are therefore essential for security of supply. Due to the short start-up times of a few minutes, highly efficient gas or gas and steam (combined cycle) power plants are predestined to absorb slumps in output and peak demand. As the most efficient form of fossil energy with an efficiency of over 63%, such plants can also be used as bridge technology an important contribution to CO2 reduction (by up to 2/3 compared to currently still operating lignite-fired power plants). This will also be considerably reduced in the scenario of a 100% renewable energy supply after the year 2050 will be the case when hydrogen and methane are used as storage media from renewable energy sources and gas turbines generate electricity on demand.

The diverse applications of gas turbines in the context of the energy system transformation, such as backup power plants or hydrogen turbines, go hand in hand with specific design requirements, in particular with regard to the efficiency of energy conversion and the reliability and flexibility of operation. These different requirements are intensively related to the coupled fluid dynamic simulation and the structural mechanical fatigue calculation. The use of integrated, multi-physical tool chains and optimization software therefore plays an important role in gas turbine design. This joint project links six different simulations—fluid dynamics, laminar convective heat transfer, 1D flux networks and turbulent convective heat transfer, heat conduction, structural mechanics, probabilistic modelling of material fatigue—which are

J. Jäschke
e-mail: jaeschke@uni-wuppertal.de

K. Klamroth
e-mail: klamroth@uni-wuppertal.de

M. Reese
e-mail: reese@uni-wuppertal.de

J. Schultes
e-mail: jschultes@uni-wuppertal.de

M. Stiglmayr
e-mail: stiglmayr@uni-wuppertal.de

P. Jaksch · A. Liefke · V. Marciniak
Siemens AG, Power and Gas, Common Technical Tools Mellinghoffer Str. 55, 45473 Mülheim an der Ruhr, Germany
e-mail: peter.jaksch@siemens.com

A. Liefke
e-mail: alexander.liefke@siemens.com

V. Marciniak
e-mail: vincent.marciniak@siemens.com

computed on a complex turbo geometry. These simulations are coupled in the multi-objective shape optimization process. See Fig. 4.1 for a schematic illustration of the multi-physical simulation/optimization cycle.

Here it is important to understand that the different design goals are closely interwoven. Efficiency of an areo-design is evaluated by computational fluid dynamics, mostly Reynolds averaged Navier Stokes, but also depends on the firing temperature. Higher firing temperature however affects the fatigue life which is evaluated by structural mechanics. Last but not least, the amount of cooling air consumption also is a driver for both, efficiency *and* fatigue life.

The challenge for the GivEn project thereby is to adjoin a highly multi-physical simulation chain with continuous coupling, to determine form gradients and form Hessians with respect to different objectives, and to make it usable in the multiobjective optimization for the turbine design process.

Coupled multi-physics simulation is an ongoing topic in turbo-machinery. For recent surveys of these fluid dynamics and heat transfer topics, see e.g. [101, 103]. The important topic of turbo-machinery life calculation is often treated separately and from a materials science point of view, see e.g. [15, 19, 80]. In contrast to the traditional separation of the mechanical and the fluid dynamics properties, the approach we follow in GivEn preserves a holistic viewpoint.

The algorithmic optimization of turbo-machinery components by now has a long history. While in the beginning genetic algorithms were used predominantly, in recent times data driven methods like Gaussian processes or (deep) neural networks predominate [1, 21, 98]. The strength of such procedures lie in global search approaches. As an alternative, gradient based optimization using adjoint equations are seen as a highly effective local search method [34, 38], see also [59] for a recent review

D. Luft · V. Schulz
Universität Trier, Fachbereich IV, Research Group on PDE-Constrained Optimization, 54296 Trier, Germany
e-mail: luft@uni-trier.de

V. Schulz
e-mail: volker.schulz@uni-trier.de

L. Mäde · S. Schmitz · J. Steiner
Siemens Gas and Power GmbH & Co. KG, Probabilistic Design, GP PGO TI TEC PRD, Huttenstr. 12, 10553 Berlin, Germany
e-mail: lucas.maede@siemens.com

S. Schmitz
e-mail: schmitz.sebastian@siemens.com

J. Steiner
e-mail: johannes.steiner@siemens.com

B. Engel
University of Nottingham, Gasturbine and Transmission Research Center (G2TRC), Nottingham NG72RD, UK
e-mail: engel.benedikt@nottingham.ac.uk

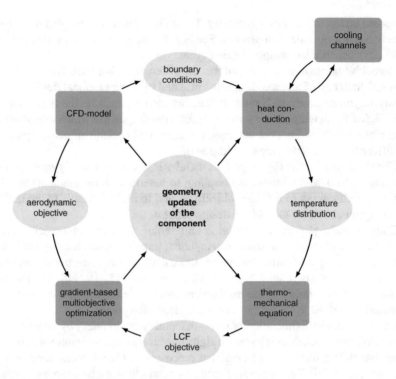

Fig. 4.1 Information flow and dependencies between project parts

including a comparison of the methods and [6] for bringing the data driven and the adjoint world together using gradient enhanced Gaussian processes [1, 33].

When combining the challenge of multi-physics and multi-objective optimization, it would be desirable to treat mechanical and fluid dynamic aspects of turbomachinery design on the same footing. A necessary prerequisite for this is the probabilistic modelling of the mechanisms of material damage, as this enables the application of the adjoint method [14, 39, 41–44, 68, 91]. This is not possible with a deterministic calculation of the lifetime of the weakest point, as taking the minimum over all points on the component is a non differential operation.

The GivEn consortium exploits these new opportunities for multi-objective and multiphysics optimization. It brings together a leading original equipment manufacturer (Siemens Power and Gas), technology developing institutions (German Aero Space Center (DLR) and Siemens CT) as well as researchers from academia (Universities of Trier and Wuppertal). Since 2017 this consortium addresses the challenges described in a joint research effort funded by the BMBF under the funding scheme "mathematics for innovation". With the present article, we review the research done so far and give an outlook on future research efforts.

This paper is organized as follows. In Sect. 4.2 we describe our research work on the different physical domains including the usage of adjoint equations, improved shape gradients and gradient based multiobjective optimization. Following the design scheme outlined in Fig. 4.1, we start with aerodynamic shape optimization in Sect. 4.2.1 using modern mesh morphing based on the Steklov-Poincaré definition of shape gradients [92–94], then proceed to heat transfer and the thermal loop in Sect. 4.2.2. Section 4.2.3 includes related probabilistic failure mechanisms. The model range from empirical models based on Weibull analysis and point processes to elaborate multi scale models. Section 4.2.4 presents shape optimization methods that are based on the probability of failure and develops a highly efficient computational framework based on conformal finite elements. Section 4.2.5 presents novel fundamental results on the existence of Pareto fronts in shape optimization along with algorithmic developments in multiobjective gradient based shape optimization including scalarization, biobjective gradient descent and gradient enhanced Gaussian processes.

In Sect. 4.3 we describe the industrial perspective from the standpoint of the DLR and Siemens. While in Sect. 4.3.1 the DLR gives a description of the interfaces with and the possible impact to the DLR's own R&D roadmap, Siemens Power & Gas in Sect. 4.3.2 relates adjoint based multiobjective optimization with adjoint multiobjective tolerance design and presents an application on real world geometries of 102 casted and scanned turbine vane geometries.

Let us note that this work is based on the papers [13, 20, 26, 41, 42, 60–62, 66, 68] that have been published with (partial) funding by the GivEn consortium so far. As this report is written after about half of the funding period of the project, we also give comments on future research plans within GivEn and beyond.

4.2 Areas of Mathematical Research and Algorithmic Development

The project GivEn researches the multiobjective free-form optimization of turbo geometries. For this purpose, the thermal and mechanical stress of the turbine blades and their aerodynamic behavior must be modelled, simulated and optimised. In the following we describe the components of the multiphysical simulation and optimization, namely aerodynamic shape optimization, heat transfer and thermal loop, probabilistic objective functionals for cyclic fatigue, shape optimization for probabilistic structure mechanics, multiobjective optimization, and probabilistic material science.

4.2.1 Aerodynamic Shape Optimization

Shape optimization is an active research field in mathematics. Very general basic work on shape calculus can be found in [27, 50, 99]. Aerodynamic investigations can be found in [87, 89]. New approaches understand shape optimization as the optimization on shape manifolds [92, 105] and thus enable a theoretical framework that can be put to good practical use, while at the same time leading to mathematical challenges, as no natural vector space structure is given. Otherwise, applications usually use finite dimensional parameterizations of the form, which severely limits the space of allowed shapes.

4.2.1.1 Aerodynamic State and Adjoint Equations

We search for stationary points of a discretized objective functional $J_D(q, x)$, where x is a set of discrete points describing the shape to be optimized. The aerodynamic state

$$q = \begin{pmatrix} \rho \\ \rho V \\ \rho E \end{pmatrix}$$

around the shape is described by the density, the velocity vector and the internal energy of the fluid. It is implicitly determined by requiring it to fulfill the Navier-Stokes Equations in a rotational frame of reference

$$\frac{\partial q}{\partial t} + \mathrm{div}(F(q)) + S(q) = 0 \tag{4.1}$$

with the fluxes being

$$F = \begin{pmatrix} \rho V \\ \rho V \otimes V + p\,\mathrm{Id} - \tau \\ \rho V H_t - \tau V + Q \end{pmatrix}$$

and accounting for source terms resulting from the rotating system

$$S = \begin{pmatrix} 0 \\ 2\rho\omega \times V - \rho|\omega|^2 r\,e_r \\ 0 \end{pmatrix}.$$

Here ω denotes the vector of rotation, the local radius is r and e_r is the radial unit vector. The heat flux density is given by $Q = -\kappa\nabla T$ and the viscous stresses are

$$\tau = 2\mu s - (\frac{2}{3}\mu - \mu_v)(\mathrm{Tr}\ s)\,\mathrm{Id}$$

with

$$s_{ij} = \frac{1}{2}\left(\frac{\partial V^j}{\partial x_i} + \frac{\partial V^i}{x_j}\right).$$

Assuming an ideal gas the rothalpy is given by $H_t = E + \frac{p}{\rho}$ and the internal energy of the gas is $E = \frac{1}{\gamma-1}\frac{p}{\rho} + \frac{1}{2}$.

The equations are Favre averaged and the result closed by means of turbulence models. By applying the divergence theorem, these equations are spatially discretized using the finite volume method. Assuming $\frac{\partial q}{\partial t} = 0$ a steady state solution method is obtained by pseudo-time marching.

The adjoint procedure to calculate gradients is derived from the Lagrange formalism: $J(q^*, x)$ can be computed under the assumption, that the solution method for Eq. 4.1 has a contractive fixed point $G(q^*, x) = q^*$. Treating this equation as constraint, and differentiating the resulting Lagrange formalism

$$L(q^*, x, \lambda) = J(q^*, x) + \lambda^T \left(G(q^*, x) - q^* \right) \qquad (4.2)$$

w.r.t q, one obtains a solution procedure for μ^*:

$$\lambda_{i+1}^T = \frac{\partial J(q^*, x)}{\partial q} + \lambda_i^T \frac{\partial G(q^*, x)}{\partial q}$$

The shape derivatives of the aerodynamic functionals are then

$$\frac{\mathrm{d}J}{\mathrm{d}x} = \frac{\partial J(q^*, x)}{\partial x} + \lambda^{*T} \frac{\partial G(q^*, x)}{\partial x} \qquad (4.3)$$

The terms can be computed using algorithmic differentiation in reverse mode. References [6, 81] describe the construction of the discrete adjoint for the turbomachinery simulation suite TRACE (cf. [10]), called adjointTRACE.

4.2.1.2   Surface Gradient

In the shape space setting, the use of volume formulations has been shown in combination with form metrics of the Steklov-Poincaré type [93, 94] were shown to be numerically very advantageous, since the volume formulation in comparison to the formally equivalent boundary formulation for canonical discretizations have better approximation properties and also weaker smoothness requirements of the functions involved. Additionally the Steklov-Poincaré type metrics require a free combination of volume and boundary formulations together with an inherently good approximation of the Shape-Hessian operators.

In order to exploit these theoretical advances for industrial applicability meeting high-end standards, the TRASOR (TRACE Shape Optimization Routine) software package for non-parametric shape optimization routines has been created. This software package is built on several solver bundles connected by an interface in Python 2.7 and 3.5. One major package bundle provided by the DLR and incorporated in TRASOR is TRACE 9.2. The TRASOR software incorporates shape gradient representations using Steklov-Poincaré-metrics (cf. [93, 94]) based on shape sensitivities derived by automatic differentiation provided by adjointTRACE

TRASOR also interfaces with FEniCS 2017.2.0 [5, 63], which is a Python based finite element software utilizing several sub-modules, such as the Unified Form Language (UFL [4]), Automated Finite Element Computing (DOLFIN [64, 65]) and PETSc [7] as a linear algebra backend, in order to solve differential equations based weak formulations. Various solver options, including CG, GMRES, PETCs's built in LU solver and preconditioning using incomplete LU and Cholesky, SOR or algebraic multigrid methods are available in FEniCS and thus applicable in TRASOR. FEniCS/PETSc also offers the possibility to parallelize finite element solving, making the Steklov-Poincaré gradient calculation scalable in processor number.

Features of the software package TRASOR include

- automatic file generation and management for TRACE and adjointTRACE
- interface between TRACE and FEniCS, including automatic FEniCS mesh generation from .cgns files
- steepest descent optimization using TRACE intern gradients
- steepest descent optimization using Steklov-Poincaré gradients calculated in FEniCS
- target parameter selection for various parameters found in TRACE, including all parameters listed in [37]
- generation of .pvd and .vtu files of gradients, sensitivities, meshes and flow simulation data for visual post processing

TRASOR features are tested on the low-pressure turbine cascade T106A designed by MTU Aero Engines (cf. [53]). The algorithm using Steklov-Poincaré gradients is outlined in Algorithm 1.

In order to exploit FEniCS it is necessary to create an unstructured computational mesh with vertices prescribed by TRACE. As FEniCS 2017.2.0 is not fully capable of supporting hexahedral and quadrilateral elements (this should be available with FEniCS 2020), hexahedral and quadrilateral elements used in TRACE are partitioned to conforming tetrahedral and triangular elements respecting the structured TRACE mesh. The conversion process including the data formats for TRACE to FEniCS mesh conversion are depicted in Fig. 4.2 (cf. [65, 86]).

To represent the TRACE generated mesh sensitivities $D_{ad}J_D(q, x)$ as a Steklov-Poincaré gradient a sufficient metric has to be chosen. In the following we describe the metric in a continuous framework, by assuming a shape continuous version $J(\Omega_{ext})$ of the discrete functional $J_D(q, x)$. It is helpful to understand that Steklov-Poincaré metrics need not have an immediate connection to the state equations of

Fig. 4.2 TRACE to FEniCS pipeline

shape optimization problems at hand, which are introduced in Sect. 4.2.1 for our case. These metrics form linear systems and serve the purpose of transforming the raw shape sensitivities to gradients with increased regularity, which makes them more suitable for mesh morphing routines used in shape optimization. Following [94], we implemented a linear elasticity model, which will function as a Steklov-Poincaré metric. So in this case, additional regularity for the resulting gradients is given by increasing smoothness through solution of the linear elasticity system

$$
\int_{\Omega_{\text{ext}}} \sigma\left(\nabla^{StP} J(\Omega_{\text{ext}})\right) : \varepsilon(U)\, dx = D\, J(\Omega_{\text{ext}})[U] \quad \forall\, U \in H_0^1(\Omega_{\text{ext}}, \mathbb{R}^d)
$$

$$
\begin{aligned}
\nabla^{StP} J(\Omega_{\text{ext}}) &= 0 && \text{on } \Gamma_{\text{Inlet/Outlet}} \\
\sigma(U) &= \lambda \operatorname{Tr}\bigl(\varepsilon(U)\bigr)I + 2\mu\varepsilon(U) \\
\varepsilon(U) &= \frac{1}{2}(\nabla U + \nabla U^\top),
\end{aligned}
\tag{4.4}
$$

where $\lambda \in \mathbb{R}$, $\mu \in \mathbb{R}_+$ are the so called Lamé parameters. If D is the entire duct including the shape of the turbine blade Ω, then $\Omega_{\text{ext}} = D \setminus \Omega$ is the external computational domain where the fluid dynamics takes place. Continuous Galerkin type elements of order one are used for target and test spaces in the FEniCS subroutine conducting the shape gradient calculation.

The following Algorithm 1 is a prototype for a shape optimization problem, including the Steklov-Poincaré gradient representation. Results of an exemplary compar-

---

1 Set flow parameters in TRACE.cgns, optimization parameters and targets in TRASOR.py
2 Build TRASOR file architecture
3 Assemble and load FEniCS data from TRACE.cgns
4 **while** $\|\nabla^{StP} J(\Omega_{\text{ext},k})\| > \varepsilon_{\text{shape}}$ **do**
5   Flow simulation and (AD) checkpoint creation using TRACE
6   Calculate mesh sensitivities by automatic differentiation using adjointTRACE
7   Pass mesh sensitivities to FEniCS setup
8   Generate Steklov-Poincaré gradient in FEniCS:
9   Calculate Lamé-Parameters
10   Solve linear elasticity problem (4.4)
11   Extract target and flow values to update/ create protocols and .pvd/.vtu files
12   Deform FEniCS mesh using FEniCS Steklov-Poincaré gradient and ALE (Arbitrary Lagrangian-Eulerian)
13   Create TRACE_deformation.dat files from FEniCS Steklov-Poincaré gradient
14   Deform TRACE mesh using PREP
15 **end**

Algorithm 1: TRASOR algorithm using Steklov-Poincaré gradients

---

ison between TRACE and Steklov-Poincaré gradients using the introduced routine are shown in Fig. 4.12. The TRACE gradient is generated by solving a linear elasticity mesh smoothing system with Dirichlet boundaries being the lattice sensitivities

$D_{ad} J(\Omega_{ext})$, whereas the Steklov-Poincaré gradient is calculated by solving the linear elasticity system (4.4) with Lamé parameters $\lambda \equiv 0$ and constant $\mu > 0$.

Further, a Steklov-Poincaré gradient representation using different bilinear forms matching the shape Hessian of the RANS flow and the target at hand are object of further studies, which might open new possibilities with superior convergence and mesh stability behavior.

4.2.2 Heat Transfer and the Thermal Loop

The numerical simulation of coupled differential equation systems is a challenging topic. The difficulty lies in the fact that the involved (P)DEs may differ in type and also in order, and thus require different types and quantities of boundary conditions [8]. The question of the correct coupling is closely related to the construction of so-called transparent boundary conditions, which are based on the coupling of interior and exterior solutions.

The numerical simulation of coupled differential equation systems by means of co-simulation has the innate advantage that one can choose the optimal solver for each sub-system, for example by employing pre-existing simulation software. Most of the work done in this field concerns transient, i.e. time dependent, problems. In our case, however, we are interested in steady state systems which rarely get special attention in current research.

Our model problem arises from the heat flow in a gas turbine blade. Since higher combustion temperatures result in better efficiency [84], engineering always strives for means to achieve these. However, this is limited by the material properties of the turbine blade, especially its melting point. One way to mitigate this, is by cooling the blade from the inside. This is done by blowing air through small cooling ducts. These ducts have a complex geometry to increase turbulence of the airflow and maximize heat transfer from the blade to the relatively cool air. For an overview, we refer to [49]. Due to the small length-scales and high turbulence, regular fluid dynamics simulation techniques are infeasible for the simulation of the airflow within the ducts. Instead, they are modeled as a one-dimensional flow with parametric models for friction and heat transfer, similar to the work in [72, 101]. This model is the result of experimental and empirical studies and is designed to be as simple as possible while still capturing the dynamics reasonably well. This results in a mostly linear model, however there is a quadratic dependence on the velocity v in the friction term.

$$w\frac{\partial v}{\partial x} = A\frac{\partial p}{\partial x} + \frac{A}{2D_h}f\rho v^2 + A\rho\omega^2 r\frac{\partial r}{\partial x}$$
$$\frac{\partial(vT_c)}{\partial x} = S$$
$$\rho = \frac{w}{vA} \tag{4.5}$$
$$p = \rho R_s T_c$$

Here, w is the mass flow through the channel which is assumed constant (i.e. only one inlet and outlet), v is the fluid velocity, ρ is the fluid's density and p and T_c denote the pressure and Temperature of the fluid, respectively. A is the cross-sectional area and D_h the hydraulic diameter of the channel, with f as the Fanning friction factor. ω and r are only relevant in the rotating case and denote the angular velocity and distance from the axis of rotation. R_s is the specific gas constant of the fluid and S denotes the heat source term from the heat flux through the channel walls.

The system (4.5) takes the form of a DAE, but can be transformed into a system of ODEs by some simple variable substitutions. Since the physical motivation behind the terms is easier to understand in the DAE form, this is omitted here.

The heat conduction within the blade material is modeled by a PDE. In the transient case, this would be a heat equation. In the stationary case, it is given by a Laplace equation. The heat transfer across the boundary is given by Robin boundary conditions, that prescribe a heat flux across the boundary depending on the temperature difference between "inside" and "outside". Temperature in (4.6) is denoted by T_m to signify that it is mathematically a different entity than the temperature in the cooling channel, denoted by T_c in (4.5). k denotes the thermal conductivity of the blade metal, while h_{int} and h_{ext} are the heat transfer coefficients of the internal and external boundary.

$$
\begin{aligned}
k\nabla^2 T_m &= 0 && \text{in } \Omega \\
-k\tfrac{\partial T_m}{\partial p} &= h_{\text{int}}(T_m - T_{\text{int}}) && \text{at } \partial\Omega_{\text{int}} \\
-k\tfrac{\partial T_m}{\partial n} &= h_{\text{ext}}(T_m - T_{\text{ext}}) && \text{at } \partial\Omega_{\text{ext}}
\end{aligned}
\tag{4.6}
$$

The coupling between Eqs. (4.5) and (4.6) is realized via the boundary condition, more specifically the internal boundary temperature T_{int} as a function of T_c, on the conduction side and the source term S in the cooling duct equations, which is a function of the values of $\partial T_m/\partial n$ on the cooling duct boundary. To ensure the coupling is proper and energy conserving, we are currently working on a port-Hamiltonian formulation of the model system.

For the numerical simulation, the coupled system is discretized using a finite elements scheme for the conduction part (4.6). This is done, because it ensures we can choose the mesh for the conduction part in a way that it is identical with the mesh used for the structural mechanics simulation described in Sect. 4.2.4, that uses the calculated temperatures as an input. For the cooling duct part (4.5), we use a finite volume scheme, as that makes it easier to have energy conservation across the boundary and provides a clear mapping of the PDE boundary to cooling channel elements. Since it is a one-dimensional system, this is just an alternative approach to obtain the standard finite difference central-differences scheme in a way that's more convenient for this application. Since we employ a central-differences scheme, we also do not have to worry about artificial diffusion.

The resulting discretized system is then solved by solving each subsystem and updating the boundary condition respectively the right hand side of the other system, alternating between the two subsystems until the solutions of two consecutive

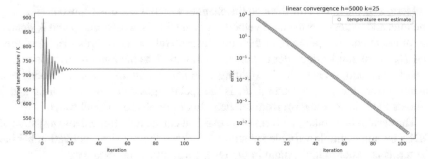

Fig. 4.3 Behavior of the cooling duct outlet temperature (left) and error-estimate (right) for a converging set of parameters ($h_{int} = h_{ext} = 5000, k = 25$)

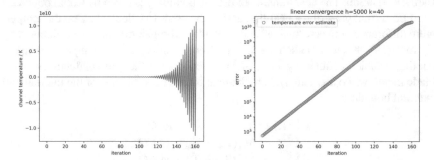

Fig. 4.4 Behavior of the cooling duct outlet temperature (left) and error-estimate (right) for a diverging set of parameters ($h_{int} = h_{ext} = 5000, k = 40$)

iterations differ by a sufficiently small margin. This back-and-forth iteration is reminiscent of a Gauß-Seidel iteration scheme, or more general, a fixed-point iteration.

Numerical tests have shown that this iterative solution indeed exhibits linear convergence as seen in Fig. 4.3, with the solution behaving like a dampened oscillation approaching the "correct" solution. These numerical tests also indicated that the convergence is not unconditional, but depends on the parameter values chosen for the system, especially the thermal conductivity k and the heat transfer coefficients h. High values of h lead to divergence and turn the aforementioned dampened oscillation into one with an exponentially increasing amplitude as seen in Fig. 4.4.

4.2.3 Probabilistic Objective Functionals for Material Failure

Since the pioneering work of Weibull [104], the probabilistic modelling of material failure has been an established field of material science, see about [9]. Applications to the *Low Cycle Fatigue* (LCF) damage mechanism can be found in [29, 76, 100]. In these studies, crack formation is modelled by percolation of intra-granular cracks or by kinetic theory for the combination of cracks. The mathematical literature mainly contains generic volume or surface target functions without direct material reference. In numerical studies, global compliance is usually chosen as the objective functional, which also does not establish a direct relationship to material failure, see e.g. [18].

The objective functional used in GivEn for the probability of failure originates [43, 44, 68] see also [39] for multi-scale modeling. A connection between probabilistic functional objectives of materials science and the mathematical discipline of shape optimization is produced for the first time in [43], see also [12–14].

The aim in this sub-area is the probabilistic modelling of material damage mechanisms and the calculation of form derivatives and form Hessian operators for the failure probabilities of thermal and mechanically highly stressed turbine blades.

The physical cause of the LCF mechanism in the foreground is the sliding of crystal dislocations along lattice planes with maximum shear stress and is therefore dependent on the random crystal orientation. For this reason, so-called intrusions and extrusions occur at the material surface, which eventually lead to crack formation [15, 79]. In an effective approach, this scattering of material properties can be empirically investigated within the framework of reliability statistics.

In the deterministic approach prevalent in mechanical engineering, life expectancy curves are used to determines the service life at each point of the blade surface. The shortest of these times is to failure over all points is then, under consideration of safety discounts, converted to the permitted safe operating time of the gas turbine. The minimum formation inherent in this process means that the target functions cannot be differentiated. Probabilistic target functions, on the other hand, can be defined according to the form and continuously adjusted.

In particular, the stability of discretization schemes must be examined in both with regard to geometric approximation of the forms as well as the solutions. The background is that H^1 solutions are insufficient for a finite probabilistic target functional, especially if notch support is also considered [67]. Suffice of this must be used a $W^{k,p}$ solution and approximation theory [17].

Next, the calculation of the shape gradients and shape Hesse operators of the functionals essentially follows [99], with open questions about the existence and properties of shape gradients for the surface- and stress-driven damage mechanism LCF still to be clarified. This program has been started within the GivEn research initiative, cf. [11]. Analogous to [43], the solution strategy is based on a uniform regularity theory for systems of elliptic PDEs, cf. [2, 3, 17]. In particular, the mathematical status of the continuously-adjusted equation deserves further attention, as this has a high regularity loss for surface-driven LCF.

In the following we present a hierarchy of probabilistic failure models that give rise to objective functionals related to reliability. We start with the simple Weibull model, proceed with a probabilistic model for LCF proposed by [68, 91] and then give an outlook on the multi-scale modeling of the scatter in probabilistic LCF, see [26].

4.2.3.1   The Weibull Model via Poisson Point Processes

Technical ceramic has multiple properties such as heat or wear resistance that make
them a widely used industrial material. Different to other industrial material, the
physical properties of ceramic materials highly depend on the manufacturing process. What determines the failure properties the most, are small inclusions that stem
from the sintering process. These make ceramic a brittle material, leading to a some-
what high possibility of failure of the component under tensile load often before the
ultimate tensile strength is reached [15].

When applying tensile load, these inclusions may become the initial point of
a crack, developing into a rupture if a certain length of the radius of the crack is
exceeded at a given level of tensile stress. Therefore, the probability of failure under
a given tensile load is the probability that a crack of critical length occurs. Or to phrase
it differently, the survival probability in this case is the probability that exactly zero
of these critical cracks occur. Thus, for a given domain $\Omega \subset \mathbb{R}^d$, $d = 2, 3$ with a
suitable counting measure $\mathscr{N}$ [54], we can express the failure probability in the
following way,

$$\text{PoF}(\Omega) = 1 - \mathbb{P}\big(\mathscr{N}(A_c(\Omega)) = 0\big), \qquad (4.7)$$

where $A_c(\Omega))$ is the set of critical cracks. The probability, that one of the inclusions
grow into a critical crack, mainly depends on the local stress tensor $\sigma_n(u)$, which
itself is determined by a displacement field $u \in H^1(\Omega, \mathbb{R}^d)$, that is the solution of
a linear elasticity equation. As there is no other indication, it is feasible to assume
that the location, size and orientation of the initial inclusions are independent of
each other and uniformly randomly distributed. Under these assumptions, it follows
that the counting measure $\mathscr{N}(\Omega)$ is a Poisson point process (PPP). Taking further
material laws into account it follows that [14]

$$\text{PoF}(\Omega|u) = 1 - \mathbb{P}(\mathscr{N}(A_c(\Omega, u)) = 0) = 1 - \exp\{-v(A_c(\Omega, u))\}, \qquad (4.8)$$

with the intensity measure of the PPP

$$v(A_c(\Omega, u)) = \frac{\Gamma(\frac{d}{2})}{2\pi^{\frac{d}{2}}} \int_{\Omega} \int_{S^{d-1}} \int_{a_c}^{\infty} dv_a(a)\, dn dx. \qquad (4.9)$$

With some reformulations we find our objective functional of Weibull type

$$J_1(\Omega, u) := v(A_c(\Omega, u)) = \frac{\Gamma(\frac{d}{2})}{2\pi^{\frac{d}{2}}} \int_{\Omega} \int_{S^{d-1}} \left(\frac{\sigma_n}{\sigma_0}\right)^m dn dx. \qquad (4.10)$$

This functional (4.10) will be one of the objective functionals in the following (multiobjective) gradient based shape optimization.

4.2.3.2 Probabilistic Models for LCF

Material parameters relevant for fatigue design, like the HCF fatigue resistance were considered as a random variable for a long time [75, 79, 97] and distributions and their sensitivities were even recorded in general design practice standards [30, 31, 52, 74]. The existence of flaws, such as crystal dislocations, non-metallic inclusions or voids in every material has early lead to the discovery of the statistical size effect [51, 70, 71, 78]. Within the last decade, a local probabilistic model for LCF based on the Poisson point process was developed by Schmitz et al. [90, 91] for predicting the statistical size effect in any structural mechanics FEA model. It approximates the material LCF life statistics with a Weibull distribution which allows developing a closed form integral solution for the distribution scale η (see Eq. (4.13)). Recently, Mäde et al. [68] presented a validation study of the combined size and stress gradient effect modeling approach within the framework of Schmitz et al. [91]. If stress gradients are present in components, they can have an increased LCF life. The benefit is proportional to the stress gradient but also material dependent [28, 95, 96, 106]. A stress gradient support factor $n_\chi = n_\chi(\chi \mid \vartheta)$, a functional of the normalized stress gradient χ and material-specific parameters ϑ [96], is introduced to quantify the effect. While the size effect as described by the surface integral (4.13) causes an actual delay or acceleration in fatigue crack initiation, researchers share the interpretation that stress gradient support effects in LCF root back to retarded propagation of meso-scale cracks in the decreasing stress field [22, 56, 58, 69, 79, 82]. In order to compare the stress gradient effect for different materials, a common detectable, "technical" crack size must be defined. Since the stress gradient $\chi(\mathbf{x})$ is, like the stress field, a local property, it was integrated into the calculation of the local deterministic life $N_{\text{det}}(\mathbf{x})$ with the Coffin-Manson-Basquin model:

$$\frac{\varepsilon_a(\mathbf{x})}{n_\chi(\chi(\mathbf{x}) \mid \vartheta)} = \frac{\sigma'_f}{E} \cdot (2N_{\text{det}}(\mathbf{x}))^b + \varepsilon'_f \cdot (2N_{\text{det}}(\mathbf{x}))^c . \qquad (4.11)$$

Here, the stress is computed with the aid of the linear elasticity equation, which this time is not a technical tool for smoothing gradients as in (4.16), but represents the physical state, namely

$$
\begin{aligned}
\nabla \cdot \sigma(u) + f &= 0 & &\text{in } \Omega \\
\sigma(u) &= \lambda(\nabla \cdot u)I + \mu(\nabla u + \nabla u^\top) & &\text{in } \Omega \\
u &= 0 & &\text{on } \partial\Omega_D \\
\sigma(u) \cdot n &= g & &\text{on } \partial\Omega_N.
\end{aligned}
$$

Here, Ω represents the component, $\lambda > 0$ and $\mu > 0$ are Lamé coefficients and $u : \Omega \to \mathbb{R}^3$ is the displacement field on Ω obtained as a reaction to the volume forces f and the surface loads g. We connect the topic of optimal probabilistic reliability to shape optimization elasticity PDE as state equation and classify Poisson point process models according to their singularity [11]. Following [68, 91], we obtain for the probability of failure at a number of use cycles n

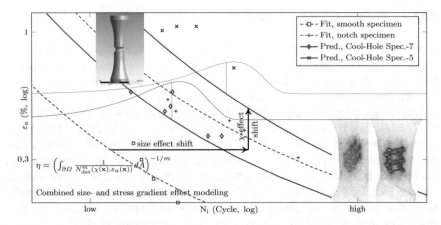

Fig. 4.5 Strain Wöhler plot of LCF test data, calibrated (dashed) and predicted (solid) median curves for smooth (□), notch (+) and cooling hole specimens (◇, ×). All Wöhler curves are interpolated median values of the Weibull LCF distributions exemplary indicated with the thin density function plot

$$\text{PoF}(\Omega, n) = 1 - e^{-n^m J_R(\Omega, u)} \tag{4.12}$$

The functional $J(\Omega, u)$ that is arising out of this framework is given by:

$$J_R(\Omega, u_\Omega) := \int_{\partial\Omega} \left(\frac{1}{N_{\text{det}}(\nabla u_\Omega(x), \nabla^2 u_\Omega(x))} \right)^m dA. \tag{4.13}$$

N_{det} denotes the deterministic numbers of life cycles at each point of the surface of the component and m is the Weibull shape parameter.

Mäde et al. have calibrated the material parameters $\vartheta, E, \sigma'_f, \varepsilon'_f, b, c$ as well as the Weibull shape parameter m with the Maximum-Likelihood method simultaneously using smooth and notch specimen data simultaneously [67, 68]. The resulting model was able to predict the LCF life distribution for certain component-similar specimens (see Fig. 4.5[1]).

In the following, we apply this model as cost functional in order to optimize the component Ω w.r.t. reliability.

4.2.3.3 Multi-scale Modeling of Probabilistic LCF

While the Weibull-based approach from the previous subsection allows a closed-form solution and therefore fast risk assessment computation times, the microstructural

---

[1] Reprinted from Comp. Mat. Sci., 142, (2018) pp. 377–388, Mäde et al., Combined notch and size effect modeling in a local probabilistic approach for LCF, Copyright (2017), with permission from Elsevier.

mechanisms of LCF suggest a different distribution shape [73]. Since this is not yet assessable by LCF experiments in a satisfying way, Engel et al. have used numerical simulations of probabilistic Schmid factors to create an LCF model considering the grain orientation distribution and material stiffness anisotropy in cylindrical Ni-base superalloy specimens [26].

Polycrystalline FEA models were developed to investigate the influence of local multiaxial stress states a result of as grain interaction on the resulting shear stress in the slip systems. Besides isotropic orientation distributions also the case of a preferential orientation distribution was analysed. From the FE analyses, a new Schmid factor distribution, defined by the quotient of $\max(\tau_{rss})$ maximum resolved shear stress at the slip systems and von Mises stress σ_{vM}, was derived as a probabilistic damage parameter at each node of the model. Qualitatively as well as quantitatively, they differ largely from the single grain Schmid factor distribution of Moch [73] and also from the maximum Schmid factor distribution of grain ensembles presented by Gottschalk et al. [45]. Experimental LCF data of two different batches presented by Engel [24, 25] showed different LCF resistances and microstructural analyses revealed a preferential grain orientation in the specimens which withstood more cycles (see Fig. 4.6). By combining the Schmid factor based LCF life model of Moch [73] and the Schmid factor distribution generated by FEA, Engel et al. were able to predict just that LCF life difference [26]. Ultimately, it was found that the microstructure-based lifing model is able to predict LCF lives with higher accuracy than the Weibull approach by considering the grain orientation and their impact on the distributions of Young's moduli and maximum resolved shear stresses. However, the application is computationally demanding and its extension to arbitrary components still has to be validated.

4.2.4   Shape Optimization for Probabilistic Structure Mechanics

This sub-area deals with two kinds of failure mechanisms, failure of brittle material under tensile load and low cycle fatigue (LCF). As explained in the preceding section, the probability of failure for both failure mechanisms can be expressed as local integral of the volume (ceramics) or surface (LCF) over a non-linear function that contains derivatives of the state u_Ω subject to the elasticity equation (Sect. 4.2.3.2).

Since problems in shape optimization generally do not result in a closed solution, the numerical solution plays a major role, e.g. in [50]. Typically, the PDEs occurring as a constraint are discretized using the finite element method. Here, a stable and accurate mesh representation of Ω is needed for stable numerical results. Deciding for a mesh always means balancing the need for accuracy of the representation on the one hand and on the other hand the time to solution. Especially in applications such as shape optimization, where usually hundreds to thousands of iterations and thus changes in the geometry and mesh are needed to find a converged solution, the

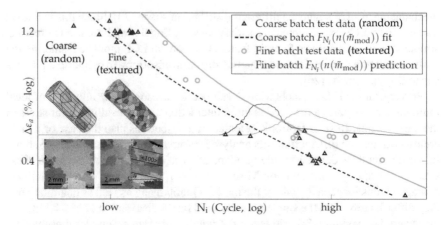

Fig. 4.6 Strain Wöhler plot of LCF test data, calibrated (dashed) and predicted (solid) median curves for specimens with isotropic and preferential grain orientation distribution (coarse and fine grain) by Engel et al. [26]. Calibration and prediction was carried out using the Schmid factor based LCF life distribution. The underlying Schmid factor distribution was derived from polycrystalline FEA simulations which considered the lattice anisotropy and orientation distribution in both specimen types

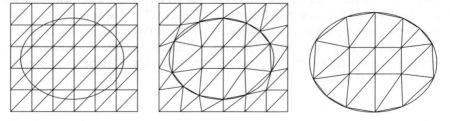

Fig. 4.7 Visualization of adaption of the grid (Reprinted from Progress in Industrial Mathematics at ECMI 2018, pp. 515–520, Bolten et al., Using Composite Finite Elements for Shape Optimization with a Stochastic Objective Functional, Copyright: Springer Nature Switzerland AG 2019.)

meshing in each iteration often becomes a bottle neck in terms of computational cost. Recent research therefore aims to find methods to move the grid points of a given representation in a stability preserving way, rather than to perform a re-meshing. These approaches result in unstructured grids. To exploit the means of high performance computing however, structured grids are way more desirable. This let us to consider an approach first developed in [46–48]. For demonstration purposes, the technique is described in two dimensions but easily extends to three dimensions.

We consider a rectangular domain $\tilde{\Omega}$ which is discretized by a regular triangular grid. We assume that all admissible shapes that occur during the optimization process lie in this domain. The regular grid on $\tilde{\Omega}$ is denoted by $\tilde{T}$, the number of elements by N^{el} and the number of nodes by N^{no}. In a second step, the boundary $\delta\Omega_0$ of the shape to be optimized Ω_0 is superimposed onto the grid, see Fig. 4.7a. The regular grid is then adapted to the boundary by moving the closest nodes to the intersections of the grid and the boundary, see Fig. 4.7b. The adapted grid is denoted by $\tilde{T}_0$. As the nodes are moved only, the connectivity of $\tilde{T}_0$ is the same as before. During the adaption process, the cells of $\tilde{T}_0$ are assigned a status as cells lying inside or outside of the component. The computations are only performed on those cells that are inside the component. When updating the grid according to the new shape of the domain Ω_1 and so forth, the adaption process starts with $\tilde{\Omega}$ again, while taking into account the information about the previous domain, the step length and search direction that led to the current domain. By this, only the nodes that lie in a certain neighborhood of the previous boundary have to be checked for adaption. This leads to a speed up in meshing compared to actual re-meshing techniques. Additionally, as the grid is otherwise regular and the connectivity is kept, the governing PDE only has to be changed in entries representing nodes on elements that have been changed, hence no full assembly is needed, which is an advantage to both re-meshing and mesh morphing techniques.

The objective functional (4.10) is discretized via finite elements. For reduction of the computational cost, the adjoint approach than leads to the derivative

$$\frac{d J_1(X, U(X))}{dX} = \frac{\partial J_1(X, U)}{\partial X} + \Lambda \left[\frac{\partial F(X)}{\partial X} - \frac{\partial B(X)}{\partial X} U \right] \qquad (4.14)$$

$$B^\top(X)\Lambda = \frac{\partial J_1(X, U)}{\partial U} \qquad (4.15)$$

$$B(X)U = F(X), \qquad (4.16)$$

with (4.15) being the adjoint equation giving the adjoint state and (4.16) is the discretized linear elasticity equation, giving the discrete displacement U. X represents the discretized domain Ω.

With (4.16) and (4.15) the derivative (4.14) is calculated on the structured mesh as visualized in Fig. 4.8a. For the optimization, more closely described in the following Sect. 4.2.5, the gradient is smoothed using a Dirchlet-to-Neumann map [88] (see Fig. 4.8b). This provides the shape gradients needed for further gradient based optimization steps in the following section.

4.2.5 Multiobjective Optimization

The engineering design of complex systems like gas turbines often requires the consideration of multiple aspects and goals. Indeed, the optimization of the reliability of a structure usually comes at the cost of a reduced efficiency and a higher volume and, hence, a higher production cost. Other relevant optimization criteria are, for example, the minimal buckling load of a structure or its minimal natural frequency [50]. In this section, we exemplify such trade-offs at the example of the mechanical integrity and the cost of a ceramic component in a biobjective PDE constrained shape optimization problem. While in the context of gas turbines this biobjective model

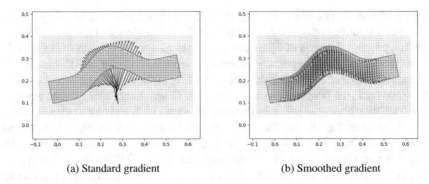

(a) Standard gradient   (b) Smoothed gradient

Fig. 4.8 Gradient in standard scalar product and smoothed gradient; $N_x = 64$, $N_y = 32$

should be extended by an aerodynamic objective function as illustrated in Fig. 4.1, the below analysis is representative in the sense that it well illustrates the advantages and difficulties in multiobjective shape optimization. Towards this end, we model the mechanical integrity $J_1(\Omega, u)$ of a component Ω as described in Sects. 4.2.3 and 4.2.4, see (4.10), while the cost $J_2(\Omega)$ is assumed to be directly proportional to the volume of the component, i.e., $J_2(\Omega) = \int_\Omega dx$.

Multiobjective shape optimization including mechanical integrity as one objective is widely considered, see, e.g., [16] for a recent example. Most of these works neither consider probabilistic effects nor use gradient information. The formulation introduced in Sects. 4.2.3 and 4.2.4 overcomes these shortcomings. It was first integrated in a biobjective model in [20], where two alternative gradient-based optimization approaches are presented. We review this approach and present new numerical results based on structured grids and advanced regularization.

4.2.5.1 Pareto Optimality

Multiobjective optimization asks for the simultaneous minimization of p conflicting objective functions $J_1, \ldots, J_p$, with $p \geq 2$. We denote by $\mathbf{J}(\Omega) = (J_1(\Omega), \ldots, J_p(\Omega))$ the outcome vector of a feasible solution $\Omega \in \mathcal{O}^{\mathrm{ad}}$ (i.e., an admissible shape). Since in general the optimal solutions of the objectives $J_1, \ldots, J_p$ do not coincide, a multidimensional concept of optimality is required. The so-called Pareto optimality is based on the component-wise order [23]: A solution $\Omega \in \mathcal{O}^{\mathrm{ad}}$ is *Pareto optimal* or *efficient*, if there is no other solution $\Omega' \in \mathcal{O}^{\mathrm{ad}}$ such that $\mathbf{J}(\Omega') \leqslant \mathbf{J}(\Omega)$, i.e., $J_i(\Omega') \leq J_i(\Omega)$ for $i = 1, \ldots, p$ and $\mathbf{J}(\Omega') \neq \mathbf{J}(\Omega)$. In other words, a solution is efficient if it can not be improved in one objective J_i without deterioration in one other objective function J_k.

4.2.5.2 Foundations for Multi-physics Multi-objective Shape Optimization

The existence of Pareto fronts for the multiobjective case are considered in a simplified analytical model replacing the RANS equations by potential theory with boundary layer losses. Pareto fronts can be replaced by scalarizations using the techniques from [12, 43] or [14, 35]. Their continuous course is investigated by variation of the scalarization and the associated optimality conditions. The convergence of the discretized Pareto optimum solutions against the continuous Pareto optimum solution shall be studied according to the approach of [50].

Optimizing the design of some component in terms of reliability, efficiency and pure performance includes the consideration of various physical systems that interact with each other. This leads naturally to a multiobjective shape optimization problem over a shape space $\mathcal{O}^{\text{ad}}$ with requirements represented by cost functionals $\mathbf{J} = (J_1, \ldots, J_p)$:

$$\begin{cases} \text{Find } \Omega^* \in \mathcal{O}^{\text{ad}} \text{ such that} \\ (\Omega^*, \boldsymbol{v}_{\Omega^*}) \text{ is Pareto optimal w.r.t. } \mathbf{J}. \end{cases}$$

The class of cost functional we use to model the requirements on the component are connected with physical state equations and arise from the probabilistic framework. They are described by

$$J_{\text{vol}}(\Omega, \boldsymbol{v}) := \int_\Omega \mathscr{F}_{\text{vol}}(x, \boldsymbol{v}, \nabla \boldsymbol{v}, \ldots, \nabla^k \boldsymbol{v}) \, dx,$$

$$J_{\text{sur}}(\Omega, \boldsymbol{v}) := \int_{\partial\Omega} \mathscr{F}_{\text{sur}}(x, \boldsymbol{v}, \nabla \boldsymbol{v}, \ldots, \nabla^k \boldsymbol{v}) \, dA.$$

Due to the fact that the event of failure, e.g. the crack initiation process takes place on the surface of the component, the physical systems need to fulfill regularity conditions in order to be includable in this setting. We describe the possible designs of the component in this shape optimization problem by Hölder-continuous functions which give us the possibility to freely morph the shapes in various designs while remaining the premised regularity conditions. In this situation uniform regularity estimates for the solutions of the physical system are needed in order to ensure the existence of a solution to this design problem in terms of Pareto optimality. The aim of this subproject is to translate a multi physical shape optimal design problem into the context of a well-posed multiobjective optimization problem.

We couple internal and external PDEs in order to describe the various forces that are inflected on the component. In this framework, using techniques based on precompactness of embedding between Hölder spaces of different index like in [12, 43], we are able to show [40] the existence of Pareto optimal shapes in terms of subsection 4.2.5.1 which form a Pareto front, see also [20] for a related result. We also prove the completeness of the Pareto front in the sense that the Pareto front coincides with the Pareto front of the closure of the feasible set (which is equivalent to the fact that every non-Pareto admissible shape is dominated by a Pareto optimal shape).

Further we investigated scalarization techniques which transform the multi-objective optimization problem into a uni-variate problem. In particular we considered the so-called achievement function and ε-constraint methods which depend, besides on the cost functional, on an additional scalarization parameter that represents the different weightings of the optimization targets, as e.g. reliability or efficiency. Hence, the shape space on which the optimization process takes place can also depend on this parameter and with it the corresponding space of optimal shapes as well. Under suitable assumptions on the continuous dependency of the scalarization method on the scalarization parameter, we are able to show a continuous dependency of the optimal shapes spaces on the parameter as well. For details we refer to the forthcoming work [40].

4.2.5.3 Multiobjective Optimization Methods

Algorithmic approaches for multiobjective optimization problems can be associated with two common paradigms: scalarization methods and non-scalarization methods. In [20], two algorithmic approaches are described: the weighted sum method as an example for a scalarization method, see, e.g., [23], and a multiobjective descent algorithm as an example for a non-scalarization method, see [32]. Gradient descent strategies were implemented for both methods to search for Pareto critical points, i.e., points for which no common descent direction for all objectives exists. In this section, we focus on weighted sum scalarizations and present new numerical results for a biobjective test case.

The weighted sum method replaces the multiobjective function $\mathbf{J}$ by the weighted sum of the objectives $J^\omega(\Omega) = \sum_{i=1}^p \omega_i J_i(\Omega)$. Here, $\omega \geqslant 0$ is a weighting vector that represents the relative importance of the individual objective functions. We assume without loss of generality that $\sum_{i=1}^p \omega_i = 1$. The resulting scalar-valued objective function J^ω can then be optimized by (single-objective) gradient descent algorithms, see e.g., [36]. If a global minimum of the weighted sum scalarization J^ω is obtained, then this solution is a Pareto optimal solution of the corresponding multiobjective optimization problem, see e.g., [23]. The converse is not true in general, i.e., not every Pareto optimal solution can be obtained by the weighted sum method. Indeed, the weighted sum method can not be used to explore non-convex parts of the Pareto front. Nevertheless, an approximation of the Pareto front can be obtained by appropriately varying the weights.

4.2.5.4 Case Study and Numerical Implementation

To illustrate the multiobjective perspective on shape optimization we exemplarily consider a biobjective instance with objectives J_1 and J_2 as introduced above (i.e., mechanical integrity and cost), and consider a simple ceramic component $\Omega \subset \mathbb{R}^2$

Fig. 4.9 Case study: general setup and starting solution

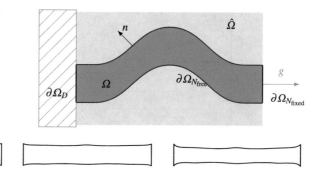

Fig. 4.10 Near Pareto critical solutions obtained by the weighted sum method

made from beryllium oxide (BeO) (with material parameter setting equal to [20], in particular Weibull's modulus $m = 5$). The volume force f is set to 1000 Pa. See Fig. 4.9 for an illustration of a possible (non-optimal) shape. This shape is used as starting solution for the numerical tests described below. The component is of length 0.6 m and is assumed to have a thickness of 0.1 m. It is fixed on the left boundary Ω_D and the tensile load is acting on the right boundary $\Omega_{N_{\text{fixed}}}$. The parts Ω_D and $\Omega_{N_{\text{fixed}}}$ are fixed, while the part $\Omega_{N_{\text{free}}}$ can be modified during the optimization process. The biobjective shape optimization problem is then given by

$$\min_{\Omega \in \mathcal{O}^{\text{ad}}} \mathbf{J}(\Omega) := (J_1(\Omega, u),\ J_2(\Omega))$$

$$\text{s.t. } u \in H^1(\Omega, \mathbb{R}^2) \text{ is the solution of a linear elasticity equation.}$$

(4.17)

The component is discretized using a regular 45×25 grid, see Sect. 4.2.4 and Fig. 4.8.

We use a gradient descent method to minimize the weighted sum objective function J^ω for different weight vectors $\omega \geqslant 0$. This is implemented using the negative gradient as search direction and the Armijo-rule to determine a step-size, see e.g. [36]. During the iterations, the component is modified by free form deformations using the method developed in Sect. 4.2.4. Since we have a regular mesh inside the component, only the grid points close to the boundary have to be adapted. When the modification during one iteration is too large, a complete remeshing is performed, still using the approach described in Sect. 4.2.4. To avoid oscillating boundaries and overfitting, we apply a regularization approach based on [94]. Numerical results for three choices of the weight vector ω are shown in Fig. 4.10, and an approximation of the Pareto front is given in Fig. 4.11. In these cases no remeshing step had to be performed, because the step length was restricted to the mesh size.

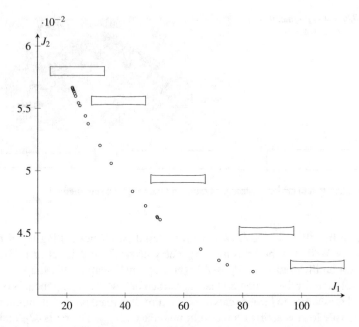

Fig. 4.11 Approximated Pareto front obtained by the weighted sum method

4.2.5.5 Gradient Enhanced Kriging for Efficient Objective Function Approximation

To cut computational time of the optimization process one can apply surrogate models to estimate expensive to compute objective functions. Optimization on the surrogate model is relatively cheap and yields new points which then in a next step are evaluated with the expensive original objective function. In the biobjective model presented above, the mechanical integrity J_1 is expensive, while the volume J_2 can be easily evaluated. We thus suggest to replace only the expensive objective J_1 by a model function.

Let $\{\Omega_1, \ldots, \Omega_M\} \subset \mathcal{O}^{\mathrm{ad}}$ be sampled shapes with responses $\{y_1, \ldots, y_M\} := \{J_1(\Omega_1), \ldots, J_1(\Omega_M)\}$. *Kriging* is a type of surrogate model that assumes that the responses $\{y_1, \ldots, y_M\}$ are realizations of Gaussian random variables $\{Y_1, \ldots, Y_M\} := \{Y(\Omega_1), \ldots, Y(\Omega_M)\}$ from a Gaussian random field $\{Y(\hat{\Omega})\}_{\hat{\Omega} \in \mathcal{O}^{\mathrm{ad}}}$. For an unknown shape Ω_0 the Kriging model then predicts

$$\hat{y}(\Omega_0) = \mathbb{E}[Y(\Omega_0) | Y(\Omega_1) = y_1, \ldots, Y(\Omega_M) = y_M],$$

i.e., the estimated objective value of Ω_0 is the conditional expectation of $Y(\Omega_0)$ under the condition that the random field is equal to the responses at the sampled shapes, or in other words the predictor is an interpolator. An advantage of this method is, that the model also provides information about the uncertainty of the prediction, denoted as $\hat{s}(\Omega_0)$, see [55] for more details.

If, as in our case, gradient information is available, one can incorporate this into the Kriging model which is then called *gradient enhanced Kriging*. One follows the same idea: the gradients $\{\bar{y}_1, \ldots, \bar{y}_{\bar{M}}\} := \{\nabla J_1(\Omega_1), \ldots, \nabla J_1(\Omega_{\bar{M}})\}$ are assumed to be realizations of the Gaussian random variables/vectors $\{\bar{Y}_1, \ldots, \bar{Y}_{\bar{M}}\}$. Adding these random variables to the ones w.r.t. the objective values enables one to predict objective values and gradients at unknown shapes Ω_0, see also [55] for more details.

In the optimization choosing the predictor $\hat{y}(\Omega_0)$ as the objective to acquire new points to evaluate with the original function may yield poor results. Since then one assumes that the prediction has no uncertainty, i.e. $\hat{s}(\Omega_0) = 0$, and areas, for which the predictor has bad values and a high uncertainty while the original function has better values than the best value at the moment, may be overlooked. Hence, one has to choose an acquisition function that incorporates $\hat{y}(\Omega_0)$ and $\hat{s}(\Omega_0)$ to balance the exploitation and exploration in the optimization.

We note that this gradient enhanced Kriging approach is a direction of ongoing development for the in house optimization process `AuoOpti` at the German Aerospace Center (DLR), see e.g. [57, 85] for design studies using the `AutoOpti` framework. In our future work, we therefore intend to benchmark the EGO-based use of gradient information with the multiobjective descent algorithms and identify their respective advantages for gas turbine design.

4.3   Applications

In this section we present the industrial implications of the GivEn consortium's research. The German Aerospace Center (DLR), Institute for Propulsion Technology, here is an important partner with an own tool development that involves an in-house adjoint computational fluid dynamics solver TRACE as well as a multiobjective optimization toolbox AutoOpti. As TRACE and AutoOpti are widely used in the German turbo-machinery industry, a spill-over of GivEn's method to the DLR assures an optimal and sustainable distribution of the research results.

In a second contribution, Siemens shows that results developed in the GivEn project can also be directly used in an industrial context, taking multiobjective tolerance design as an example.

4.3.1   German Aerospace Center (DLR)

Industrial turbomachinery research at DLR includes several activities that benefits from the insights gained in this project. These activities primarily pursue two goals:

(i) Assessment of technology potential for future innovations in the gas turbine industry.
(ii) Development of efficient design and optimization tools that can be used, for instance, to perform (i).

4.3.1.1 Challenges of Industrial Turbomachinery Optimization

The gain in aerodynamic performance of both stationary gas turbines and aircraft engines that has been achieved over the last decades, leaves little room for improvement if solely aerodynamics is considered. More precisely, aerodynamic performance enhancements that neglect the issues of manufacturing, structural dynamics or thermal loads, will typically not find their way into application. One of the reasons for this is the fact that real engines currently designed already have small "safety" margins. Summarizing, one can conclude that aerodynamic performance, structural integrity as well as manufacturing and maintenance costs have become competing design goals. Therefore, the design of industrial turbomachinery has become a multi-disciplinary multi-objective optimization problem. Accordingly, DLR is highly interested in advances concerning both simulation tools for coupled problems and multi-disciplinary optimization (MDO) techniques.

Multiobjective optimizations based on high-fidelity simulations are firmly established in the design of turbomachinery components in research and industry [59, 102]. As explained above, current developments increasingly demand the tighter coupling of simulations from multiple disciplines. Reliable evaluations of such effects require the simultaneous consideration of aerodynamics, aeroelasticity and aerothermodynamics. Moreover, optimization should account for the influence of results from these disciplines on component life-times.

Gradient-free optimization methods, typically assisted by surrogate modeling, prevail in today's practical design processes [77, 83]. A tendency towards optimizing with higher level of detail and optimizing multiple stages simultaneously leads to higher dimensional design spaces, making gradient-free methods increasingly expensive and gradient-based optimization the better suited approach.

4.3.1.2 Expected Impact of GivEn Results

The methods described in the preceding sections describe how derivatives for a fully coupled aerothermal design evaluation process can be computed efficiently and how a gradient based optimization procedure can be constituted for the design criteria of efficiency and component life-time. The process developed in the frame of this project serves as a research tool and a base to adopt the methods for other applications, resulting in different levels of simulation fidelity, different sets of disciplines as well as different objectives.

To illustrate this on exemplary computational results, a comparison of a TRACE gradient and a Steklov-Poincaré gradient for the isentropic total pressure loss coef-

(a) Steklov-Poincaré gradient   (b) TRACE Mesh Smoothing gradient

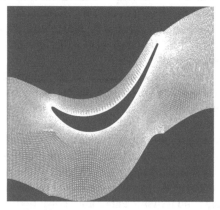

(c) FEniCS computational mesh of the T106A

Fig. 4.12 Comparison of Steklov-Poincaré gradient (upper left) with TRACE Mesh Smoothing gradient (upper right) on a FEniCS computational mesh of the T106A (lower center)

ficient in relative frame of reference based on dynamic pressure is portrayed in Fig. 4.12. The TRACE gradient generated by DLR software represents techniques used in the industry today, whereas the Steklov-Poincaré gradient stems from the current research frontier in Shape Optimization.

We can see additional gain of regularity in the gradient through Steklov-Poincaré representation, in particular the pronounced rise in sensitivity at the trailing edge is handled by redistributing sensitivities at the pressure side in a smooth manner, thus guaranteeing better stability of the mesh morphing shape optimization routine. This clearly shows the potential benefits practitioners in the industry could reap from trickle down of current research through the GivEn project.

The partners from DLR consider the goals of this project as an important milestone that could enable researchers to tackle, among others, the following problems:

– Concurrent optimization of turbine aerodynamics and cycle performance with the goal of reducing cooling air mass flows. Such optimization should take into account the redistribution and mixing of hot and cold streaks to be able to predict aerodynamic loads in downstream stages.

- Assessment of the technology potential of operating the burner at partial admission (or even partial shutdown) conditions in order to achieve good partial load performance while avoiding a significant increase in emissions.
- Assessment of potentials of thermal clocking. The idea is to be able to reduce cooling air if the relatively cold wake behind cooled vanes is used in downstream stages. Such optimizations will be based on unsteady flow predictions.

The necessary changes to create adjoints for existing evaluation processes stretches from parametrization to the simulation and post-processing codes for the different disciplines involved. Moreover, DLR expects to benefit from the coupling strategies developed here. These should be both sufficiently accurate and apt for an appropriate strategy to define coupled adjoint solvers. A particularly important milestone that this project is to achieve is the establishment of a life-time prediction that is suitable for gradient-based optimization.

DLR will not only apply these advances in aerothermal design problem but also hopes that general conclusions can be drawn that carry over to other multidisciplinary design problems that involve coupled simulations of flows and structures.

Theoretical concepts are shared in joint seminars in the early stages of the project to lay the foundation of common implementation of prototypes in later phases of the project. The role of these prototypes is to explore the implementation of the methods and simultaneously are used to communicate about changes to existing design evaluation chains and requirements for practical optimizations.

4.3.2 Siemens

Due to their low computational cost compared to other approaches, adjoint methods have become an interesting optimization approach for industry applications. Classical approaches such as parametric designs require a good understanding of the dependencies and impact of parameters and an optimization target, in order to identify a rather small but also relevant parameter space. In contrast, adjoint methods do not require to restrict the parameter space in advance. Shape optimization of gas turbine components, e.g. blades, require a multi objective shape optimization process based on multi-physical simulations. The upfront definition of a set of parameters is a demanding challenge and restricts the accessible design space. The application of methods developed in the GivEn project provides the opportunity to extend the possible design space. As a first example for an industrial application of adjoint methods within the GivEn Project, an application on the effects of manufacturing variations is presented.

This new application can have a major impact in the very competitive market of gas turbine for power generation at relatively low cost and time horizon. The following sections describe some aspects of the development of such a method using the example of a turbine vane.

4.3.2.1 Effects of Manufacturing Variations

The vanes and blades of a gas turbine are designed to deflect and expand the flow leaving the combustion chamber and generate torque at the rotor shaft. It implies that these components must operate at very high temperature while maintaining good aerodynamics and mechanical properties. Therefore, the material chosen for such hot gas section parts usually belongs to the Ni-base superalloy family and the manufacturing process involves precision casting which is a very complex and expensive technique. During the manufacturing process, the shape accuracy of each casting is assessed by geometrical measurements. If all of the measured coordinates lie within a specified tolerance band, the part is declared compliant and otherwise scrapped. In order to reduce uneconomic scrapping, the re-evaluation of acceptable tolerances is a constantly present challenge. In the past, point-based measurements and subsequent combination with expert judgement were conducted. Nowadays, advanced 3D scanners enable the acquisition of highly detailed geometric views of each part in a short amount of time which are better suited to characterize the analyze geometrical deviations. They furthermore allow to focus on the *effects* of geometric variations on the component's functions rather than on the geometric variations themselves. Nonetheless, evaluating the characteristics of each component produced using traditional computational methods—CFD and FEM—would not be industrially feasible due to the computational resource required. In this context, the capability of adjoint codes to take into account very efficiently the effects of a great number of parameters on one objective function is a crucial asset. It allows to benefit from the high resolution scans and model the manufacturing variations on the surface directly as the displacement of each point on the surface.

However, since adjoint equations are linearized with respect to an objective function, the accuracy of the predictions provided with the help of an adjoint code will be limited by the magnitude of the deformations. In other words, the deformations must be small enough such that a linear approximation of the effects on the objective function is appropriate. Therefore the usefulness of adjoint codes in an industrial context must be investigated using *real* manufacturing deviations. To this aim, 102 scans of heavy-duty turbine vanes have been used for the validation of the adjoint codes. Since the role of a turbine vane consists in fulfilling aerodynamic and mechanical functions, the objectives chosen for the investigation must be chosen within these two families.

4.3.2.2 Validation of the Tools

In a context of a more volatile energy production induced by renewable sources, gas turbines must be started or stopped very often and produce energy the most efficiently possible.

For turbine vanes, starting or stopping a machine more often implies that important temperature gradients will be experienced also more often leading potentially to LCF problems. The occurrence of LCF issues can be numerically estimated by computing

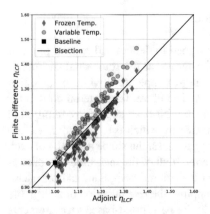 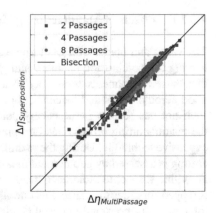

Fig. 4.13 Weibull scale parameter (η_{LCF}, right) and isentropic efficiency (η, left) obtained by adjoint codes versus finite differences

the maximal number of start/stop cycles above which a crack appears on the vane surface. For this assessment, the probabilistic Weibull-based approach by Schmitz et al. has been chosen. It was combined with the primal and adjoint equations and has been implemented into a Siemens in-house software. In Fig. 4.13, left, a comparison of the gradients of the Weibull scale parameter η_{LCF} computed with the adjoint code and the finite difference for each of the 102 vanes considered is presented. Actually two sets of data are present: one assuming that the temperature field on the vane is independent of the actual geometry—the so-called frozen temperature hypothesis— while the other considers on the contrary a variable temperature field. In both cases, all points are very close to the first bisectional curve, meaning that for each of the 102 vanes, the manufacturing deviations have a linear impact on the Weibull scale parameter. In other words, the magnitude of the real manufacturing deviations are small enough so their effects on LCF can be successfully taken into account by an adjoint code. A more detailed investigation can be found in the publication of Liefke et al. [60].

The efficiency of a gas turbine depends of course on the aerodynamic characteristic of the vanes. However the behaviour of each single vane makes only sense if they are integrated into a row together. When considering manufacturing variations, an additional difficulty appears namely the interaction of the manufacturing deviations of adjacent blades with each other. In this case, it would be interesting to compare the gradients obtained with the adjoint code versus the finite difference, not only for a purely axis-symmetrical situation (e.g. one passage) but also for several different arrangements (e.g. more than one passage). Figure 4.13, right, presents the gradients of the stage isentropic efficiency for such a configuration using DLR's CFD suite TRACE [6]. It can be seen that independently of the arrangement considered—2,4 or 8 passages—there is a very good agreement between the gradients predicted by the adjoint code and those obtained with the help of the finite difference. The work of Liefke et al. [61] summarizes the complete investigation of this case.

4.3.2.3   Industrial Perspectives

The previous section demonstrated that adjoint codes can be successfully used in an industrial context to quantify the impact of manufacturing variations on low-cycle fatigue and isentropic efficiency. It could possible to extend this approach to additional physical phenomena such as high-cycle fatigue or creep, given that the equations modelling these phenomena can be differentiated and of course that the impact of the manufacturing variation remains linear.

In addition to the direct and short-term benefits for the manufacturing of turbomachine components, the results presented in this section also demonstrate that adjoint codes can be successfully deployed and used in an industrial context. The confidence and experience gained will pave the way for other new usage within Siemens. Especially, the tools and concepts developed within the GivEn project will contribute to the creation of more rapid, efficient and robust multidisciplinary design optimization of compressors and turbines either based fully on adjoint codes or in combination with surrogate models.

Acknowledgements The authors were partially supported by the BMBF collaborative research project GivEn under the grant no. 05M18PXA.

References

1. Adams, B.M., Bohnhoff, W.J., Dalbey, K.R., Eddy, J.P., Eldred, M.S., Gay, D.M., Haskell, K., Hough, P.D., Swiler, L.P.: DAKOTA, a multilevel parallel object-oriented framework for design optimization, parameter estimation, uncertainty quantification, and sensitivity analysis: version 5.0 user's manual. Sandia National Laboratories, Tech. Rep. SAND2010-2183 (2009)
2. Agmon, S., Nirenberg, A., Douglis, L.: Estimates near the boundary for solutions of elliptic partial differential equations satisfying general boundary conditions I. Commun. Pure Appl. Math. **12**, 623–727 (1959)
3. Agmon, S., Nirenberg, A., Douglis, L.: Estimates near the boundary for solutions of elliptic partial differential equations satisfying general boundary conditions II. Commun. Pure Appl. Math. **17**, 35–92 (1964)
4. Alnæs, M.S.: UFL: a finite element form language. In: Logg, A., Mardal, K.-A., Wells, G.N. (eds.) Automated Solution of Differential Equations by the Finite Element Method, Lecture Notes in Computational Science and Engineering, vol. 84, chapter 17. Springer (2012)
5. Alnæs, M.S., Blechta, J., Hake, J., Johansson, A., Kehlet, B., Logg, A., Richardson, C., Ring, J., Rognes, M.E., Wells, G.N.: The FEniCS project version 1.5. Arch. Numer. Softw. **3**(100) (2015)
6. Backhaus, J., Schmitz, A., Frey, C., Mann, S., Nagel, M., Sagebaum, M., Gauger, N.R.: Application of an algorithmically differentiated turbomachinery flow solver to the optimization of a fan stage. In: AIAA-Paper 2017-3997 (2017)
7. Balay, S., Abhyankar, S., Adams, M., Brown, J., Brune, P., Buschelman, K., Dalcin, L., Dener, A., Eijkhout, V., Gropp, W., et al.: PETSc Users Manual. Argonne National Laboratory (2019)
8. Bartel, A., Günther, M.: PDAEs in refined electrical network modeling. SIAM Rev. **60**(1), 56–91 (2018)
9. Batendorf, S.B., Crosse, J.G.: A statistical theory for the fracture of brittle structures subject to nonuniform polyaxial stress. J. Appl. Mech. **41**, 459–465 (1974)

10. Becker, K., Heitkamp, K., Kügeler, E.: Recent progress in a hybrid-grid CFD solver for turbomachinery flows. In: Proceedings Fifth European Conference on Computational Fluid Dynamics ECCOMAS CFD00 2010 (2010)
11. Bittner, L.: On shape calculus with elliptic PDE constraints in classical function spaces. arXiv preprint arXiv:2001.06297 (2020)
12. Bittner, L., Gottschalk, H.: Optimal reliability for components under thermomechanical cyclic loading. Control Cybern. **45**, 2–35 (2016)
13. Bolten, M., Gottschalk, H., Hahn, C., Saadi, M.: Numerical shape optimization to decrease failure probability of ceramic structures. Comput. Visual Sci. (2019)
14. Bolten, M., Gottschalk, H., Schmitz, S.: Minimal failure probability for ceramic design via shape control. J. Optim. Theory Appl. **166**(3), 983–1001 (2015)
15. Bäker, M., Harders, H., Rösler, J.: Mechanisches Verhalten der Werkstoffe, 3rd edn. Vieweg + Teubner, Wiesbaden (2008)
16. Chirkov, D.V., Ankudinova, A.S., Kryukov, A.E., Cherny, S.G., Skorospelov, V.A.: Multi-objective shape optimization of a hydraulic turbine runner using efficiency, strength and weight criteria. Struct. Multidiscipl. Optim. **58**, 627–640 (2018)
17. Ciarlet, P.: Mathematical Elasticity - Volume I: Three-Dimensional Elasticity, Studies in Mathematics and Its Applications, vol. 20. North-Holland, Amsterdam (1988)
18. Conti, S., Held, H., Pach, M., Rumpf, M., Schultz, R.: Risk averse shape optimization. SIAM J. Contr. Optim. **49**, 927–947 (2011)
19. Cui, W.: A state-of-the-art review on fatigue life prediction methods for metal structures. J. Marine Sci. Techn. **7**(1), 43–56 (2002)
20. Doganay, O.T., Gottschalk, H., Hahn, C., Klamroth, K., Schultes, J., Stiglmayr, M.: Gradient based biobjective shape optimization to improve reliability and cost of ceramic components. Optim. Eng. (2019)
21. Dorfner, C., Nicke, E., Voss, C.: Axis-asymmetric profiled endwall design by using multi-objective optimisation linked with 3D RANS-flow-simulations. In: ASME Turbo Expo 2007: Power for Land, Sea, and Air, pp. 107–114. American Society of Mechanical Engineers Digital Collection (2007)
22. Dowling, N.E.: Notched member fatigue life predictions combining crack initiation and propagation. Fatigue Fract. Eng. Mater. Struct. **2**, 129–138 (1979)
23. Ehrgott, M.: Multicriteria Optimization. Springer, Berlin, Heidelberg (2005)
24. Engel, B.: High temperature low-cycle fatigue of the Ni-base superalloy RENÉ80. In: LCF8 (2017)
25. Engel, B.: Einfluss der lokalen Kornorientierung und der Korngrösse auf das Verformungs- und Ermüdungsverhalten von Nickelbasis Superlegierungen. Ph.D. thesis, Technische Universität Kaiserslautern (2018)
26. Engel, B., Mäde, L., Lion, P., Moch, N., Gottschalk, H., Beck, T.: Probabilistic modeling of slip system-based shear stresses and fatigue behavior of coarse-grained Ni-base superalloy considering local grain anisotropy and grain orientation. Metals **9**(8), 813 (2019)
27. Eppler, K.: Efficient Shape Optimization Algorithms for Elliptic Boundary Value Problems. Technische Universität Berlin, Habilitation (2007)
28. Eriksson, R., Simonsson, K., Leidermark, D., Moverare, J.: Evaluation of notch effects in low cycle fatigue of alloy 718 using critical distances. MATEC Web Conf. **165**, 15001 (2018)
29. Fedelich, B.: A stochastic theory for the problem of multiple surface crack coalescence. Int. J. Fract. **91**, 23–45 (1998)
30. Fiedler, M., Varfolomeev, I., Wächter, M.: Rechnerischer Bauteilfestigkeitsnachweis unter expliziter Erfassung nichtlinearen Werkstoff-Verformungsverhaltens. Technical report, Forschungskuratorium Maschinenbau e.V. (2015)
31. Fiedler, M., Wächter, M., Varfolomeev, I., Vormwald, M., Esderts, A.: Rechnerischer Festigkeitsnachweis für Maschinenbauteile unter expliziter Erfassung nichtlinearen Werkstoff-Verformungsverhaltens, 1st edn. FKM-Richtlinie. VDMA-Verlag GmbH, Frankfurt (2019)
32. Fliege, J., Svaiter, B.F.: Steepest descent methods for multicriteria optimization. Math. Meth. Oper. Res. **51**(3), 479–494 (2000)

33. Forrester, A., Sobester, A., Keane, A.: Engineering Design Via Surrogate Modelling: A Practical Guide. Wiley (2008)
34. Frey, C., Nürnberger, D., Kersken, H.: The discrete adjoint of a turbomachinery RANS solver. In: Proceedings of ASME-GT2009 (2009)
35. Fujii, N.: Lower semicontinuity in domain optimization problems. J. Optim. Theory Appl. **59**(3), 407–422 (1988)
36. Geiger, C., Kanzow, C.: Numerische Verfahren zur Lösung unrestringierter Optimierungsaufgaben. Springer, Berlin, Heidelberg (1999)
37. German Aerospace Center: Variable definitions valid for TRACE SUITE 9.2 (2019). http://www.trace-portal.de/userguide/trace/variableNames.pdf
38. Giles, M.B., Pierce, N.A.: An introduction to the adjoint approach to design. Flow Turbul. Combust. **65**(3–4), 393–415 (2000)
39. Gottschalk, H., Krause, R., Rollmann, G., Seibel, T., Schmitz, S.: Probabilistic Schmid factors and scatter of LCF life. Mater. Sci. Engin. Techn. **46**, 156–164 (2015)
40. Gottschalk, H., Reese, M.: An analytical study in multi physics and multi criteria shape optimization. In preparation (2020)
41. Gottschalk, H., Saadi, M.: Shape gradients for the failure probability of a mechanic component under cyclic loading: a discrete adjoint approach. Comput. Mech. **64**(4), 895–915 (2019)
42. Gottschalk, H., Saadi, M., Doganay, O.T., Klamroth, K., Schmitz, S.: Adjoint method to calculate the shape gradients of failure probabilities for turbomachinery components. In: ASME Turbo Expo 2018: Turbomachinery Technical Conference and Exposition. American Society of Mechanical Engineers Digital Collection (2018)
43. Gottschalk, H., Schmitz, S.: Optimal reliability in design for fatigue life, part i: existence of optimal shapes. SIAM J. Contr. Optim. **52**, 2727–2752 (2014)
44. Gottschalk, H., Schmitz, S., Rollmann, G., Krause, R.: A probabilistic approach to low cycle fatigue. In: Proceedings of ASME Turbo Expo 2013, pp. GT2013–94899 (2012)
45. Gottschalk, H., Schmitz, S., Seibel, T., Rollmann, G., Krause, R., Beck, T.: Probabilistic Schmid factors and scatter of low cycle fatigue (LCF) life. Mat.-wiss u. Werkstofftech. **45**(2), 156–164 (2015)
46. Hackbusch, W., Sauter, S.: Adaptive composite finite elements for the solution of PDEs containing nonuniformely distributed micro-scales. Matem. Mod. **8**, 31–43 (1996)
47. Hackbusch, W., Sauter, S.: Composite finite elements for problems containing small geometric details. Comput. Visual. Sci. **1**, 15–25 (1997)
48. Hackbusch, W., Sauter, S.: Composite finite elements for the approximation of PDEs on domains with complicated micro-structures. Num. Math. **75**, 447–472 (1997)
49. Han, J.-C.: Recent studies in turbine blade cooling. Int. J. Rotating Mach. **10**(6), 443–457 (2004)
50. Haslinger, J., Mäkinen, R.A.E.: Introduction to Shape Optimization - Theory, Approximation and Computation. SIAM, Advances in Design and Control (2003)
51. Hempel, M.: Stand der Erkenntnisse über den Einfluß der Probengröße auf die Dauerfestigkeit. Draht **8**, 385–394 (1957)
52. Hertel, O., Vormwald, M.: Statistical and geometrical size effects in notched members based on weakest-link and short-crack modelling. Eng. Fract. Mech. **95** (2012)
53. Hoheisel, H., Kiock, R., Lichtfuss, H.J., Fottner, L.: Influence of free stream turbulence and blade pressure gradient on boundary layer and loss behaviour of turbine cascades. In: ASME 1986 International Gas Turbine Conference and Exhibit, pp. V001T01A102–V001T01A102. American Society of Mechanical Engineers (1986)
54. Kallenberg, O.: Random Measures. Akademie Verlag, Berlin (1982)
55. Keane, A., Forrester, A., Sobester, A.: Engineering Design via Surrogate Modelling: A Practical Guide. American Institute of Aeronautics and Astronautics, Inc., September (2008)
56. Kontermann, C., Almstedt, H., Scholz, A., Oechsner, M.: Notch support for LCF-loading: a fracture mechanics approach. Procedia Struct. Integr. **2**, 3125–3134 (2016)
57. Kröger, G., Voß, C., Nicke, E.: Axisymmetric casing optimization for transonic compressor rotors. In: ISROMAC 13, April 4–9, 2010, Honolulu, USA (2010)

58. Lazzarin, P., Tovo, R., Meneghetti, G.: Fatigue crack initiation and propagation phases near notches in metals with low notch sensitivity. Int. J. Fatigue **19**(8–9), 647–657 (1997)
59. Li, Z., Zheng, X.: Review of design optimization methods for turbomachinery aerodynamics. Prog. Aerosp. Sci. **93**, 1–23 (2017)
60. Liefke, A., Jaksch, P., Schmitz, S., Marciniak, V., Janoske, U., Gottschalk, H.: Towards multi-disciplinary turbine blade tolerance design using adjoint methods. In: ASME Turbo Expo 2020: Power for Land, Sea, and Air. American Society of Mechanical Engineers Digital Collection (2020)
61. Liefke, A., Marciniak, V., Backhaus, J., Frey, C., Gottschalk, H., Janoske, U.: Aerodynamic impact of manufacturing variation on a nonaxisymmetric multi-passage turbine stage with adjoint CFD. In: ASME Turbo Expo 2019: Power for Land, Sea, and Air. American Society of Mechanical Engineers Digital Collection (2019)
62. Liefke, A., Marciniak, V., Janoske, U., Gottschalk, H.: Using adjoint CFD to quantify the impact of manufacturing variations on a heavy duty turbine vane. In: ECCM-ECFD 2018. Eccomas - European Community on Computational Methods in Applied Sciences (2018)
63. Logg, A., Mardal, K.-A., Wells, G.N. (eds.): Automated Solution of Differential Equations by the Finite Element Method. Lecture Notes in Computational Science and Engineering, vol. 84. Springer (2012)
64. Logg, A., Wells, G.N.: DOLFIN: automated finite element computing. ACM Trans. Math. Softw. **37**(2) (April 2010)
65. Logg, A., Wells, G.N., Hake, J.: DOLFIN: a C++/Python finite element library. In: Logg, A., Mardal, K.-A., Wells, G.N. (eds.) Automated Solution of Differential Equations by the Finite Element Method, Lecture Notes in Computational Science and Engineering, vol. 84, chapter 10. Springer (2012)
66. Luft, D., Schulz, V., Welker, K.: Efficient techniques for shape optimization with variational inequalities using adjoints. arXiv preprint arXiv:1904.08650 (2019)
67. Mäde, L., Gottschalk, H., Schmitz, S., Beck, T., Rollmann, G.: Probabilistic LCF risk evaluation of a turbine vane by combined size effect and notch support modeling. In: Turbo Expo: Power for Land, Sea, and Air, volume 7A: Structures and Dynamics, 06 2017. V07AT32A004
68. Mäde, L., Schmitz, S., Gottschalk, H., Beck, T.: Combined notch and size effect modeling in a local probabilistic approach for LCF. Comput. Mater. Sci. **142**, 377–388 (2018)
69. Manson, S., Hirschberg, M.: Low cycle fatigue of notched specimens by consideration of crack initation and propagation. NASA Technical Note TN D-3146, National Aeronautics and Space Administration, Lewis Research Center Cleveland, Ohio, 06 1967
70. Markovin, D., Moore, H.F. (eds.): The effect of size of specimens on fatigue strength of three types of steel. In: Proceedings of ASTM, vol. 44. ASTM (1944)
71. Massonet, C. (ed.): The effect of size, shape and grain size on the fatigue strength of medium carbon steel. In: Proceedings of ASTM, vol. 56. ASTM (1956)
72. Meitner, P.L.: Computer code for predicting coolant flow and heat transfer in turbomachinery. Technical Report, 89-C-008 (1990)
73. Moch, N.: From Macroscopic Models of Damage Accumulation to Probability of Failure of Gas Turbines. Ph.D. thesis, Bergische Universität Wuppertal (2018)
74. Monnot, J., Heritier, B., Cogne, J.: Effect of Steel Manufacturing Processes on the Quality of Bearing Steels, chapter Relationship of Melting Practice, Inclusion Type, and Size with Fatigue Resistance of Bearing Steels, pp. 149–164. ASTM International, West Conshohocken (1988)
75. Murakami, Y., Usuki, H.: Quantitative evaluation of effects of non-metallic inclusions on fatigue strength of high strength steels. II: Fatigue limit evaluation based on statistics for extreme values of inclusion size. Int. J. Fatigue **11**(5), 299–307 (1989)
76. Olschewski, J., Ziebs, J., Fedelich, B.: Modellierung des Schädigungsverhaltens der Legierung in 738LC unter mehrachsiger thermisch-mechanischer Beanspruchung. Abschlussbericht SFB 339, Projekt C2:C21–63 (1997)
77. Peter, J., Marcelet, M.: Comparison of surrogate models for turbomachinery design. WSEAS Trans. Fluid Mech. **3**(1), 10–17 (2008)

78. Phillipp, H.A.: Einfluß von Querschnittsgröße und Querschnittsform auf die Dauerfestigkeit bei ungleichmäßig verteilten Spannungen. Forschung im Ingenieurwesen **13**, 99–111 (1942)
79. Radaj, D., Vormwald, M.: Ermüdungsfestigkeit, 3rd edn. Springer, Berlin, Heidelberg, New York (2007)
80. Rao, J.S., Narayan, R., Ranjith, M.C.: Lifing of turbomachinery blades: a process driven approach. In: ASME Turbo Expo 2008: Power for Land, Sea, and Air, pp. 151–158. American Society of Mechanical Engineers Digital Collection (2008)
81. Sagebaum, M., Özkaya, E., Gauger, N.R., Backhaus, J., Frey, C., Mann, S., Nagel, M.: Efficient algorithmic differentiation techniques for turbo-machinery design. In: AIAA-Paper 2017-3998 (2017)
82. Sakane, M., Ohnami, M.: Notch effect in low-cycle fatigue at elevated temperatures-life prediction from crack initiation and propagation considerations. J. Eng. Mater. Technol. **108**(4), 279–284 (1986)
83. Samad, A., Kim, K.-Y.: Surrogate based optimization techniques for aerodynamic design of turbomachinery. Int. J. Fluid Mach. Syst. **2**(2), 179–188 (2009)
84. Saravanamuttoo, H.I.H., Rogers, G.F.C., Cohen, H.: Gas Turbine Theory. Pearson Education (2001)
85. Schaefer, P., Hofmann, W.H., Gieß, P.-A.: Multiobjective optimization for duct and strut design of an annular exhaust diffuser. In: ASME Turbo Expo 2012: Turbine Technical Conference and Exposition, pp. 1679–1689. American Society of Mechanical Engineers Digital Collection (2012)
86. Schloe, N.: meshio - I/O for mesh files (2019). https://github.com/nschloe/meshio
87. Schmidt, S.: Efficient Large Scale Aerodynamic Design Based on Shape Calculus. Ph.D. thesis, Universität Trier (2010)
88. Schmidt, S., Schulz, V.: Impulse response approximations of discrete shape Hessians with application in CFD. SIAM J. Contr. Optim. **48**(4), 2562–2580 (2009)
89. Schmidt, S., Schulz, V.: Shape derivatives for general objective functions and the incompressible Navier-Stokes equations. Control Cybern. **39**(3), 677–713 (2010)
90. Schmitz, S., Rollmann, G., Gottschalk, H., Krause, R.: Risk estimation for LCF crack initiation. In: Proc. ASME. 55263, 7A: Structures and Dynamics, vol. GT2013-94899, p. V07AT27A007. ASME, June (2013)
91. Schmitz, S., Seibel, T., Beck, T., Rollmann, G., Krause, R., Gottschalk, H.: A probabilistic model for LCF. Comput. Mater. Sci. **79**, 584–590 (2013)
92. Schulz, V.: A Riemannian view on shape optimization. Found. Comput. Math. **14**, 483–501 (2014)
93. Schulz, V., Siebenborn, M.: Computational comparison of surface metrics for PDE constrained shape optimization. Comput. Meth. Appl. Math. **16**, 485–496 (2016)
94. Schulz, V., Siebenborn, M., Welker, K.: Efficient PDE constrained shape optimization based on Steklov-Poincaré-type metrics. SIAM J. Optim. **26**, 2800–2819 (2016)
95. Siebel, E., Meuth, H.: Die Wirkung von Kerben bei schwingender Beanspruchung. VDI Zeitschrift **91**(13), 319–323 (1949)
96. Siebel, E., Stieler, M.: Ungleichförmige Spannungsverteilung bei schwingender Beanspruchung. VDI-Zeitschrift **97**(5), 121–126 (1955)
97. Siebert, C.A.: The effect of surface treatments on the fatigue resistance of a hardened SAE 1065 steel. Technical report, Engineering Research Institute, The University of Michigan Ann Arbor, 1956. Project 1685
98. Siller, U., Voß, C., Nicke, E.: Automated multidisciplinary optimization of a transonic axial compressor. In: 47th AIAA Aerospace Sciences Meeting Including The New Horizons Forum and Aerospace Exposition, p. 863 (2009)
99. Sokolowski, J., Zolesio, J.-P.: Shape Optimization - Shape Sensitivity Analysis. Springer, Berlin, Heidelberg (1992)
100. Sornette, D., Magnin, J.-P., Brechet, Y.: The physical origin of the Coffin-Manson law in low-cycle fatigue. Europhys. Lett. **20**, 433–438 (1992)

101. Sultanian, B.: Gas Turbines: Internal Flow Systems Modeling, vol. 44. Cambridge University Press (2018)
102. Thévenin, D., Janiga, G. (eds.): Optimization and Computational Fluid Dynamics. Springer Science & Business Media (2008)
103. Tyacke, J., Vadlamani, N.R., Trojak, W., Watson, R., Ma, Y., Tucker, P.G.: Turbomachinery simulation challenges and the future. Prog. Aerosp. Sci. 100554 (2019)
104. Weibull, W.: A Statistical Theory of the Strength of Materials. Handlingar / Ingeniörsvetenskapsakademien, Generalstabens litografiska anstalts förlag (1939)
105. Welker, K.: Efficient PDE Constrained Shape Optimization in Shape Spaces. Ph.D. thesis, Universität Trier (2016)
106. Wundt, B.M.: Effects of Notches on Low-Cycle Fatigue. A Literature Survey. American Society for Testing and Materials (1972)

Chapter 5
Using the Stein Two-Stage Procedure to Calculate Uncertainty in a System for Determining Gas Qualities

Leonid Kuoza

Abstract This paper examines the measurement uncertainty calculation for results from network state reconstruction and gas quality tracking in gas transport and distribution networks. The Monte Carlo method is used for this purpose. It analyses how many Monte Carlo runs are needed to determine the measurement uncertainty with sufficient accuracy and reliability. The analysis shows that the number of Monte Carlo runs depends on the stochastic properties of the calculation results, particularly on the excess of probability distribution. It shows how Stein's two-stage procedure can be adapted to determine the measurement uncertainty reliably.

Keywords Measurement technique · Gas quality tracking · Measurement uncertainty · Monte Carlo method · Stein's two-stage procedure

5.1 Introduction

Gas is a powerful source of energy and a major contributor to a reliable energy supply in Germany. In Germany, gas is billed according to its energy content. The quality of natural gas and its energy content vary, however, due to its natural origin. In addition, biogas—biomethane and hydrogen—are increasingly being fed into the natural gas network.

Therefore, to reliably calculate the energy content, as a basis for usage-based billing, the calorific values must be allocated precisely for every consumer, in accordance with calibration laws and regulations.

Measuring the calorific values directly at the exit or handover points, with a calorimeter or gas chromatograph, is complex and expensive. The gas conditioning used to maintain a maximum 2% deviation of the calorific value at the exit point, compared to the volume-weighted input calorific values, is equally complex and expensive. This conditioning is needed particularly when biogas is input, to comply

L. Kuoza (✉)
PSI Gas & Oil, Ruhrallee 201, 45136 Essen, Germany
e-mail: lkuoza@psi.de

© Springer Nature Switzerland AG 2021   107
S. Göttlich et al. (eds.), *Mathematical Modeling, Simulation and Optimization for Power Engineering and Management*, Mathematics in Industry 34,
https://doi.org/10.1007/978-3-030-62732-4_5

with the 2% rule. It is less expensive and more effective to use a system that enables reconstruction of the flow conditions and tracking of gas qualities, and thus exact allocation of the billing-relevant parameters, although enough measurements also have to be made here in order to obtain correct results. Computer-supported systems for tracking calorific values are a worthwhile alternative [2]. We differentiate here between reconstruction systems, which meet the requirements of PTB-A7.64, and allocation systems pursuant to DVGW worksheet G685 [5].

The gas quality values of reconstruction systems at all exit nodes (which are compliant with PTB-A7.64) are calibrated measured values. In accordance with calibration laws, they require a conformity assessment and conformity testing by a certified testing organization. The important thing in this case is that calibrated measuring devices are used to measure all input and output volumes and all gas qualities to be tracked at the input points and that the measurements are transferred through secure and/or redundant data channels. The permissible measurement errors are specified in the calibration laws. The permissible error for calorific values is less than 0.8% of the scale end value of the measuring device ($14\,\mathrm{kWh\,N^{-1}\,m^{-3}}$ for gas quality reconstruction systems), the permissible error for standard densities is 0.5% of the measured value and the permissible error for CO_2 content is $0.5\,\mathrm{Mol\%}$.

If the output volumes are not measured sufficiently, allocation methods could be used as a replacement, according to DVGW publication G685. Compliance with the 2% rule is important here. This means, the calorific value at the exit nodes may only deviate from the volume-weighted or arithmetic mean value of the input calorific values by a maximum of two percent.

In both cases, a mathematical model of pipeline networks is defined and resolved in the form of differential equations [7]. The flow data gained through this are used to track the movement of "gas blocks" through the gas network. As a result, it is possible to obtain information about gas qualities at every point and every time in the gas network.

Each measured value is an approximation of the actually measured parameter. The uncertainty with which a measured value can be recorded must be identified together with the value itself. Evaluation of the calculation results is also needed when using reconstruction systems to calculate billing-relevant gas qualities. Therefore, the uncertainty of each billing value must be identified. The uncertainty of calculation results can theoretically be calculated in two ways: analytically or through the Monte Carlo method. Under the analytic procedure, the uncertainty of the input data is reproduced, taking the sensitivities into account. Under the Monte Carlo method, many calculations are made to determine the sampling distribution of the calculation results. Due to the high complexity of the mathematical models and the non-continuous changes to gas qualities in the pipelines, the sensitivity analysis has proven to be unsuitable for gas quality tracking. As such, the "Guide to the Expression of Uncertainty in Measurement" (GUM) suggests using the Monte Carlo method to determine the measurement uncertainties for such complex models [1, 3].

The uncertainty of the calculation results must not exceed the error limits.

5.2 Monte Carlo Method for Gas Quality Reconstruction

When applying the Monte Carlo method to gas quality reconstruction, we presume that the uncertainties of the input data are known. The input data include measurement values for flows, pressures and gas qualities, as well as network parameters regarding pipe lengths, pipe diameters and pipe roughness. The input data are changed at random, in accordance with their uncertainties, and the network state is calculated. This process is repeated multiple times. This creates a series of values for each billing parameter—a sample. These values are then used to determine the uncertainties of the billing parameters.

In this context, the following questions must be answered:

- How many Monte Carlo runs are needed for the results to be sufficiently precise?
- What does "sufficiently precise" mean in this context?

5.2.1 Simple Examples

In the following simple examples, the key characteristics of the Monte Carlo method are illustrated in the context of our task at hand.

The following designations are used:

- $E(\cdot)$, $V(\cdot)$, $\sigma(\cdot)$ and $u(\cdot)$: Expected value, variance, standard deviation and uncertainty of a random parameter
- $P(\cdot)$: Probability of an event
- $Q(p)$: Quantile for probability p
- $N(\mu; \sigma)$: Normal distribution

5.2.1.1 One Input and One Output

First, a trivial network in which one input is linked with one output (Fig. 5.1) and the gas quality is unvarying. The gas quality at the output is always equal to the gas quality at the input.

The following is assumed:

- The calorific value at the input amounts to 10 kWh per cubic metre
- With an uncertainty of 0.1 kWh per cubic metre ($\sigma \approx 0.05$)
- Normal distribution of the observed (measured) calorific value

Fig. 5.1 Trivial network

$$CV_{\text{in,obs}} = N(10; 0.05) \tag{5.1}$$

The value 10.03 kWh per cubic metre is measured. Under the Monte Carlo method, CV_{in} would be manipulated as if it had a normal distribution with a mean value of 10.03 and an assumed standard deviation of 0.05.

$$CV_{\text{in}} = N(10.03; 0.05) \tag{5.2}$$

In this trivial network, the result of the simulation is simply an identity function:

$$CV_{\text{out}} = f(CV_{\text{in}}) = CV_{\text{in}} \tag{5.3}$$

Therefore, the resulting calorific value will also have a like normal distribution:

$$CV_{\text{out}} = N(10.03; 0.05) \tag{5.4}$$

An important conclusion can be drawn from these trivial considerations: The expected value of the result corresponds to the value that we calculate from the reconstruction calculation, without manipulation of the input data, and is equally distorted by measurement errors. Therefore, the Monte Carlo method cannot deliver a more precise result than the reconstruction calculation. This statement also applies to more complex gas networks. Accordingly, the mean value of the results is irrelevant and is ignored. At the same time, the Monte Carlo method provides information about the uncertainty of the calculation results, provided we know the uncertainty of the measured values.

5.2.1.2 Two Inputs and One Output

The next example examines a case in which gas is mixed. There are two inputs and one output (Fig. 5.2). Both the calorific values and the flows at the inputs have some uncertainties.

In this case, the calorific value at the output is calculated as follows:

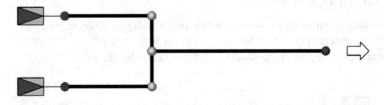

Fig. 5.2 Gas mixture

$$CV = CV_1 \frac{Q_1}{Q_1 + Q_2} + CV_2 \frac{Q_2}{Q_1 + Q_2} = \frac{\sum_i CV_i Q_i}{\sum_i Q_i} \qquad (5.5)$$

Under the Monte Carlo method, we add a random variable with an expected value of null to flows and calorific values.

$$CV_i \rightarrow CV_i + \delta_i \qquad (5.6)$$
$$Q_i \rightarrow Q_i + \varepsilon_i \qquad (5.7)$$

Since these random parameters are small in comparison to the original values, we linearise the upper equation. This results in the following equation for the calorific value.

$$
\begin{aligned}
CV &= \frac{\sum_i CV_i Q_i}{\sum_i Q_i} \left(1 + \sum_i \delta_i \frac{Q_i}{\sum_j CV_j Q_j} + \sum_i \varepsilon_i \left(\frac{CV_i}{\sum_j CV_j Q_j} - \frac{1}{\sum_j Q_j} \right) \right) \\
&= \frac{\sum_i CV_i Q_i}{\sum_i Q_i} + \sum_i \delta_i \frac{Q_i}{\sum_j Q_j} + \sum_i \frac{\varepsilon_i}{\sum_j Q_j} \left(CV_i - \frac{\sum_j CV_j Q_j}{\sum_j Q_j} \right)
\end{aligned}
\qquad (5.8)
$$

As expected, the expected value of the random quantity agrees with the upper equation.

$$E(CV) = \frac{\sum_i CV_i Q_i}{\sum_i Q_i} \qquad (5.9)$$

Assuming that the measurement errors are independent of one another, the variance can be calculated as follows:

$$V(CV) = \sum_i V(\delta_i) \left(\frac{Q_i}{\sum_j Q_j} \right)^2 + \sum_i \frac{V(\varepsilon_i)}{\left(\sum_j Q_j \right)^2} \left(CV_i - \frac{\sum_j CV_j Q_j}{\sum_j Q_j} \right)^2 \qquad (5.10)$$

To understand the importance of this equation, let us examine the case of two inputs and assume that the flows are identical and that the random quantities used have same distribution.

$$Q_i = Q \qquad (5.11)$$
$$V(\delta_i) = V(\delta) \qquad (5.12)$$
$$V(\varepsilon_i) = V(\varepsilon) \qquad (5.13)$$

Then

$$V(CV) = \frac{V(\delta)}{2} + \frac{V(\varepsilon)}{8Q^2} (CV_1 - CV_2)^2 \qquad (5.14)$$

Apparently, the variance of the calorific value result can be both less than and greater than the variance of the calorific value measurement. If the uncertainty of the flow measurements is small enough, then the variance of the result is also small. Therefore, the averaging of independent random values makes the variance smaller. If the uncertainty of the flow measurements becomes large, the variance of the result increases. This increase also depends on the difference between the calorific values of the mixed flows.

The following question is also relevant: Under which conditions is the result variance less than the measurement variance?

This condition is:

$$\frac{V(\varepsilon)}{8Q^2}(CV_1 - CV_2)^2 < \frac{V(\delta)}{2} \tag{5.15}$$

or

$$\frac{u(\varepsilon)}{Q} < 2\frac{u(\delta)}{|CV_1 - CV_2|} \tag{5.16}$$

5.2.2   Example with Reversal of the Flow Direction

In large networks, it is a frequent occurrence for gas in certain pipelines to stop moving. This occurs when there are two areas in the network that are both well balanced. As much gas is taken out as gas is fed in each of these areas. In such cases, a minor balance difference causes a flow. The gas flows in one direction or the other, depending whether the balance difference is positive or negative. If the gas quality also differs between these two areas, the gas qualities at the output points between these areas can vary more widely.

In this case, we are interested in examining how a situation like this in gas transport influences the number of Monte Carlo runs needed to calculate the uncertainties of the calculation results.

To do so, we use the following simple example and analyse it numerically, using a pseudo-random number generator.

A pipeline with a very small flow $100 - 99 = 1$ is examined, as depicted in Fig. 5.3. Under the Monte Carlo method, flows are manipulated at random; each value has an uncertainty of 2%. Consequently, the gas in the pipeline can flow in one direction and, in other cases, in the other direction. The result can be a mixture of two different gases or not.

This case can be analysed through numerical tests.

We determine the following values for the calorific value at the output:

- Value without input data manipulation: $11\,kWh/m^3$
- Expected value: $11.1\,kWh/m^3$
- Standard deviation $0.23\,kWh/m^3$
- Uncertainty (half the confidence interval): $0.45\,kWh/m^3$
- Excess: 1.9

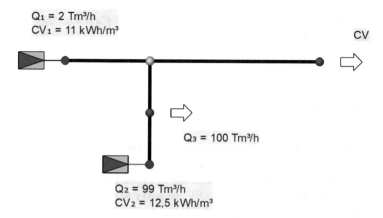

Fig. 5.3 Variable flow direction

In Sect. 5.3.4, it is shown that of the parameters of this distribution the excess has the greatest impact on the number of Monte Carlo runs.

5.3 Accuracy and Reliability

In this section, the meaning of the accuracy of the uncertainty estimate for the gas quality is defined and its dependency on number of Monte Carlo runs is examined.

5.3.1 Standard Deviation and Uncertainty

The uncertainty is interpreted as half of the interval for a confidence level of 95%:

$$u = \frac{1}{2} (Q(\alpha + 0.95) - Q(\alpha)) \tag{5.17}$$

In the symmetric case:

$$u = \frac{1}{2} (Q(0.975) - Q(0.025)) \tag{5.18}$$

If the form of the distribution of the parameter x is known, then the uncertainty is linearly dependent on the standard deviation:

$$u = k\sigma(x) \tag{5.19}$$

The factor depends on the form of distribution. If the parameter has a standard distribution, $k = 1.96$.

$$u = 1.96\sigma(x) \tag{5.20}$$

5.3.2 Estimated Uncertainty

Assume that the random quantity x represents the result of the reconstruction calculation, for example, the calorific value at the output. As part of the Monte Carlo method, we generate a sample x_i. The result is estimations for variance, standard deviation and uncertainty. These estimates are also random quantities: If we repeat the Monte Carlo method, we get different estimates.

Consequentially, the accuracy of the estimate has a probabilistic, rather than a deterministic character. Expressed informally, we can define the accuracy requirements as follows: The estimated uncertainty, at a reliability of $p = 95\%$, should have a relative error rate r of 5% at most. It is further assumed that our random quantity has a truly unknown uncertainty γ, either defined through a standard deviation and a coverage factor or as half of a confidence interval.

The estimation u has a reliability of p and a relative accuracy of r if

$$P\left(u(1 - r) \leq \gamma \leq u(1 + r)\right) \geq p \tag{5.21}$$

We could also rewrite the condition for $r \ll 1$ as follows:

$$P\left(\left|\frac{u}{\gamma} - 1\right| < r\right) \geq p \tag{5.22}$$

5.3.3 Number of Monte Carlo Runs

We will attempt to determine the number of Monte Carlo runs that guarantee the required accuracy. The following designations are used:

- x_i and s^2: sample of the random quantity x and variance of the sample.

An unbiased estimation of the variance is always used. The variance of a sample is then calculated as follows:

$$s^2 = \frac{1}{n - 1} \sum_i \left(x_i - \frac{1}{n} \sum x_i\right)^2 \tag{5.23}$$

The variance s^2 is also a random quantity. If x has an excess of null, then the following applies:

$$E(s^2) = \sigma^2 \tag{5.24}$$

$$V(s^2) = \frac{2\sigma^4}{n-1} \tag{5.25}$$

It is assumed that s^2 has a standard distribution

$$s^2 \sim N\left(\sigma^2; \sigma^2\sqrt{\frac{2}{n-1}}\right) \tag{5.26}$$

or to put it differently:

$$\left(\frac{s^2}{\sigma^2} - 1\right)\sqrt{\frac{n-1}{2}} \sim N(0; 1) \tag{5.27}$$

Then let us find the number n for which the following is true:

$$\begin{aligned}
P\left(\left|\frac{s}{\sigma} - 1\right| > r\right) &\approx P\left(\left|\frac{s^2}{\sigma^2} - 1\right| > 2r\right) \\
&= P\left(|N(0; 1)| > r\sqrt{2(n-1)}\right) \\
&= 1 - p
\end{aligned} \tag{5.28}$$

At an accuracy p of 95% and a relative error r of 5%, this results in the following formula for n:

$$n > \frac{1}{2r^2}Q_{N(0;1)}\left(1 - \frac{1-p}{2}\right)^2 + 1 = 770 \tag{5.29}$$

5.3.4   Excess Not Equal to Null

Now the case is examined where the kurtosis is not equal to three. The distribution of the result value now has an excess not equal to null. We can no longer calculate the sampling variance as before. Once again the unbiased estimate from Eq. (5.23) for the sampling variance is examined.

In this case, the variance of s^2 is calculated as follows:

$$V(s^2) = \frac{\sigma^4}{n}\left((\kappa - 1) + \frac{2}{n-1}\right) \tag{5.30}$$

The kurtosis of the distribution is $\kappa \geq 1$. Now let us repeat the same procedure, to estimate the number of Monte Carlo runs:

$$s^2 \sim N\left(\sigma^2; \sigma^2 \sqrt{\frac{1}{n}\left((\kappa - 1) + \frac{2}{n-1}\right)}\right) \sim N\left(\sigma^2; \sigma^2 \sqrt{\frac{\kappa - 1}{n}}\right) \qquad (5.31)$$

or to put it differently:

$$\left(\frac{s^2}{\sigma^2} - 1\right)\sqrt{\frac{n}{\kappa - 1}} \sim N(0; 1) \qquad (5.32)$$

We are looking for the number of n such that the following is true:

$$P\left(\left|\frac{s}{\sigma} - 1\right| > r\right) \sim P\left(\left|\frac{s^2}{\sigma^2} - 1\right| > 2r\right)$$
$$= P\left(|N(0; 1)| > 2r\sqrt{\frac{n}{\kappa - 1}}\right) \qquad (5.33)$$
$$= 1 - p$$

This results in the following formula for n:

$$n > \frac{\kappa - 1}{4r^2} Q_{N(0;1)}\left(1 - \frac{1-p}{2}\right)^2 \qquad (5.34)$$

For $\kappa = 3$, the almost same formula as before, Eq. (5.29), results. As such, the number of runs required to achieve the desired accuracy depends on the kurtosis/excess of the distribution of result values. This parameter is not known in advance, however. Therefore, the number of Monte Carlo runs is determined dynamically in the following.

5.4 Stein's Two-Stage Procedure

One procedure for the dynamic determination of the number of Monte Carlo runs is Stein's two-stage procedure. Other methods include the GUM S1 adaptive schema [1] and the modified GUM S1 adaptive schema [8]. In the following, let us examine Stein's two-stage procedure in more detail.

Originally, Stein's two-stage procedure [6] was developed as a statistical test for a linear hypothesis (or as a special case thereof: student's test for $E(x) = 0$), whose power is independent of the variation of random quantity. This method enables determination of the expectation of a random quantity with sufficient accuracy, independently of its variance. Let us assume that we have a sample x_i of independent, normally distributed random quantities, with the mean value μ and a standard deviation σ that are unknown. We will attempt to determine the value μ with an accuracy of δ and a reliability of $1 - \alpha$.

We select an initial number of runs n_1 (first stage) and determine the sampling variance:

$$s^2 = \frac{1}{n_1 - 1} \sum_i (x_i - \bar{x})^2 \tag{5.35}$$

The number of additional runs n_2 (second stage) is then determined as follows:

$$n_2 = \max\left(\left[\frac{s^2 Q_{t(n_1-1)}(1 - \alpha/2)^2}{\delta^2}\right] - n_1 + 1, 0\right) \tag{5.36}$$

By convention:

- $t(n_1 - 1)$: t distribution with $n_1 - 1$ degrees of freedom
- $[\cdot]$ Round up to an integer.

As long as $n_2 > 0$, additional Monte Carlo runs are performed. Finally, the mean value $\bar{x}$ is calculated for $n_1 + n_2$ values. The probability is then $1 - \alpha$ that this value will lie within the interval $[\mu - \delta, \mu + \delta]$.

The requirement that the random parameter have a normal distribution can be relaxed. The method is robust against moderate deviations of the distribution from the normal distribution.

5.4.1   Determining the Uncertainty (Batch Design)

In contrast to the mean value, the uncertainty and/or variance of a sample can no longer be seen as the total of the independent, normally distributed random variables. Therefore, the schema as described above cannot be used directly.

Batch design is used to overcome this problem [4]. The full series of values is divided into blocks with a uniform number of values—in other words, into batches. The examined value, such as the uncertainty or variance, is determined for each batch.

$$s_i^2 = \frac{1}{m - 1} \sum_{j=1}^{m} (x_{ij} - \bar{x})^2 \tag{5.37}$$

The mean value is then calculated.

$$s_b^2 = \frac{1}{n} \sum_{i=1}^{n} s_i^2 \tag{5.38}$$

This makes it possible to use Stein's two-stage procedure for parameters s_i^2. Stein's two-stage procedure is now used to determine the necessary number of batches.

5.4.2 Dependency on the Batch Size

It is apparent that the estimate s_b^2 meets the expectations for the variance of the random quantity.

$$E(s_b^2) = E(s^2) = \sigma^2 \tag{5.39}$$

Now let us answer the question as to what the variance of s_b^2 looks like and how it relates to the batch size:

$$V(s_i^2) = \frac{\sigma^4}{m}\left((\kappa - 1) + \frac{2}{m-1}\right) \tag{5.40}$$

$$V(s_b^2) = \frac{\sigma^4}{nm}\left((\kappa - 1) + \frac{2}{m-1}\right) \tag{5.41}$$

We examine the difference between the variance s_b^2 and the variance of the conventional estimate s^2, which were calculated for the same number of values $N = nm$: $\Delta = V(s_b^2) - V(s^2)$.

$$\Delta = \frac{2\sigma^4}{nm}\left(\frac{1}{m-1} - \frac{1}{nm-1}\right) = \frac{2\sigma^4}{N}\frac{1}{m} + o\left(\frac{1}{Nm}\right) \tag{5.42}$$

Here, $o\left(\frac{1}{Nm}\right)$ denotes terms that approach zero quicker than $\frac{1}{Nm}$ for $Nm \to \infty$. We can see:

- The variance of the batch-based estimates is greater than the variance of the classic estimate
- The smaller the difference, the larger the batch
- The difference does not depend on the excess of the random quantity

The following estimate applies to $\kappa = 3$:

$$V(s_b^2) \approx V(s^2)\left(1 + \frac{1}{m}\right) \tag{5.43}$$

This means we should take a sufficiently large batch. Larger batches have a disadvantage, however: Batches always have to be calculated all the way through. Even if the desired accuracy is achieved within the batch, the calculations cannot be stopped. If the size of the batch is 50, we cannot stop after 620 runs, but instead have to carry out 30 more calculations.

Moreover:

- The larger the batch, the fewer degrees of freedom the t distribution has in the formula for n_2.
- And the smaller the number of degrees of freedom, the larger is the quantile of their distribution.

5.4.3   Comparison with Earlier Estimation

Now let us try once again to derive an analytical estimate for the necessary number of runs for calculation results with standard distribution. This time we will use Stein's two-stage formula with batch design. Let us examine the sampling variance of the batch size m.

$$s^2 \sim N\left(\sigma^2; \sigma^2\sqrt{\frac{2}{m-1}}\right) = \sigma^2\left(1 + N\left(0; \sqrt{\frac{2}{m-1}}\right)\right) \qquad (5.44)$$

Now let us examine the distribution of the uncertainty of the random quantity:

$$u = k\sqrt{s^2} \sim k\sigma\left(1 + N\left(0; \frac{1}{\sqrt{2(m-1)}}\right)\right) = N\left(k\sigma; \frac{k\sigma}{\sqrt{2(m-1)}}\right) \qquad (5.45)$$

The variance of the uncertainty is

$$V(u) = \frac{(k\sigma)^2}{2(m-1)} \qquad (5.46)$$

Stein's two-stage formula for $\delta = rk\sigma$ (relative accuracy of r) then looks like this:

$$n = \left(\frac{(k\sigma)^2 Q_{t(n_1-1)}(1-\alpha/2)^2}{(rk\sigma)^2 2(m-1)} + 1\right)m = \frac{Q_{t(n_1-1)}(1-\alpha/2)^2}{2r^2}\frac{m}{m-1} + m \qquad (5.47)$$

This formula is very similar to the previously derived formula Eq. (5.29). However, the number of necessary runs is estimated a bit more pessimistically. For $N = 800$ and $m = 25$, we arrive at 892 runs. This time, however, the method guarantees that the uncertainty is reliable and estimated with sufficient accuracy, independently of how the calculation result is distributed.

5.4.4   Numerical Tests

Experiments were carried out that confirm and illustrate the properties of the Monte Carlo method and Stein's two-stage procedure derived here. A pseudo-random number generator was used to test the procedure for two examples: Result with standard distribution and result with an excess not equal to zero. The examples correspond to simple topologies, as described above—with one input and one output and with two inputs and two outputs, with an option to reverse the flow direction. In both cases, the influence of the number of runs in the first stage of Stein's two-stage procedure and the influence of the batch size were examined, as shown in Table 5.1.

Table 5.1 Influence of the number of runs in the first stage of Stein's two-stage procedure and batch size

Distribution	Number of runs in stage 1	Batch size	Number of runs (Expected value)	Standard deviation
Result with standard distribution	300	25	853	69
	400	25	850	58
	500	25	850	43
	600	25	850	47
	700	25	849	45
	800	25	850	38
	900	25	902	8
	800	10	893	43
	795	15	863	40
	800	20	854	38
	800	25	850	38
	780	30	851	41
	770	35	852	41
	800	40	856	39
Example with reverse flow	800	25	1715	85
	900	25	1711	80
	1000	25	1712	76
	1100	25	1713	73
	1200	25	1713	69
	1300	25	1712	67
	1400	25	1712	64
	1200	10	1870	68
	1200	20	1742	68
	1200	30	1692	68
	1200	40	1676	68
	1200	50	1670	68
	1200	60	1671	67
	1190	70	1676	68

5.5 Summary and Outlook

Stein's two-stage procedure can be used to determine the measurement uncertainty precisely and reliably. The number of Monte Carlo runs needed to estimate the measurement uncertainties is highly dependent on the distribution of the results. The excess of the distribution plays a decisive role.

Stein's two-stage procedure has two parameters: the number of Monte Carlo runs in stage 1 and the size of the batch. Analytical considerations and numerical tests show that the number of Monte Carlo runs in stage 1 influences the variance of the number of Monte Carlo runs and the size of the batch determines the expected value. If the batches are much too small or much too large, the number of necessary Monte Carlo runs increases.

The examination here concentrated largely on the question of the reliability and accuracy of measurement uncertainty calculation. It did not aim to reduce the number of Monte Carlo runs and thus conserve computing capacity. This aspect could be examined as a further development of this work, for example, using the genetic algorithms which are used for rare event simulation.

References

1. BIPM, IEC, IFCC, ILAC, ISO, IUPAC, IUPAP and OIML: Evaluation of Measurement Data–Supplement 1 to the 'Guide to the Expression of Uncertainty in Measurement'—Propagation of distributions using a Monte Carlo method Joint Committee for Guides in Metrology. Tech. Rep. 101, Bureau International des Poids et Mesures, JCGM (2008)
2. Herr, E., Scheibe, D., Schröder, P., Voss, K.F., Weimann, A.: Rechnergestützte Zuordnung von an den Einspeisepunkten eines Ferngasnetzes vorgenommenen Brennwertmessungen zu den an den Übergabestationen entnommenen Gasmengen [Computer-supported allocation of calorific measurements taken at the input points of a gas grid to the gas volumes discharged at the transfer stations]. Gas/Erdgas (124) (1983)
3. Kessel, R., Sommer, K.D.: Uncertainty evaluation for quality tracking in natural gas grids. In: MEASUREMENT 2013, Proceedings of the 9th International Conference, Smolenice, Slovakia (2013)
4. Müller, M., Rink, C.: On the convergence of the Monte Carlo block design. Metrologia **46**, 404–8 (2009)
5. Sarge, S.: Ermittlung von Gasbeschaffenheitskenngrößen für Erdgas im geschäftlichen Verkehr mittels Gasnetz-Zustandsrekonstruktion [Determination of gas quality parameters for natural gas in commercial practice using gas network state reconstruction]. PTB-Mitteilungen **1**(127) (2017)
6. Stein, C.: A two-sample test for a linear hypothesis whose power is independent of the variance. Ann. Math. Stat. **16**, 243–258 (1945)
7. Weimann, A.: Modellierung und Simulation der Dynamik von Gasverteilnetzen im Hinblick auf Gasnetzführung und Gasnetzüberwachung [Modelling and simulation of the dynamism of gas distribution networks with regard to gas network management and gas network monitoring]. Ph.D. thesis, Technische Universität München (1978)
8. Wübbeler, G., Harris, P.M., Cox, M.G., Elster, C.: A two-stage procedure for determining the number of trials in the application of a Monte Carlo method for uncertainty evaluation. Metrologia (47) (2010)

Chapter 6
Energy-Efficient High Temperature Processes via Shape Optimization

Christian Leithäuser and René Pinnau

Abstract We consider mathematical models and optimization techniques for a melting furnace in phosphate production. In this high temperature process radiation plays a predominant role. The main design goals are a reduction of the energy consumption as well as the product quality. In particular, we are going to focus on shape optimization techniques for an improved design of the melting furnace, which will rely on a hierarchy of models incorporating the multi-physics of the process.

6.1 Motivation

The transition towards sustainable energy in Germany can only succeed if fossil fuel consumption is significantly reduced. Around 40% of Germany's heat consumption is incurred in the industrial sector and 90% of it is used for process heat. High-temperature processes are always associated with extreme thermal losses, so that there is a powerful lever in this area for saving energy, CO2 and costs.

Mathematical modeling, simulation and optimization (MMSO) can help to utilize this huge potential by ensuring that energy is only used where necessary and that thermal losses are minimized. Optimizing the shape of industrial equipment is particularly important in high-temperature processes where radiation plays a major role. On the one hand, because thermal losses mainly take place over the surface of the melting furnace and on the other hand, because radiation and thus the energy distribution can be controlled by reflections on the boundary.

In our collaborative project, methods for the design of high-temperature processes are developed. These are derived and validated for a typical high-temperature process from chemical industry. Here, we aim for an energy saving of at least 20%. The

C. Leithäuser
Fraunhofer ITWM, Kaiserslautern, Germany
e-mail: christian.leithaeuser@itwm.fraunhofer.de

R. Pinnau (✉)
Department of Mathematics, TU Kaiserslautern, Kaiserslautern, Germany
e-mail: pinnau@mathematik.uni-kl.de

© Springer Nature Switzerland AG 2021
S. Göttlich et al. (eds.), *Mathematical Modeling, Simulation and Optimization for Power Engineering and Management*, Mathematics in Industry 34,
https://doi.org/10.1007/978-3-030-62732-4_6

relevant contribution to the energy transition results from the simple transferability of the developed methods to a large number of other processes from all areas in high-temperature industry. After the end of project, the methods are further developed at Fraunhofer ITWM into a modular tool for shape optimization of high-temperature processes.

6.1.1 Industrial Background

To increase the efficiency of industrial high-temperature processes through modeling, simulation and optimization (MMSO), we work together with our industrial partner ICL in Ladenburg, Germany. The goal is to optimize their energy consuming furnace process, which is used for the production of poly-phosphates. To model this process we use a complex multi-physics approach, including heat transport and radiation, turbulent flow, chemical reactions and phase transitions (for details see Sect. 6.2). Based on this model, both the furnace geometry as well as the positioning of the gas burner need to be optimized by new methods of shape optimization to get a better process efficiency and thus significant energy savings.

Because the heat transfer at high temperatures takes place mainly via radiation, the treatment of radiation problems is given special importance in this collaborative project, which poses special challenges for both, the simulation and optimization methods. In order to enable optimization of the complex model, methods such as space mapping based on model hierarchies are used. The developed methods can later be applied to a large number of similar high-temperature processes, particularly from chemical industry and glass production.

Among other things, poly-phosphates are used in food industry as complexing agents or stabilizers. For the production, phosphoric acid and caustic soda are first neutralized in a special container. The suspension solution then flows into the furnace, where the salt solution is heated and dehydrated at high temperature. Through endothermic chemical reactions, the desired end product sodium hexametaphosphate (sodium poly-phosphate) is finally formed in the furnace. The required high temperatures are generated by gas burners, with the flame aimed directly at the melt. The typical residence time of the melt in the furnace at high temperatures is rather large. In order to increase energy efficiency, on the other hand, a short residence time and a furnace geometry with less temperature losses over the surface are desirable.

This is where the collaborative project comes in: a coupled multi-physics simulation makes it possible to monitor the course of the reaction and by using shape optimization and the optimized positioning of the gas burner, the energy efficiency of the furnace is significantly increased without reducing the quality of the end product. The optimization ensures that the melt running through the furnace receives only the heat input in each area that is necessary for the reaction to run optimally. In particular, the heat propagation driven by radiation can be controlled by the design of the furnace shape. This theoretically enables energy savings of up to 50%. In the project, we aim at energy savings of at least 20% for the chamber furnace process, along with correspondingly reduced pollutant emissions.

6.1.2 Mathematical Background

The joint project combines methods from the following mathematical areas:

- Coupled multi-physics systems,
- Modeling, simulation and optimization under uncertainties,
- Model reduction (here based on asymptotic model reduction and moment approaches),
- Shape optimization for large scale systems.

The different expertise of the project partners allows to raise synergies and to work effectively on this industrial problem. A main focus is on the appropriate model for the high temperature process which is dominated by radiative effects. Here, we can exploit the whole model hierarchy which ranges from the radiative heat transfer model [22], over moment methods [9] to diffusive-type approximations [16]. These were already successfully employed for optimization in glass production [15, 24] and for the simulation of gas turbines [30, 31].

Based on these models we are going to use well-known tools from shape calculus [32, 33] to derive the for the first time the shape gradients for these type of model. Shape calculus has been already successfully used in aerodynamics [27], electro-magnetics [10], polymer distributors [14, 17–19] and the design of microchannel reactors [4].

A final mathematical focus of the project is on the computation of adjoint sensitivities in heat transfer problems involving fluid flows and radiation. Based on algorithmic differentiation this allows to compute very accurate gradients for optimal design applications [1, 11, 25]. In particular, we rely here on the open-source multiphysics suite SU2 [6], which finally allows to validate the shape gradients against adjoint calculations and finite difference approximations.

In this article, we focus in Sect. 6.2 on the multiphysics problem for the full three dimensional melting furnace including all relevant processes like flow, radiation, reactions and vaporization. This model can finally be validated against temperature measurements which are provided by our industrial partner ICL. Further, we describe in Sect. 6.3 how shape calculus can be used for the optimal shape design of radiation dominated processes. Concluding remarks on the sustainability of the mathematical tools in industry and future research directions are given in Sect. 6.4.

6.2 The Multiphysics Problem

The solution of this advanced industrial task is based on the interplay of different mathematical fields. In the following we describe the appropriate models which are used for the numerical simulation and the optimization. Further, we discuss the numerical solution techniques.

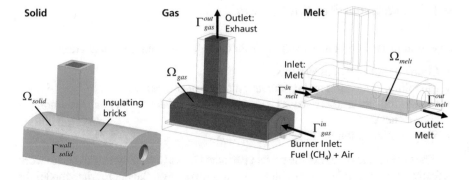

Fig. 6.1 Basic geometric setup of a furnace geometry

6.2.1 Geometric Setup

The geometry of a simplified furnace is shown in Fig. 6.1. Let $\Omega \subset \mathbb{R}^3$ denote the whole domain. This domain decomposes into three main parts: We denote the solid part consisting of different kinds of insulating bricks by Ω_{solid}. The gaseous phase Ω_{gas} contains the flow of a mixture of air, fuel, vaporized water and combustion products as well as the flame which generates the energy. The melt Ω_{melt} is sitting at the bottom of the furnace. This is where the vaporization of water and chemical reactions take place. We assume that the interface $\Gamma_{\text{gas,melt}}$ between Ω_{gas} and Ω_{melt} is geometrically fixed. This assumption of course also holds for the two other interfaces $\Gamma_{\text{solid,gas}}$ and $\Gamma_{\text{solid,melt}}$ with the solid part. The outer wall of Ω_{solid} is denoted by $\Gamma_{\text{solid}}^{\text{wall}}$. The gas burner is the inlet $\Gamma_{\text{gas}}^{\text{in}}$ of Ω_{gas} the exhaust which is connected with the chimney is the outlet $\Gamma_{\text{gas}}^{\text{out}}$. For the melt domain Ω_{melt} the inlet $\Gamma_{\text{melt}}^{\text{in}}$ and outlet $\Gamma_{\text{melt}}^{\text{out}}$ are sitting on the two far sided of the furnace. Furthermore, we define $\Omega_{\text{flame}} \subset \Omega_{\text{gas}}$ which represents the geometric extend of the flame. We use Ω_{flame} to define the flame as an energy source.

6.2.2 Energy Equation

The energy equation for the temperature T is active in the whole domain Ω

$$\rho\, \mathbf{u} \cdot \nabla h(T) - \nabla \cdot (\kappa \nabla T) = S_{\text{flame}} + S_{\text{rad}} + S_{\text{rad}}^{\text{wall}} \quad \text{in } \Omega, \tag{6.1}$$

with sensible enthalpy [2]

$$h(T) = \int_{T_{\text{ref}}}^{T} C_p\, dT, \tag{6.2}$$

Fig. 6.2 The geometric
approximation of the flame
Ω_{flame} as a subset of the
domain Ω_{gas}

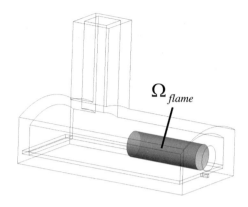

Ω_{flame}

density ρ, specific heat capacity C_p, conductivity κ and reference temperature T_{ref}. Note that all equations in this section are given in dimensional form. The flow velocity **u** in the convection term couples the energy equation with the Navier-Stokes equation (cf. Sect. 6.2.4). Naturally, **u** $= 0$ holds on the solid part Ω_{solid}. The coupling with the radiative transfer (cf. Sect. 6.2.3) comes through the source terms for the internal radiation S_{rad} and the wall radiation $S_{\text{rad}}^{\text{wall}}$ on the right hand side.

Furthermore, the source term S_{flame} originates from our simple flame model. We define it by

$$
S_{\text{flame}} = \begin{cases} \dfrac{\dot{Q}_{\text{flame}}}{\text{vol}(\Omega_{\text{flame}})} & \text{in } \Omega_{\text{flame}} \\ 0 & \text{otherwise} \end{cases}
\tag{6.3}
$$

where $\dot{Q}_{\text{flame}}$ is the rate of heat flow which originates from burning the fuel and $\text{vol}(\Omega_{\text{flame}})$ is the volume of Ω_{flame}. So our simple flame model equally distributes the heat flow over the subset $\Omega_{\text{flame}} \subset \Omega_{\text{gas}}$. The shape of the flame should roughly be resembled by Ω_{flame} (see Fig. 6.2). Of course the simple flame model is a major simplification. The more accurate alternative would be to use a full combustion model which would extensively increase the computational costs. The simple flame model makes sure that the total amount of energy fed into the system is correct and we believe that it is a reasonable choice to help to understand key aspects of the process. Furthermore, boundary conditions for the temperature are set at the inlets and the outer wall of Ω.

6.2.3   Radiative Transfer Equation

The radiative transfer equation is active in the gaseous and in the melt region, but the absorption coefficient varies significantly. Currently we either use the SP_1-approximation [22] or the method of discrete ordinates [5, 22, 23] to solve the radiative transfer. Of course the SP_1-approximation is much faster to solve, while the method of discrete ordinates is more accurate (compare also the detailed discussion in Sect. 6.3).

The following partial differential equation for the incident radiation φ is called SP_1-approximation:

$$- \nabla \cdot (D \nabla \varphi) = -S_{\mathrm{rad}}. \quad \text{in } \Omega_{\mathrm{gas}} \cup \Omega_{\mathrm{melt}} \tag{6.4}$$

The diffusion constant is given by

$$D = \frac{1}{3(\sigma_a + \sigma_s(1 - g))} \tag{6.5}$$

and is derived from the absorption coefficient σ_a, the scattering coefficient σ_s and the anisotropy factor g. The absorption coefficient in the gaseous phase is mainly influenced by its amount of water and also the amount of CO_2. The source term which couples radiative transfer with the energy equation is given by

$$S_{\mathrm{rad}} = \sigma_a \varphi - 4\sigma_a n^2 \sigma T^4, \tag{6.6}$$

where n is the refractive index of the medium and σ is the Stefan-Boltzmann constant. The first term models the absorption of radiation by the surrounding medium while the second term is the Stefan–Boltzmann law [22] describing the radiation of a black body in terms of its temperature.

The boundary condition on

$$\Gamma_{\mathrm{rad}} := \Gamma_{\mathrm{solid,gas}} \cup \Gamma_{\mathrm{solid,melt}} \cup \Gamma_{\mathrm{gas}}^{\mathrm{in}} \cup \Gamma_{\mathrm{gas}}^{\mathrm{out}} \cup \Gamma_{\mathrm{melt}}^{\mathrm{in}} \cup \Gamma_{\mathrm{melt}}^{\mathrm{out}}$$

is given by

$$- D \, \partial_{\mathbf{n}} \varphi = q_{\mathrm{rad}}^{\mathrm{wall}} \quad \text{on } \Gamma_{\mathrm{rad}}, \tag{6.7}$$

where $\partial_{\mathbf{n}}$ is the derivative in direction of the outward pointing normal vector $\mathbf{n}$. With the assumption that the walls are a diffuse gray surface the radiative heat flux at the wall $q_{\mathrm{rad}}^{\mathrm{wall}}$ can be computed by (see [2])

$$q_{\mathrm{rad}}^{\mathrm{wall}} = -\frac{\varepsilon_{\mathrm{wall}}}{2(2 - \varepsilon_{\mathrm{wall}})}(4n^2 \sigma T^4 - \varphi), \tag{6.8}$$

where $\varepsilon_{\mathrm{wall}}$ is the wall emissivity and n is the refractive index of the medium. The wall emissivity depends on the surface material of the wall. On the in- and outlets $\varepsilon_{\mathrm{wall}} = 1$ is assumed. The wall radiation term in the energy equation (6.1) is then defined by

$$S_{rad}^{wall} = \begin{cases} -q_{rad}^{wall} & \text{on } \Gamma_{rad} \\ 0 & \text{otherwise.} \end{cases} \tag{6.9}$$

6.2.4   The Flow Model

With the gaseous and the melt phase there are two different flow regimes which are coupled through their interface. We write $\mathbf{u}_g$, p_g for velocity and pressure in Ω_{gas} and $\mathbf{u}_m$, p_m for velocity and pressure in Ω_{melt}. However, since the corresponding domains are disjunct we also define

$$\mathbf{u} := \begin{cases} \mathbf{u}_g & \text{in } \Omega_{gas}, \\ \mathbf{u}_m & \text{in } \Omega_{melt}, \\ 0 & \text{in } \Omega_{solid} \end{cases} \tag{6.10}$$

and for p respectively.

The flow is modeled using the stationary Navier-Stokes equation [8]

$$\begin{aligned} -\eta \Delta \mathbf{u} + \rho(\mathbf{u} \cdot \nabla)\mathbf{u} + \nabla p &= 0 & \text{in } \Omega_{gas} \cup \Omega_{melt}, \\ \nabla \cdot \mathbf{u}_g &= 0 & \text{in } \Omega_{gas} \cup \Omega_{melt}, \end{aligned} \tag{6.11}$$

with viscosity η and density ρ. Note that for simplicity an incompressible setup was used here. However, since flow velocities and temperatures are typically high, the gaseous phase should be treated as a compressible fluid. The gaseous mixture can then be modeled as an ideal gas and both viscosity and density are temperature dependent, which leads to a coupling with the energy equation. The flow in Ω_{gas} is turbulent due to its high Reynolds number. We solve the turbulent problem by means of the Reynolds-averaged Navier–Stokes equations (RANS) [7] and the k-ω-turbulence model [37]. As for the whole multiphysics forward problem we make use of the implementation in ANSYS Fluent [2].

Mass flow boundary conditions are used at the inflows, pressure boundary conditions at the outflows and wall boundary conditions at the inner furnace walls.

On the interface $\Gamma_{gas,melt}$ we realize a one-way coupling by assuming that the speed of the gas phase is much higher than the speed the melt phase. With this assumption we use the boundary condition

$$\mathbf{u}_g = 0 \quad \text{on } \Gamma_{solid,melt} \tag{6.12}$$

for the gas phase. And for the melt phase we match the tangential stresses $\eta \mathbf{n} \times \mathbf{u}$ on both sides. Therefore, let η_g denote the viscosity of the gaseous phase and η_m the viscosity of the melt phase. Furthermore, let $\mathbf{n}$ denote w.l.o.g. the outward pointing normal vector with respect to Ω_{melt} which forms an orthonormal system with the tangents $\boldsymbol{\tau}_1$ and $\boldsymbol{\tau}_2$. Then, the boundary condition

$$\eta_m \, \partial_\mathbf{n}(\boldsymbol{\tau}_1 \cdot \mathbf{u}_m) = \eta_g \, \partial_\mathbf{n}(\boldsymbol{\tau}_1 \cdot \mathbf{u}_g) \qquad \text{on } \Gamma_{\text{gas,melt}}$$
$$\eta_m \, \partial_\mathbf{n}(\boldsymbol{\tau}_2 \cdot \mathbf{u}_m) = \eta_g \, \partial_\mathbf{n}(\boldsymbol{\tau}_2 \cdot \mathbf{u}_g) \qquad \text{on } \Gamma_{\text{gas,melt}} \qquad (6.13)$$
$$\mathbf{n} \cdot \mathbf{u}_m = 0 \qquad \text{on } \Gamma_{\text{gas,melt}}$$

must hold.

6.2.5 Reaction and Vaporization

With some reasonable simplifications we can assume that a suspension of monosodium phosphate ($Na\,H_2\,PO_4$), disodium phosphate ($Na_2\,H\,PO_4$) and water (H_2O) enters the furnace. Inside the furnace polyphosphate ($Na_6\,P_4\,O_{13}$) is produced by the reaction [38, Chap. 6]

$$2\,Na_2\,H\,PO_4 + 2\,Na\,H_2\,PO_4 \longrightarrow Na_6\,P_4\,O_{13} + 3\,H_2O, \qquad (6.14)$$

while all the water is vaporized.

For the processes under consideration a large amount of the applied energy is used to vaporize this water from the melt and to run the exothermic reactions which transform the melt into the final product. It is important to include these effects into the model to get the correct balance of energy. The direct approach would mean to add a transport equation for the concentration for every species involved.

While this approach is feasible it would make the model quite complex. Instead we want to follow a different approach: we use an effective specific heat capacity to include the effects of vaporization and reaction energy. To derive this effective specific heat capacity we consider a small sample of melt in detail while applying heat. We assume that the sample is small enough such that the temperature and the concentration of every species is uniform, i.e., we do not have any spatial dependence. This can be modeled as a system of ODEs where the variables only depend on the time t [3]. The reaction rate is modeled using an Arrhenius law (see [3])

$$k(T) = A e^{-\frac{E_A}{RT}}. \qquad (6.15)$$

Figure 6.3 shows results for such a simplified setup with Arrhenius constants $A = 1000$ and $E_a = 50\,000\,\text{J}\,\text{mol}^{-1}$. We see how the species composition, the density and the specific heat change with increasing temperature. The amount of energy necessary to reach a specific temperature level is given by Q_{actual}. However, Q_{actual} is not continuous at a temperature of $100°$ because of the vaporizing water. To circumvent this we use a smoothed version Q_{smooth} of Q_{actual} which is differentiable. The derivative of Q_{smooth} is the effective specific heat capacity $C_p^{\text{eff}}(T)$ which we use in the multiphysics simulation to model the melt.

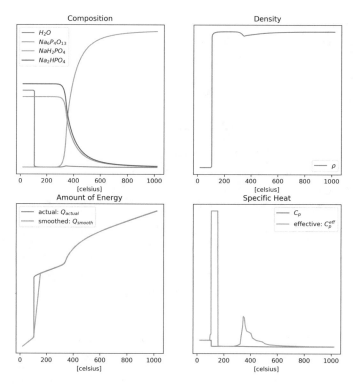

Fig. 6.3 Material composition $\mu_i N_i(T)$, density, amount of energy and specific heat over temperature T for the solution of the ODE system

6.2.6   Simulation Results

The multiphysics problem is used for forward simulations of the furnace and for the validation of the results. Optimization results obtained with reduced models are finally validated with the multiphysics problem. The multiphysics problem is solved using ANSYS Fluent [2]. Exemplary results for a simplified furnace are shown in Fig. 6.4 with process parameters from Table 6.1.

6.3   Shape Optimization

The full multiphysics problem is clearly much too expensive for a direct optimization approach. Nevertheless, it is well suited for validation purposes or as a fine model in the so-called space mapping algorithm. In the following, we describe how modern optimization algorithms can be employed for the automated design of the melting furnace.

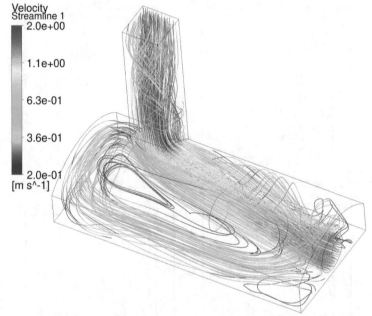

(a) Streamlines through the gaseous phase of the furnace colored by velocity.

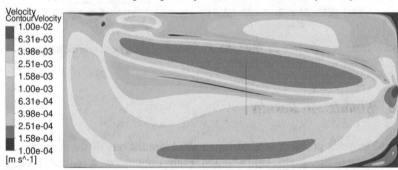

(b) Velocity at the surface of the melt.

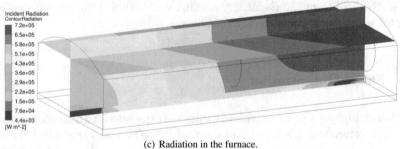

(c) Radiation in the furnace.

Fig. 6.4 Demonstrational simulation results for the 3D furnace

Table 6.1 Input parameters for the test case of the 3D furnace model

	Gaseous phase Ω_{gas}	Melt phase Ω_{melt}
Mass flow rate at inflow $\dot{m}_{gas}^{in}/\dot{m}_{melt}^{in}$	$0.8\,\mathrm{kg\,s^{-1}}$	$0.3\,\mathrm{kg\,s^{-1}}$
Temperature at inflow $T_{gas}^{in}/T_{melt}^{in}$	$27\,^{\circ}\mathrm{C}$	$100\,^{\circ}\mathrm{C}$
Pressure at outflow $p_{gas}^{out}/p_{melt}^{out}$	0	0
Rate of heat flow of flame $\dot{Q}_{flame}$	$1e6\,\mathrm{W}$	–
Density ρ	$1.2\,\mathrm{kg\,m^{-3}}$	$\rho_{\phi}(T)$
Viscosity η	$1.8e-5\,\mathrm{Pa\,s}$	$0.05\,\mathrm{Pa\,s}$
Specific heat capacity C_p	$1006\,\mathrm{J\,kg^{-1}\,K^{-1}}$	$C_{p,\phi}^{eff}(T)$
Conductivity κ	$0.024\,\mathrm{W\,m^{-1}\,K^{-1}}$	$10\,\mathrm{W\,m^{-1}\,K^{-1}}$
Absorption coefficient σ_a	$0.4\,\mathrm{m^{-1}}$	$40\,\mathrm{m^{-1}}$
Scattering coefficient σ_s	0	0
Anisotropy factor g	–	–
Refractive index n	1	1
Wall emissivity ε_{wall}	1	1

6.3.1 The Radiation Model

In this high temperature process radiation plays a dominant role, such that the full Radiative Heat Transfer Equation (RHTE) for grey matter in a homogeneous a medium would be an appropriate model [22]. Since we are considering a 24/7 process, it is sufficient to study a stationary model describing the multi-physics processes in the atmosphere and in the phosphate melt (compare also Sect. 6.2).

The full stationary, convective Radiative Heat Transfer (RTE) system on a bounded domain $\Omega \subset \mathbb{R}^3$ reads

$$-\nabla \cdot (k \nabla T(x)) + c\, u(x) \cdot \nabla T(x) = -\int_{S^2} \kappa (a\, T^4(x) - I(x, \omega))\mathrm{d}\omega \qquad \text{in } \Omega, \tag{6.16a}$$

$$\forall \omega \in S^2: \ \omega \cdot \nabla_x I(x, \omega) = \kappa (a\, T^4(x) - I(x, \omega)) \qquad \text{in } \Omega, \tag{6.16b}$$

where the unknowns are the specific radiative intensity $I(x, \omega)$ at position $x \in \Omega$ traveling in direction $\omega \in S^2$ on the unit sphere S^2, as well as the temperature $T(x)$ for $x \in \Omega$. The convective flow u is either pre-computed or can be calculated from a coupled Boussinesq model.

The model parameters are the convection coefficient c, the absorption coefficient κ, the heat conduction coefficient k and the Stefan-Boltzmann constant a. The total thermal radiation $B(T) = aT^4$ is defined by Stefan's law, which is dominant for high temperatures. For more details on the model and its parameters we refer to [22].

The model is supplemented with Robin boundary conditions (6.16c) for the temperature and a semi-transparent boundary condition (6.16d) for the incident radiation (see also [16]):

$$k n \cdot \nabla T = h(T_{out} - T) \qquad \qquad \text{on } \partial\Omega, \tag{6.16c}$$

$$I(\omega, x) = \rho(n \cdot \omega)I(\omega', x) + (1 - \rho(n \cdot \omega))aT_{out}^4, \quad n \cdot \omega < 0, \qquad x \in \partial\Omega, \tag{6.16d}$$

where $\omega' = \omega - 2(n \cdot \omega)n$ is the specular reflection of ω and n the outward unit normal along the boundary. The convective heat transfer coefficient h and the ambient temperature T_{out} are given constants. The reflectivity $\rho(n \cdot \omega) \in [0, 1]$ can be calculated from Fresnel's law (see, e.g., [16]).

This model has a five dimensional phase space for the three spatial directions and the two angular directions. In the case of a frequency dependent model, we would even need to add another dimension. Thus, this model is unsuitable for direct optimization due to its high numerical complexity. Hence, we are considering a simplified model for the radiation, which is given by the SP_1 approximation [16] and has been already successfully employed for optimization purposes in glass production [15, 24]. This yields a stationary, nonlinear heat equation coupled with the elliptic SP_1 approximation, which reads in its scaled version [24]:

$$-\varepsilon^2 k \, \Delta T + \varepsilon c \, u \cdot \nabla T = -\left(\kappa 4\pi aT^4 - \kappa\phi\right) + f_1 \quad \text{in } \Omega,$$

$$-\frac{\varepsilon^2}{3\kappa}\Delta\phi = \kappa 4\pi aT^4 - \kappa\phi + f_2 \qquad \text{in } \Omega, \tag{6.17a}$$

where the unknowns are the temperature T and the radiative flux density ϕ. The model parameters are the heat conductivity k, the convection coefficient c, and a scaling parameter $\varepsilon > 0$ describing the optical thickness of the medium. Further, κ is the absorption coefficient, a the Stefan-Boltzmann constant and f_1, f_2 are heat and radiation sources. The derivation of this model is based on a formal Neumann-series expansion [12, 16].

We assume that the boundary of the domain $\Gamma = \partial\Omega$ splits into three parts: for the temperature we prescribe an outgoing heat flux modeling the exhaust by a Neumann boundary condition at Γ_{out}. Further, we have an incoming heat flux at Γ_{in}. At the remaining part of boundary we employ Newton's cooling law [22]. The boundary condition for the radiative flux density ϕ is an approximation of the kinetic semi-transparent boundary condition (6.16d) of the RHTE [22]. The reflectivity constant χ can be calculated from the reflectivity $\rho(n \cdot \omega)$ (for details see [16]).

Finally, we supplement the SP_1-system (6.17a) with the boundary conditions

$$\varepsilon k \, n \cdot \nabla T \quad = h(T_{out} - T) \qquad \text{on } \Gamma \backslash (\Gamma_{out} \cup \Gamma_{in}),$$

$$\varepsilon k \, n \cdot \nabla T \quad = g_{in} \qquad \text{on } \Gamma_{in},$$

$$\varepsilon k \, n \cdot \nabla T \quad = g_{out} \qquad \text{on } \Gamma_{out}, \qquad\qquad (6.17b)$$

$$\frac{\varepsilon}{3\kappa} n \cdot \nabla \phi \quad = \chi (4\pi a T_{out}^4 - \phi) \qquad \text{on } \Gamma.$$

6.3.2 The Optimal Design Problem

As a first shape optimization problem we have the objective to reach a desired temperature at the bottom of the furnace, which is in the temperature range such that all chemical reactions can take place. In this way we can avoid the solution of the full reactions described in Sect. 6.2.5.

The design variable is the shape of the domain Ω, in which our state equations are solved. We prescribe the desired temperature profile T_d at the bottom $\Gamma_B \subset \Gamma$, and consider the minimization problem

$$\min_{\Omega} J(\Omega, T(\Omega)) = \min_{\Omega} ||T(\Omega) - T_d||^2_{L^2(\Gamma_B)}$$

subject to the state system (17).

Remark 1 One might also consider different cost functionals, e.g., different norms or additional regularization terms like the length of the exhaust.

6.3.3 Basic Concepts in Shape Optimization

The main tool in shape optimization is the computation of the shape derivative for the shape design problem at hand. We define the space of admissible shapes

$$\Xi \subset 2^{\mathbb{R}^d} := \{\Omega \subset D \subset \mathbb{R}^d : \Omega \text{ is a Lipschitz domain}\}$$

for $d = 2, 3$, and where D is a smooth, bounded domain containing all of these shapes.

In order to optimize a given cost functional with respect to the geometry one perturbs its shape $\Omega \in \Xi$ by the action of a compactly supported velocity field $V \in C_0^1(\bar{D}, \mathbb{R}^3)$. We transform the domain using the so called *speed method* [32] for $t \in [0, \tau]$, with $\tau > 0$ small enough, by the solution of the ODE

$$\frac{d}{dt} x(t) = V(x(t)), \qquad x(0) = X. \qquad\qquad (6.18)$$

This defines a family of diffeomorphic perturbations by

$$\mathcal{T}_t : \mathbb{R}^d \to \mathbb{R}^d, \quad X \mapsto \mathcal{T}_t(X) := x(t)$$

with $\mathcal{T}_t(x) = \mathcal{T}(t, x) \in C([0, \tau); C^k(\bar{D}, \mathbb{R}^d))$. Here, k defines the regularity of the boundary of the domain D. The corresponding family of perturbed shapes is given by $\Omega_t = \mathcal{T}_t(\Omega)$. The local existence of a diffeomorphic solution to system (6.18) is proven in [32, p. 50].

This allows us to define the shape derivative [32, p. 54].

Definition 1 (*Shape Derivative*) Let $J(\Omega)$ be a shape functional and $\Omega_t = \mathcal{T}_t(\Omega)$ be a family of perturbations given by a flow which is generated by a smooth vector field $V \in C_0^1(D; \mathbb{R}^d)$ (see Eq. (6.18)). Then, the Eulerian semiderivative is defined as the limit

$$\mathrm{d}J(\Omega)[V] = \lim_{t \downarrow 0} \frac{J(\Omega) - J(\Omega_t)}{t}. \tag{6.19}$$

If this limit exists and is linear and continuous in the direction V, we call J shape differentiable with shape derivative $\mathrm{d}J$.

Remark 2 In particular, in recent years the volume formulation, or weak form of the shape derivative, is used by many authors (see, e.g., [26, 35]), which often also yields a faster convergence of finite element approximations [13].

Remark 3 Also higher order derivatives of the shape functional can be defined. A characterization of shape derivatives using Riemannian manifolds as shape spaces yields symmetric Riemannian shape Hessians, which can used for the derivation of sufficient optimality conditions and second order optimization algorithms [29, 36].

Since the definition of the shape derivative is hard to evaluate directly, different techniques have been developed to calculate the shape derivative $\mathrm{d}J$ and here we use in particular the material derivative free Lagrangian approach (see, e.g., [33] for an overview).

This allows us to derive for the first time the shape derivative for the radiation model (6.17a). Starting from its weak form we first get the weak formulation of the adjoint equations and then the final form of the shape derivative. For the sake of a smoother presentation we omit the lengthy formulas here and refer instead to [21].

Since the shape derivative cannot be directly used in numerical algorithms, we project it with the help of an inner product into a desired Hilbert space $\mathcal{H}$ and get in this way the final shape gradient (see [34] for more details). Here, we use the inner product induced by the linear elasticity equation (compare, e.g., [28]). Then, the shape gradient can be computed from the solution of

$$(\nabla J(\Omega), V)_{\mathcal{H}} = \mathrm{d}J(\Omega)[V] \quad \text{for all} \ V \in \mathcal{H}$$

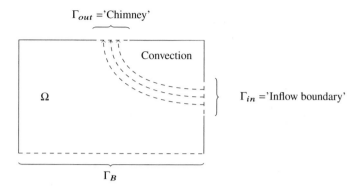

Γ_{out} ='Chimney'

Convection

Ω

Γ_{in} ='Inflow boundary'

Γ_B

Fig. 6.5 Cut through the furnace in length direction

6.3.4   Numerical Results

To validate the feasibility of our optimization approach we consider a two dimensional cut along the length axis of the melting furnace (see Fig. 6.5). Then, the shape optimization can be restricted to two spatial dimensions. The goal is to reach a desired a temperature distribution at the bottom of the furnace and the initial shape of the furnace is half of the unit square. For the numerical simulations we use a one-way coupling with the Navier-Stokes equations and precompute the convective flow u in Eq. (6.17a). Here, we neglect volume sources and energy enters the system only via the inflow Neumann boundary condition.

Remark 4   Although the derivation of the shape derivative also holds for three spatial dimensions, we consider here only the two dimensional cut. During the last year of the project it is planned to use methods from algorithmic differentiation to get the shape derivative of the full model presented in Sect. 6.2. This will then allow for the solution of the full 3d problem.

For the discretization we use the finite element library DOLFIN in Python, which is provided by the framework FEniCS version 2017.2 [20]. In particular, we use $P1$ elements for the radiative flux density ϕ and $P2$ elements for the material temperature T to avoid spurious oscillations at the boundary. For the convection and pressure field, we use the Crouzeix–Raviart finite element for the Navier-Stokes equation. Remeshing is done with gmsh 3.0.6.

The minimization problem is solved for the parameters stated in Table 6.2 by a descent method with step-size control method given by Algorithm 1.

The convective flow field for the initial configuration can be seen in Fig. 6.6, while the corresponding temperature distributions is depicted in Fig. 6.7.

Our optimal shape design algorithm terminates after twelve iterations with the optimal shape shown in Fig. 6.8, in which one also sees the corresponding temperature distribution within the furnace.

Table 6.2 Parameter used for the simulation

Description	Variable	Value
Stefan-Boltzmann constant	a	5.67e-8
Optical thickness	ε	3
Reference value in space	x_{ref}	6
Reference value temperature	T_{ref}	1000
Reference value absorption	κ_{ref}	0.056
Reference value heat conduction	k_{ref}	0.018
Physical value heat conduction	k	11000
Source[a]	f_2	0 for case (a)
		200 for case (b)
Source[a]	f_1	0
Regularization constant	μ	2e5 for case (a)
		8e5 for case (b)
Reflectivity parameter	χ	0.01554
Outflow coefficient	g_{out}	−12000
Inflow coefficient	g_{in}	6000
Navier-Stokes inflow vector	u_{in}	$[-0.4, 0]^T$
Convection coefficient	c	1.8e6
Reynolds number	Re	1080

[a] Scaled density of 2d normal distribution

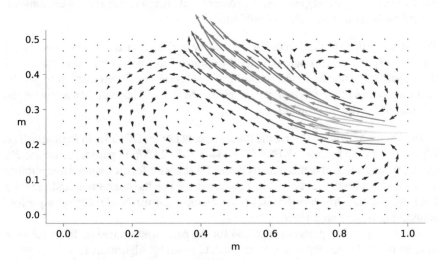

Fig. 6.6 Convective flow field calculated from the Navier-Stokes system

Result: Local optimum $\tilde{\Omega}$ of functional J or -1 for error.
1 Initialize $\Omega_0 := [0, 1] \times [0, 0.5]$ with crossed triangles;
2 **for** $i = 0$ **to** *max iterations* **do**
3 $\quad$ Solve state equation on Ω_i;
4 $\quad$ Solve adjoint equation on Ω_i;
5 $\quad$ Calculate gradient V_i;
6 $\quad$ Update $\Omega_{i+1} = \Omega_i + hV_i$ using Armijo rule.;
7 $\quad$ **if** *Armijo rule failed because of mesh quality* **then**
8 $\quad\quad$ Remesh;
9 $\quad\quad$ Solve state equation on Ω_i;
10 $\quad\quad$ Solve adjoint equation on Ω_i;
11 $\quad\quad$ Calculate gradient $\tilde{V}_i$;
12 $\quad\quad$ Update $\Omega_{i+1} = \Omega_i + h\tilde{V}_i$ using Armijo rule.;
13 $\quad\quad$ **if** *Armijo rule failed because of mesh quality* **then**
14 $\quad\quad\quad$ Return -1
15 $\quad\quad$ **end**
16 $\quad$ **end**
17 $\quad$ **if** $\frac{\|V_i\|}{\|V_0\|} < TOL$ **then**
18 $\quad\quad$ Return $\tilde{\Omega} = \Omega_i$
19 $\quad$ **end**
20 **end**

Algorithm 1: The descent algorithm with step-size control.

Fig. 6.7 Temperature distribution in the initial configuration

Since our objective in the cost functional was a desired temperature profile at the bottom, we depict in Fig. 6.9 the temperature along the bottom of the furnace. Note, that the optimal shape allows for an increase of the boundary temperature by ca. 70 degree Kelvin. This is mainly achieved by a smaller furnace with better reflectivity on the upper ceiling.

To underline the performance of the algorithm we show in Fig. 6.10 the relative norm of the shape gradients during the iterations.

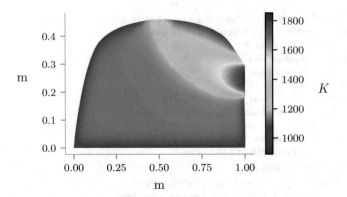

Fig. 6.8 Optimized shape with final temperature distribution

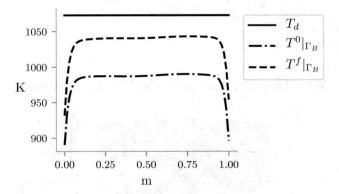

Fig. 6.9 Temperature profiles at the bottom Γ_B with T_d desired temperature profile, $T^0|_{\Gamma_B}$ temperature profile of initial geometry, $T^f|_{\Gamma_B}$ temperature profile of final geometry

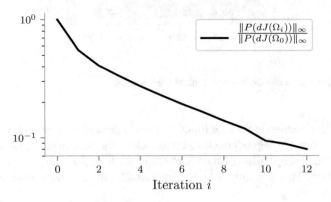

Fig. 6.10 Relative $||.||_\infty$–norm of the shape gradient

6.4   Conclusions

The ICL melting furnace must be rebuilt at regular intervals. Implementation at ICL is therefore expected within the first year after the end of the project. The Fraunhofer ITWM actively promotes the shape optimization of high-temperature processes and uses the extensive Fraunhofer network of industrial customers, thereby increasing the awareness of the topic. The first, industry-funded, follow-up projects to increase energy efficiency are expected in the first two years after the end of the project.

At the same time, the Fraunhofer ITWM, with the support of the university partners, is expanding the methodology into a modular shape optimization tool, which helps to increase the efficiency of high-temperature processes. This extends the usability to a variety of application areas, for example in chemical industry, glass production or crystal growth. Significant energy savings in these industrial areas with huge savings potential are expected in the first five years after the end of the project. This multiplier effect leads to a widespread use of the results and thus a significant contribution to the energy transition and the economic attractiveness of Germany as a business location.

Acknowledgements This project has been supported by the Federal Ministry of Education and Research (BMBF) under grant numbers 05M18UKA and 05M18AMA. The authors are indebted to the other project members for their input and support in finalizing this article. In particular, they thank N. Dietrich, N. Gauger, Th. Marx, E. Özkaya, R. Sanchez (all at TU Kaiserslautern), as well as R. Feßler, N. Siedow (Fraunhofer ITWM, Kaiserslautern) and R. Tänzler (ICL Ladenburg).

References

1. Albring, T., Sagebaum, M., Gauger, N.R.: A consistent and robust discrete adjoint solver for the SU2 framework – validation and application. In: New Results in Numerical and Experimental Fluid Mechanics X, pp. 77–86. Springer (2016)
2. ANSYS® Fluent, Release 19.3: Help System, Fluent Theory Guide. ANSYS, Inc. (2019)
3. Arima, T.: Numerical methods for chemically reacting fluid flow computation under low-mach number approximation. Tokyo J. Math. **29**(1), 167–198 (2006). https://doi.org/10.3836/tjm/1166661873
4. Blauth, S., Leithäuser, C., Pinnau, R.: Model hierarchy for the shape optimization of a microchannel cooling system (2019)
5. Chui, E., Raithby, G.: Computation of radiant heat transfer on a nonorthogonal mesh using the finite-volume method. Numer. Heat Transf. **23**(3), 269–288 (1993)
6. Economon, T.D., Palacios, F., Copeland, S.R., Lukaczyk, T.W., Alonso, J.J.: SU2: an open-source suite for multiphysics simulation and design. Aiaa J. **54**(3), 828–846 (2016)
7. Ferziger, J.H., Peric, M.: Numerische Strömungsmechanik. Springer (2008)
8. Foias, C., Manley, O., Rosa, R., Temam, R.: Navier-Stokes Equations and Turbulence. Encyclopedia of Mathematics and its Applications. Cambridge University Press (2001). https://doi.org/10.1017/CBO9780511546754
9. Frank, M.: Approximate models for radiative transfer. Bulletin of the Institute of Mathematics Academia Sinica **2**, 409–432 (2007)

10. Gangl, P., Langer, U., Laurain, A., Meftahi, H., Sturm, K.: Shape optimization of an electric motor subject to nonlinear magnetostatics. SIAM J. Sci. Comput. **37**(6), B1002–B1025 (2015). https://doi.org/10.1137/15100477X
11. Griewank, A., Walther, A.: Evaluating Derivatives: Principles and Techniques of Algorithmic Differentiation, vol. 105. Siam (2008)
12. Herty, M., Pinnau, R., Thömmes, G.: Asymptotic and discrete concepts for optimal control in radiative transfer. ZAMM Zeitschrift für Angewandte Mathematik und Mechanik **87**(5), 333–347 (2007)
13. Hiptmair, R., Paganini, A., Sargheini, S.: Comparison of approximate shape gradients. BIT Numer. Math. **55**(2), 459–485 (2015). https://doi.org/10.1007/s10543-014-0515-z
14. Hohmann, R., Leithäuser, C.: Shape optimization of a polymer distributor using an Eulerian residence time model. SIAM J. Sci. Comput. **41**(4), B625–B648 (2019). https://doi.org/10.1137/18M1225847
15. Kossiga, A., Béchet, F., Siedow, N., Lochegnies, D.: Influence of radiative heat transfer model on the computation of residual stresses in glass tempering process. Int. J. Appl. Glass Sci. (2017). https://doi.org/10.1111/ijag.12335
16. Larsen, E.W., Thömmes, G., Klar, A., Seaïd, M., Götz, T.: Simplified P_n approximations to the equations of radiative heat transfer and applications. J. Comput. Phys. **183**, 652–675 (2002)
17. Leithäuser, C., Pinnau, R., Feßler, R.: Approximate controllability of linearized shape-dependent operators for flow problems. ESAIM: COCV **23**(3), 751–771 (2017). https://doi.org/10.1051/cocv/2016012
18. Leithäuser, C., Pinnau, R., Feßler, R.: Designing polymer spin packs by tailored shape optimization techniques. Optim. Eng. **19**(3), 733–764 (2018). https://doi.org/10.1007/s11081-018-9396-3
19. Leithäuser, C., Pinnau, R., Feßler, R.: Shape design for polymer spin packs: modeling, optimization and validation. J. Math. Ind. **8** (2018). https://doi.org/10.1186/s13362-018-0055-2
20. Logg, A., Wells, N.G., J, H.: DOLFIN: a C++/Python Finite Element Library, chap. 10. Springer (2012)
21. Marx, T., Dietrich, N., Pinnau, R.: Shape optimization for the SP1–model for convective radiative heat transfer. In: Pinnau, R., Gauger, N., Klar, A. (eds.) Modeling, Simulation and Optimization in the Health- and Energy-Sector. Springer (2020)
22. Modest, M.: Radiative Heat Transfer. Academic Press, San Diego (2003)
23. Murthy, J., Mathur, S.: Finite volume method for radiative heat transfer using unstructured meshes. J. Thermophys. Heat Transf. **12**(3), 313–321 (1998)
24. Pinnau, R.: Analysis of optimal boundary control for radiative heat transfer modeled by the SP_1-system. Commun. Math. Sci. **5**, 951–969 (2007)
25. Sagebaum, M., Gauger, N.R., Naumann, U., Lotz, J., Leppkes, K.: Algorithmic differentiation of a complex c++ code with underlying libraries. In: ICCS, pp. 208–217 (2013)
26. Schmidt, S.: Weak and strong form shape hessians and their automatic generation. SIAM J. Sci. Comput. **40**(2), C210–C233 (2018). https://doi.org/10.1137/16M1099972
27. Schmidt, S., Schulz, V.: Shape derivatives for general objective functions and the incompressible Navier-Stokes. Control Cybern. **39** (2010)
28. Schulz, V., Siebenborn, M.: Computational comparison of surface metrics for PDE constrained shape optimization. Comput. Meth. Appl. Math. **16**, 485–496 (2016)
29. Schulz, V., Siebenborn, M., Welker, K.: Efficient PDE constrained shape optimization based on Steklov-Poincaré-type metrics. SIAM J. Optim. **26**, 2800–2819 (2016)
30. Seaid, M., Frank, M., Klar, A., Pinnau, R., Thömmes, G.: Efficient numerical methods for radiation in gas turbines. J. Comput. Appl. Math. **170**(1), 217–239 (2004)
31. Seaid, M., Klar, A., Pinnau, R.: Numerical solvers for radiation and conduction in high temperature gas flows. Flow Turbul. Combust. **75**, 173–190 (2005). https://doi.org/10.1007/s10494-005-8589-y
32. Sokolowsky, J., Zolesio, J.P.: Introduction to Shape Optimization. Springer (1992)

33. Sturm, K.: Minimax Lagrangian approach to the differentiability of nonlinear PDE constrained shape functions without saddle point assumption. SIAM J. Control Optim. **53**, 2017–2039 (2015)
34. Sturm, K., Eigel, M.: Reproducing kernel Hilbert spaces and variable metric algorithms in PDE constrained shape optimisation. Optim. Methods Softw. (2016). https://doi.org/10.1080/10556788.2017.1314471
35. Sturm, K., Laurain, A.: Distributed shape derivative via averaged adjoint method and applications. ESAIM Math. Model. Numer. Anal. **50** (2015). https://doi.org/10.1051/m2an/2015075
36. Welker, K.: Efficient PDE constrained shape optimization in shape spaces. Ph.D. thesis (2016)
37. Wilcox, D.C., et al.: Turbulence Modeling for CFD, vol. 2. DCW industries La Canada, CA (1998)
38. Zobel, D.: Verfahrensentwicklung und Technische Sicherheit in der Anorganischen Phosphorchemie. Expert Verlag (2018)

Chapter 7
Power-to-Chemicals: A Superstructure Problem for Sustainable Syngas Production

Dominik Garmatter, Andrea Maggi, Marcus Wenzel, Shaimaa Monem, Mirko Hahn, Martin Stoll, Sebastian Sager, Peter Benner, and Kai Sundmacher

Abstract A novel benchmark superstructure is defined for the production of syngas from renewable energy, H_2, CO_2 and biogas. Fixed bed reactors (FBR) for the dry reforming (DR) or the reverse water gas shift (RWGS) are used to convert the reactants into raw syngas, which is then purified in a sequence of pressure and/or temperature swing adsorption (PSA/TSA) steps. We discuss the simulation of the resulting model equations. The complexity of the dynamic process model is tackled by model order reduction and parallel-in-time integration using PARAREAL in order to allow a faster model evaluation in particular for the multi-query context

D. Garmatter (✉) · M. Stoll
Technische Universität Chemnitz, Straße der Nationen 62, 09111 Chemnitz, Germany
e-mail: dominik.garmatter@mathematik.tu-chemnitz.de

M. Stoll
e-mail: martin.stoll@mathematik.tu-chemnitz.de

A. Maggi (✉) · M. Wenzel · S. Monem · P. Benner · K. Sundmacher
Max Planck Institute for Dynamics of Complex Technical Systems, Sandtorstr. 1, 39106 Magdeburg, Germany
e-mail: maggi@mpi-magdeburg.mpg.de

M. Wenzel
e-mail: wenzel@mpi-magdeburg.mpg.de

S. Monem
e-mail: monem@mpi-magdeburg.mpg.de

P. Benner
e-mail: benner@mpi-magdeburg.mpg.de

K. Sundmacher
e-mail: sundmacher@mpi-magdeburg.mpg.de

S. Monem · M. Hahn · S. Sager · P. Benner · K. Sundmacher
Otto von Guericke University Magdeburg, Universitätsplatz 2, 39106 Magdeburg, Germany
e-mail: mirhahn@ovgu.de

S. Sager
e-mail: sager@ovgu.de

© Springer Nature Switzerland AG 2021
S. Göttlich et al. (eds.), *Mathematical Modeling, Simulation and Optimization for Power Engineering and Management*, Mathematics in Industry 34,
https://doi.org/10.1007/978-3-030-62732-4_7

arising in a design optimization process. Future challenges regarding mixed-integer optimization and more complex model dynamics are discussed.

Keywords Sustainable syngas production · Superstructure · Simulation · Parallel in time · Model order reduction

7.1 Introduction

Large reductions in CO_2 emissions are necessary to minimize the effects of human made climate change in the future. For this reason, the world economy needs to be gradually shifted away from fossil fuels towards a more sustainable basis for the production of heat, power and value-added products. In this context, CO_2 has become an interesting alternative carbon source to fossil fuels since it is abundantly available from point sources (e.g. cement plants) and from air. One very active area of research is the production of liquid fuels (e.g. gasoline, diesel, jet fuel) from renewable resources and energy: CO_2 emissions could be drastically reduced if liquid fuels were produced on the basis of CO_2 instead of fossil fuels. Liquid fuels can be readily produced by Fischer-Tropsch synthesis [1] or via methanol (among other routes). Both processes require syngas, a mixture of hydrogen (H_2) and carbon monoxide (CO). For liquid fuel production, the H_2/CO ratio of 2:1 is desired. Two promising chemical reactions to produce syngas from CO_2 are the reverse water-gas shift (RWGS) and the dry reforming (DR) reactions, see, e.g., [2, 3]. In the RWGS reaction, H_2 is used to convert CO_2 into CO according to the following equation:

$$CO_2 + H_2 \rightleftharpoons CO + H_2O. \tag{7.1}$$

In the DR reaction, methane (CH_4) is used instead of H_2 to convert CO_2 and the reaction proceeds according to:

$$CO_2 + CH_4 \rightleftharpoons 2H_2 + 2CO. \tag{7.2}$$

As both reactions are thermodynamically limited (i.e. the reactants are not fully converted into the products), purification is generally necessary before the syngas can be used for liquid fuel production. This can be achieved by pressure or temperature swing adsorption (PSA or TSA, respectively), in which an adsorbent is used to selectively remove chemical components from a gas mixture.

All possible process pathways for syngas production based on the given reactions and separation techniques can be included in a superstructure, which is introduced in the following.

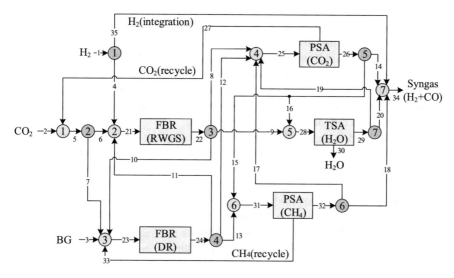

Fig. 7.1 Superstructure for sustainable production of syngas from H_2, CO_2 and BG via RWGS, DR in FBRs with subsequent product separation in different PSA and TSA units. Green and red circles denote mixers and splitters, respectively. Numbers indicate process streams

7.1.1 Reactor-Separator-Recycle Superstructure

The superstructure depicted in Fig. 7.1 is based on the assumption that clean H_2, CO_2 and biogas (BG, a mixture of CO_2 and CH_4), delivered by streams 1, 2 and 3, are available from sustainable sources such as water electrolysis based on renewable power, direct air-capture and biomass. The reactor network representation corresponds to a superstructure of fixed-bed reactors (FBR) where the RWGS and DR reactions are taking place inside the reactors that process the inputs from streams 21 and 23. In the separation network, the undesired components H_2O, CO_2 and CH_4 are separated from the raw syngas via pressure swing adsorption (CO_2 and CH_4) or temperature swing adsorption (H_2O) fed by stream 25, 31 and 28, respectively. Unreacted CO_2 and CH_4 can be recycled to the reactors (streams 27 and 33). On the other hand, H_2O is discarded since it can not be used as a reactant (stream 30). Pure hydrogen can be added by stream 35 after the separators to adjust the syngas H_2/CO ratio.

7.1.2 Contribution

The main contribution is a mathematically rigorous description of the novel superstructure presented here. The governing equations of the various units being either ordinary differential equations (ODEs) or partial differential equations (PDEs) are described and the algebraic equations resulting from the connections between the units are derived. After discretizing the model equations in space, a differential-algebraic equation (DAE) system is obtained. In a numerical investigation, the DAE system is first solved and discussed for an exemplary configuration of the super-structure. Afterwards, two state-of-the-art methods for mathematical modelling of time-dependent problems are discussed that aim at speeding up the solution time of the system, PARAREAL and model order reduction.

PARAREAL, see, e.g., [4, 5], aims at decomposing the global time domain into several smaller domains and, given initial values on these subdomains, the global problem splits up into local subproblems that, in each iteration, can be solved in parallel. The initial values are generated using a cheap but possibly inaccurate coarse time integrator and the subproblems are then solved in parallel using an expensive but accurate fine time integrator. Initial values for the next iteration are generated in a correction step utilizing the information from the fine integrator. While PARAREAL decomposes the time domain to speed up the solution time, model order reduction (MOR) aims at reducing the dimension of the state of a system such that only a small surrogate model has to be solved which is computationally less demanding, see, e.g., [6, 7]. The techniques of proper orthogonal decomposition (POD) and discrete empirical interpolation (DEIM) are reviewed and applied to the governing equations of the units (reactors and separators). Numerical results will show the efficiencies of both approaches.

Finally, challenges and future research directions are discussed. This will include: more complex superstructures, enhanced dynamics in the governing equations, opti-mizing the superstructure with respect to suitable objective functions resulting in a mixed-integer nonlinear program, improvements and/or combinations of MOR and PARAREAL in the context of (large-scale) DAE systems.

7.2   Mathematical Model

In this section, the models used to describe the different process units of the super-structure, i.e., the FBR reactor and the PSA and TSA units, are shortly derived and explained, see, e.g., [8] for further details. A general form of the interconnection of the process streams is proposed, which is necessary to solve the simulation problem.

7.2.1   Connectivity of Unit Operations

Mixers and splitters are used before and after each process unit to allow connections between all possible process units. The mixers and splitters are described by mass balances (algebraic equations). It is assumed that no heat or power is generated or lost in the mixers and splitters, that no chemical reaction occurs and that they have no storage capacity (zero volume). Exemplary, the total and component mass balance equations of mixer 1 in Fig. 7.1 are given by

$$0 = N^{(2)} + N^{(27)} - N^{(5)} \quad \text{and} \quad \mathbf{0} = N^{(2)}\mathbf{x}^{(2)} + N^{(27)}\mathbf{x}^{(27)} - N^{(5)}\mathbf{x}^{(5)},$$

where $N^{(k)}$ denotes the total molar flow of stream k and $\mathbf{x}^{(k)}$ is a vector of mole fractions of stream k given by

$$\mathbf{x}^{(k)} = \left(x_{H_2}^{(k)}, x_{H_2O}^{(k)}, x_{CO}^{(k)}, x_{CO_2}^{(k)}, x_{CH_4}^{(k)} \right)^{\mathsf{T}},$$

such that $0 = N^{(2)}\mathbf{x}^{(2)} + N^{(27)}\mathbf{x}^{(27)} - N^{(5)}\mathbf{x}^{(5)}$ actually contains 5 component mass balance equations. Since the splitters only serve to split streams and do not separate chemical components, the composition of the exiting streams is always identical to the inlet stream. This can be expressed via additional constraints $x_\alpha^{(in)} = x_\alpha^{(out,n)}$, where $x_\alpha^{(in)}$ denotes the mole fraction of component $\alpha \in S := \{H_2, H_2O, CO, CO_2, CH_4\}$ in the inlet stream and $x_\alpha^{(out,n)}$ then denotes the mole fraction of component α in one of the n outlet streams of the specified splitter (a set of such constraints is required for each splitter). Thus, no component mass balance equations are needed for the splitters. As a consequence, mixers and splitters can be described solely by algebraic equations and one can express them in matrix-vector form as

$$\mathbf{A}\mathbf{b} = \mathbf{0}, \tag{7.3}$$

where $\mathbf{A}$ is a connectivity matrix that is fixed by the superstructure and takes values in $\{-1, 0, 1\}$ and $\mathbf{b}$ is a vector that is composed of

$$\mathbf{b} = \left(N^{(1)}, N^{(1)}\mathbf{x}^{(1)}, \mathbf{x}^{(1)}, N^{(2)}, N^{(2)}\mathbf{x}^{(2)}, \mathbf{x}^{(2)}, \ldots, N^{(k)}, N^{(k)}\mathbf{x}^{(k)}, \mathbf{x}^{(k)} \right)^{\mathsf{T}}. \tag{7.4}$$

It is important to understand that Fig. 7.1 represents a superstructure, where a vector of splitting factors is assigned to each splitter and thus determines which streams of the superstructure are active. As a result, the connectivity matrix $\mathbf{A}$ depends on these splitting factors. Furthermore, note that the variables collected in $\mathbf{b}$ change in time such that one has to ensure that these algebraic equations are fulfilled at every point in time. Finally, the inlet streams of CO_2, H_2 and biogas (see streams 1,2 and 3 in Fig. 7.1, respectively) are

$$x_{H_2}^{(1)} = 1 \quad \text{and} \quad x_\alpha^{(1)} = 0, \quad \forall \alpha \in S \setminus \{H_2\},$$
$$x_{CO_2}^{(2)} = 1 \quad \text{and} \quad x_\alpha^{(2)} = 0, \quad \forall \alpha \in S \setminus \{CO_2\},$$
$$x_{CO_2}^{(3)} = 0.5, \quad x_{CH_4}^{(3)} = 0.5 \quad \text{and} \quad x_\alpha^{(3)} = 0, \quad \forall \alpha \in S \setminus \{CH_4, CO_2\}.$$

Now that the connections between the process units are established, the behavior of the reactors and separators are described by dynamic mathematical models that relate the input to the output of these unit operations.

7.2.2 Reactors

For the modeling of reactors and separators, isothermal and isobaric operations are assumed. Therefore, the total molar concentration remains constant along the axial coordinate. These simplifying assumptions allow to exclude energy and momentum balance equations from the model. Before the governing equations of the reactors are discussed, the reaction rates are introduced.

7.2.2.1 Reaction Rates

The rate at which a component $\alpha \in \mathcal{S}$ is generated or consumed by a chemical reaction is denoted by $\sigma_\alpha(\tilde{\mathbf{x}})$, where $\tilde{\mathbf{x}} = \{\tilde{x}_{H_2}, \tilde{x}_{H_2O}, \tilde{x}_{CO}, \tilde{x}_{CO_2}, \tilde{x}_{CH_4}\}^\top$ denote the mole fractions inside the respective reactor. For the RWGS, the reaction rate is given as [9]

$$\sigma_\alpha(\tilde{\mathbf{x}}) = \nu_{\alpha,RWGS} \frac{b \dfrac{p_{RWGS}}{\tilde{x}_{H2}}\left(\dfrac{\tilde{x}_{H2}\tilde{x}_{CO2}}{K_{eq,WGS}} - \tilde{x}_{CO}\tilde{x}_{H2O}\right)}{\left(1 + \gamma_1\tilde{x}_{CO} + \gamma_2\tilde{x}_{H2} + \gamma_3\dfrac{\tilde{x}_{H2O}}{\tilde{x}_{H2}}\right)^2}$$

and for the DR it reads [10]

$$\sigma_\alpha(\tilde{\mathbf{x}}) = \nu_{\alpha,DR} \frac{k p_{DR}^2 \tilde{x}_{CH4}\tilde{x}_{CO2} - k p_{DR}^4 (\tilde{x}_{CO})^2 (\tilde{x}_{H2})^2 / K_{eq,DR} p_{ref}^2}{(1 + K_1 p_{DR}\tilde{x}_{CH4} + K_2 p_{DR}\tilde{x}_{CO})(1 + K_3 p_{DR}\tilde{x}_{CO2})},$$

where $\nu_{\alpha,RWGS}$ and $\nu_{\alpha,DR}$ are the stoichiometric coefficients for the component $\alpha \in \mathcal{S}$. The remaining parameters including exemplary parameter values and the units of measurement of the respective parameters are collected in Table 7.3.

7.2.2.2 Fixed Bed Reactor

A fixed bed reactor is a tube filled with solid particles. A gaseous stream reacts along its length according to a specified rate. When z denotes the spatial coordinate, the change in the mole fractions $\tilde{\mathbf{x}}$ inside the reactor can be described for each component $\alpha \in \mathcal{S}$ by the PDE

$$\begin{aligned}
\frac{\partial \tilde{x}_\alpha(t, z)}{\partial t} &= D\frac{\partial^2 \tilde{x}_\alpha(t, z)}{\partial z^2} - v(z)\frac{\partial \tilde{x}_\alpha(t, z)}{\partial z} \\
&\quad + \frac{\rho_{cat}}{c_{tot}}\frac{1-\varepsilon}{\varepsilon}\left(\sigma_\alpha(\tilde{x}_\alpha(t, z)) - \tilde{x}_\alpha(t, z)\sum_{\alpha\in\mathcal{S}}\sigma_\alpha(\tilde{x}_\alpha(t, z))\right),
\end{aligned} \tag{7.5}$$

which has to be fulfilled for all $(t, z) \in (0, T] \times (0, L)$, where $L > 0$ specifies the length of the reactor and $T > 0$ the final time. Furthermore, $v(z)$ is the interstitial flow velocity, D is the axial dispersion coefficient, and the remaining parameters are again described in Table 7.3.

The chemical reaction occurring within a FBR may lead to an increasing or decreasing molar flowrate. Consequently, as the pressure is fixed, the velocity field is in general not constant. Thus, the flow velocity $v(z) = v(z, \tilde{x}_\alpha(t, z))$ depends implicitly on time through the mole fraction $\tilde{x}_\alpha(t, z)$. The change in the flow velocity can be derived from the total mass balance and is given by the ODE

$$\frac{dv(z)}{dz} = \frac{\rho_{cat}}{c_{tot}} \frac{1 - \varepsilon}{\varepsilon} \sum_{\alpha \in S} \sigma_\alpha(\tilde{x}_\alpha(t, z)), \quad \forall z \in (0, L]. \tag{7.6}$$

Since the flow velocity can be calculated from the inlet molar flow rate $N^{(in)}$ and the reactor dimensions, the initial condition is given by

$$v(0) = \frac{N^{(in)}}{A_{cross} \varepsilon c_{tot}}, \tag{7.7}$$

where A_{cross} is the cross-sectional area of the reactor tube. Since the RWGS, according to (7.1), preserves the total mole flowrate due to its stoichiometry, the derivative in (7.6) is zero such that the velocity field in the RWGS-FBR is constant. Finally, the initial and boundary conditions required for (7.5) are assigned:

$$\begin{aligned} &\tilde{x}_\alpha(0, z) = \tilde{x}_{\alpha, init}, \\ &\tilde{x}_\alpha(t, 0) = x_\alpha^{(in)}(t), \quad \tilde{x}_\alpha(t, L) = x_\alpha^{(out)}(t), \\ &\left. \frac{\partial \tilde{x}_\alpha}{\partial z} \right|_{t, z=-\gamma} = 0, \quad \left. \frac{\partial \tilde{x}_\alpha}{\partial z} \right|_{t, z=L+\gamma} = 0. \end{aligned} \tag{7.8}$$

To enforce the Neumann conditions, an expansion of the model domain by non-reactive segments γ is performed before and after the actual control volume [11]. The outlet flowrate of the reactor is calculated as

$$N^{(out)} = v(L) A_{cross} \varepsilon c_{tot}. \tag{7.9}$$

Parameters describing the physical dimensions of each unit such as length, diameter, catalyst density, as well as those related to the fluid properties, e.g., feed flowrate, temperature and pressure, dispersion coefficient, and velocity, are unit-dependent. For the sake of readability, this is not reflected in the notation within this section.

7.2.2.3 Pressure and Temperature Swing Adsorption

In pressure and temperature swing adsorption (PSA and TSA) units, a tubular column
filled with solid adsorbent is used to remove certain chemical components which tend
to adsorb (i.e. attach) more readily than other components to the solid adsorbent.
Thereby, a separation of chemical components can be achieved. Since the solid
adsorbent can only take up a limited amount of chemical components, PSA and TSA
cannot be run continuously in real operation. The adsorbent needs to be regenerated,
i.e., the adsorbed chemical components need to be released, before it can be used
again. Thus, PSA and TSA are cyclic processes and at least two columns (one in
separation mode, one in regeneration mode) are required for each separation task for
a quasi-continuous operation. The general concept of PSA and TSA is sketched in
Fig. 7.2.

In this work, however, PSA and TSA are not modeled as a cyclic process since this
would lead to a complex overall process behavior. As a first estimate, we describe
both processes in a continuous way that is similar to the FBR reactor. This is motivated by the fact that on a large enough time scale or using many PSA/TSA units
simultaneously, the cyclic behavior is negligible and the process can be described as
quasi-continuously operating. Thus, we assume that the capacity of the adsorption
bed does not decrease during operation, i.e., the adsorbent material is assumed to be
always in its regenerated state. As isothermal and isobaric operation are assumed,
the change in the mole fractions in an adsorption column can be described for each
component $\alpha \in \mathcal{S}$ by the PDE [12]

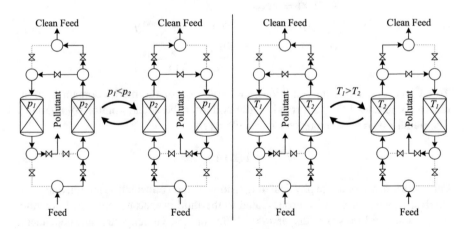

Fig. 7.2 General structure of a PSA unit (left) and a TSA unit (right) consisting of two adsorption
columns each

$$\frac{\partial \tilde{x}_\alpha(t, z)}{\partial t} = D \frac{\partial^2 \tilde{x}_\alpha(t, z)}{\partial z^2} - v(z) \frac{\partial \tilde{x}_\alpha(t, z)}{\partial z}$$
$$- \frac{\rho_{ads}}{c_{tot}} \frac{1 - \varepsilon}{\varepsilon} \left(\frac{\partial q_\alpha(t, z)}{\partial t} - \tilde{x}_\alpha(t, z) \sum_{\alpha \in S} \frac{\partial q_\alpha(t, z)}{\partial t} \right), \tag{7.10}$$

which has to be fulfilled for all $z \in (0, L)$, $t \in (0, T]$ and the adsorbed phase concentration of species α is denoted by $q_\alpha(t, z)$. The change of the adsorbed phase concentration can be described by an ODE [12]

$$\frac{\partial q_\alpha(t, z)}{\partial t} = \frac{15 D_{eff}}{R_p^2} q_\alpha^*(t, z), \tag{7.11}$$

where $q_\alpha^*(t, z)$ can be described by a linear Henry adsorption isotherm [13] as

$$q_\alpha^*(t, z) = \frac{c_{tot}}{\rho_{ads}} k_{H,\alpha} \tilde{x}_\alpha(t, z), \tag{7.12}$$

with $k_{H,\alpha}$ being the Henry adsorption constant of component $\alpha \in S$ and the remaining parameters being collected in Table 7.3. Inserting (7.11) and (7.12) into (7.10) yields the final PDEs for the PSA/TSA units

$$\frac{\partial \tilde{x}_\alpha(t, z)}{\partial t} = D \frac{\partial^2 \tilde{x}_\alpha(t, z)}{\partial z^2} - v(z) \frac{\partial \tilde{x}_\alpha(t, z)}{\partial z} \tag{7.13}$$
$$- \frac{15 D_{eff}}{R_p^2} \frac{1 - \varepsilon}{\varepsilon} \left(k_{H,\alpha} \tilde{x}_\alpha(t, z) - \tilde{x}_\alpha(t, z) \sum_{\alpha \in S} k_{H,\alpha} \tilde{x}_\alpha(t, z) \right),$$

which have to be fulfilled for all $z \in (0, L)$, $t \in (0, T]$. Note that while CO_2 and CH_4 can be effectively separated by PSA, a TSA unit is more suitable for H_2O removal. However, regardless of which method is used, Eq. (7.13) is applicable.

As in fixed bed reactors, the flow velocity depends implicitly on the mole fraction and the spatial change in the flow velocity can be modeled by

$$\frac{dv(z)}{dz} = -\frac{15 D_{eff}}{R_p^2} \frac{1 - \varepsilon}{\varepsilon} \sum_{\alpha \in S} k_{H,\alpha} \tilde{x}_\alpha(t, z). \tag{7.14}$$

The required initial and boundary conditions for (7.13) and (7.14) are analogous to (7.7) and (7.8) of the FBR. Furthermore, the outlet flow rate of the purified stream can be calculated as in (7.9). The outlet flow rate of the separated stream is calculated from the total mass balance of the PSA/TSA unit according to

$$N^{(sep)} = N^{(in)} - N^{(pur)}, \tag{7.15}$$

where $N^{(in)}$ is the inlet stream to the PSA/TSA unit, $N^{(sep)}$ is the separated stream (either CO_2, H_2O or CH_4) and $N^{(pur)}$ is the purified stream. The separated stream is assumed to contain only the separated component. Exemplarily, the total mass balance for the CH_4 removal PSA unit in Fig. 7.1 is given by $N^{(33)} = N^{(31)} - N^{(32)}$ and the composition of the separated stream 33 is fixed to $x_{CH_4}^{(33)} = 1$ and $x_\alpha^{(33)} = 0$ for $\alpha \in S \setminus \{CH_4\}$. Using the same logic, the separated stream flow rate and composition can be calculated for the PSA and TSA units for the removal of CO_2 and H_2O, respectively.

7.3 Forward Simulation

7.3.1 Discretization and DAE-System

The upwind, cell-centered finite volume scheme with equally spaced faces is selected for the discretization along the spatial coordinates of the reactor and separator units. The variables are assumed to be constant inside each cell and the derivatives at the faces are substituted by first order Taylor expansions. Let Δz denote the length of one of the N equally spaced finite volumes and let $\tilde{x}_{\alpha,i}(t) \approx \tilde{x}_\alpha(t, i\Delta z)$, where i refers to the index of the volume. For ease of notation, the dependencies of $\tilde{x}_\alpha$ on (t, z) and of $\tilde{x}_{\alpha,i}$ on (t) are suppressed. In the same fashion, $v_i(t) \approx v(i\Delta z, \tilde{x}_\alpha(t, i\Delta z))$, and the dependencies of v are also suppressed.

With this, the discretization of the FBR equations (7.5) and (7.6) for a generic inner cell i yields

$$\frac{d\tilde{x}_{\alpha,i}}{dt} = -v_i \frac{\tilde{x}_{\alpha,i} - \tilde{x}_{\alpha,i-1}}{\Delta z} + D \frac{\tilde{x}_{\alpha,i+1} + \tilde{x}_{\alpha,i-1} - 2\tilde{x}_{\alpha,i}}{\Delta z^2}$$
$$+ \frac{\rho_{cat}}{c_{tot}} \frac{1-\varepsilon}{\varepsilon} \left(\sigma_\alpha(\tilde{x}_{\alpha,i}) - \tilde{x}_{\alpha,i} \sum_{\alpha \in S} \sigma_\alpha(\tilde{x}_{\alpha,i}) \right), \tag{7.16}$$

$$v_i = v_{i-1} + \Delta z \frac{\rho_{cat}}{c_{tot}} \frac{1-\varepsilon}{\varepsilon} \sum_{\alpha \in S} \sigma_\alpha(\tilde{x}_{\alpha,i}). \tag{7.17}$$

For the inlet ($i = 1$) and the outlet ($i = N$) of the reactor (including the added non-reactive segments γ), the boundary conditions (7.8) are implemented during discretization. The PSA/TSA equations (7.13) and (7.14) can be discretized accordingly. Applying the discretization to all reactors and separators results in a semi-explicit autonomous DAE-system of index 1, see, e.g., [14] for an overview of the topic,

$$\dot{u}(t) = d(u(t), v(t)),$$
$$0 = a(u(t), v(t)), \tag{7.18}$$

for $t \in (0, T]$ with consistent initial values $u(0) = u_0$, $v(0) = v_0$. Here,

- $u(t)$ collects all differential variables, i.e., each of the 5 process units (2 FBR reactors, 2 PSA and one TSA unit) is discretized in space and each unit then contributes a purely time-dependent differentiable variable for each discretization point,
- $v(t)$ collects all algebraic variables, i.e., all variables contributing to **b** from (7.4) and thus describes the influence of the mixers and splitters,
- $d(u(t), v(t))$ collects all differential equations, i.e., the 5 ODE-systems resulting from the spatial discretization of the 5 process units,
- and $a(u(t), v(t))$ collects all algebraic equations, i.e., the equations of the mixers and splitters corresponding to (7.3).

7.3.2 Simulation Results

An exemplary forward simulation is performed to illustrate the behavior of the reactors and separators. For this purpose, the splitting factors are chosen such that the sequence of RWGS-FBR, DR-FBR, PSA(CH_4), PSA(CO_2) and TSA(H_2O) is obtained. Each unit is discretized with 50 equally-spaced finite volumes in the axial coordinate. The unit lengths are 7 m for FBR and 5 m for PSA/TSA, excluding the non-reactive segments γ. For all the units, a diameter of 3.6 cm is selected. The H_2, CO_2 feed-stream flowrates are 1 mol/s, BG is 2 mol/s. The resulting DAE system of index 1 is solved within MATLAB® R2018b using the ode23t solver. For the time horizon [0, 1]s, the computational time was 26.9 s and 27.3 s for active and inactive recycle streams, respectively. While the solver determines consistent initial values for the algebraic variables, an initialization vector of the correct order of magnitude is required. Figure 7.3a and b show the spatial change in composition in the DR-FBR reactor and the PSA(CH_4)-separator at $t = 1$ s, respectively, for active recycles.

Figure 7.3a shows the depletion of the reactants CO_2 and CH_4 along the axial coordinate and the generation of H_2 and CO. Water does not take part in the DR reaction. The slight decrease in mole fraction of water is explained by the increase

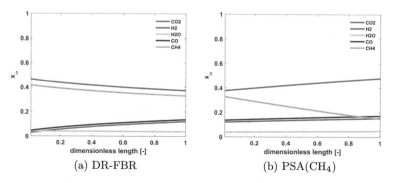

(a) DR-FBR   (b) PSA(CH_4)

Fig. 7.3 Change in mole fractions along the axial coordinate

of the total molar flowrate during the DR reaction, since two moles of reactants form four moles of products. In Fig. 7.3b, the mole fraction of methane decreases due to its preferential adsorption along the bed and as a consequence, the mole fractions for the other components must increase.

7.3.3 PARAREAL *for DAEs*

As seen in the previous section, the forward simulation is already computationally demanding and this demand will only increase when more complex variants of the model problem are considered in the future as discussed in Sect. 7.4. As a result, speed-up possibilities have to be found and this section investigates PARAREAL [4, 5].

The main idea of PARAREAL is to decompose the global time domain $[0, T]$ into N_t smaller subdomains. Given initial values on each of these subdomains, the global time-dependent problem in each iteration of the method splits up into N_t many local problems on these subdomains, which can then be solved in parallel. The initial values can be generated using a *coarse integrator*, which should be *cheap* but can be inaccurate, and the subproblems are then solved in parallel using a *fine integrator*, which has to be *accurate* and is thus more expensive. Afterwards, the next iterate of PARAREAL is generated via a correction step, where the fine solutions of the subproblems are used to correct the coarse sequential integrator.

In order to conveniently describe PARAREAL, problem (7.18) is reformulated into a single system

$$f(\dot{y}(t), y(t)) = 0, \quad \text{and} \quad y(0) = y_0, \tag{7.19}$$

where $y(t)$ collects $u(t)$, $v(t)$ and f collects $\dot{u}(t) - d(u(t), v(t))$ as well as $-a(u(t), v(t))$. The equidistant time decomposition $0 := t_0 < t_1 \cdots < t_{N_t} := T$ results in the coarse time step-size $\delta_{t,c} = \frac{T}{N_t}$. $G(t_{n+1}, t_n, \delta_{t,c}, y_n^k)$ denotes the coarse integrator, that integrates on the subdomain $[t_n, t_{n+1}]$ with step-size $\delta_{t,c}$ and provides an inaccurate approximation to $y(t_{n+1})$, the solution of (7.19), using the initial values y_n^k, where k indicates the iteration number of PARAREAL. In the same fashion, the fine integrator $F(t_{n+1}, t_n, \delta_{t,f}, y_n^k)$, that provides a more accurate approximation to $y(t_{n+1})$ using the initial values y_n^k and the fine step-size $\delta_{t,f} \le \delta_{t,c}$ is introduced. With this notation at hand, PARAREAL is described in Algorithm 1.

Algorithm 1 is terminated after a fixed amount of K iterations, but one can also terminate as soon as the relative change in the iterate $\left\| y^{k+1} - y^k \right\| / \left\| y^k \right\|$ is below some tolerance. Note that in line 10, the values of the coarse integrator $G_{n+1}^k = G(t_{n+1}, t_n, \delta_{t,c}, y_n^k)$ have been calculated in the previous iteration already and thus can be reused. The coarse integrator G operates on the introduced time-grid $t_0, \ldots, t_{N_t}$ and, as it should be fast, can for example be a Runge-Kutta method (RKM) of low order. As DAEs require the invertibility of the Runge-Kutta matrix, see, e.g., [15, Chap. VI.1], the implicit Euler method is chosen. The fine integrator F

Algorithm 1 PARAREAL(y_0, N_t, K, G, F)

1: $k = 0, y_0^0 = y_0$
2: **for** $n = 0, \ldots, N_t - 1$ **do**
3:     $y_{n+1}^0 = G(t_{n+1}, t_n, \delta_{t,c}, y_n^0)$ ▷ first Initial values
4: **end for**
5: **for** $k = 1, \ldots, K$ **do**
6:     **parfor** $n = 0, \ldots, N_t - 1$ **do**
7:         $F_{n+1}^k = F(t_{n+1}, t_n, \delta_{t,f}, y_n^k)$ ▷ parallel step
8:     **end parfor**
9:     **for** $n = 0, \ldots, N_t - 1$ **do**
10:         $y_{n+1}^{k+1} = G(t_{n+1}, t_n, \delta_{t,c}, y_n^{k+1}) + F_{n+1}^k - G_{n+1}^k$ ▷ correction step
11:     **end for**
12: **end for**
13: **return** y^K

---

should be a time-integration method of higher order, and sticking to RKMs a possible choice is the Lobatto IIIC method. In addition, F can operate on a time-grid that is a refinement of $t_0, \ldots, t_{N_t}$, such that multiple steps of F are required to compute F_{n+1}^k in line 7 of Algorithm 1. This increases the cost of the parallel step but also the accuracy of the fine integrator F.

7.3.3.1 PARAREAL: Numerical Results

Regarding the splitting factors, the unit lengths, the amount of discretization points and other model parameters, the same setup as in Sect. 7.3.2 is used. For the time horizon [0, 1], a coarse step size of $\delta_{t,c} = \frac{1}{20}$ is chosen, resulting in 20 subintervals such that 20 cores are used for the parallel computations. For the fine step size, two setups are introduced: the first one using $\delta_{t,f} = \delta_{t,c}$ and the second one using $\delta_{t,f} = \delta_{t,c}/4$ such that in the second setup, the fine integrator F has to perform 4 steps per subinterval. Algorithm 1 was implemented in MATLAB® 2019b with its parallel computing toolbox using the implicit Euler method as coarse integrator G and the Lobatto IIIC method as the fine integrator F. Both RKMs were adapted to DAEs as described in [15, Chap. VI.1], and the nonlinear systems were solved with Newton's method, where the derivatives were generated via the automatic differentiation library ADiMat [16].

PARAREAL is now applied to the problem in question and is stopped as soon as the relative iteration error is below 10^{-4} or after 10 iterations. Additionally, the *fine solution* is computed via the fine integrator $F(t_f, t_0, \delta_{t,f}, y_0)$ as well as the *reference solution* via the internal MATLAB function ode23t. The quality of the PARAREAL solution is verified by measuring the relative error between the PARAREAL iterate and the fine solution over the course of the iteration (err_fine), as well as the relative error between each PARAREAL iterate and the reference solution at the last timestep (err_ref). Additionally, the speedup between PARAREAL and the fine solution is computed over the course of the iteration and the results are depicted in Table 7.1.

Table 7.1 Results of the PARAREAL experiment for the two different fine time-step sizes using the basic model setup from Sect. 7.3.2

# Iter	$\delta_{t,f} = \delta_{t,c}$			$\delta_{t,f} = \delta_{t,c}/4$		
	Speedup	err_fine	err_ref	Speedup	err_fine	err_ref
1	5.578	1.114e-02	1.029e-04	7.959	1.412e-02	1.031e-04
2	2.749	4.787e-03	7.265e-05	3.999	6.241e-03	7.299e-05
3	1.811	2.112e-03	2.149e-05	2.689	3.151e-03	2.197e-05
4	1.366	1.030e-03	1.386e-05	2.023	1.814e-03	1.420e-05
5	1.096	5.112e-04	4.993e-06	1.621	1.016e-03	4.833e-06
6	0.916	2.487e-04	6.712e-06	1.352	5.653e-04	6.721e-06
7	0.786	1.207e-04	6.500e-06	1.162	3.156e-04	6.565e-06
8	0.689	5.897e-05	6.283e-06	1.018	1.735e-04	5.757e-06
9	0.612	2.882e-05	6.586e-06	0.901	9.511e-05	8.877e-06
10	n/a	n/a	n/a	0.812	5.284e-05	5.061e-06

To keep the time comparison for the speedup fair, MATLAB is only allowed to use 1 computational thread inside the RKMs.

It can be seen that in both setups the speedup decays over the course of the PARAREAL iterations as does the error towards the fine solution, which was to be expected. Furthermore, it can be seen that both relative errors are already very small after only 1 PARAREAL iteration and that the first setup terminates due to the relative iteration error after 9 iterations. Therefore, it can be concluded that in this experiment, already 1 or 2 PARAREAL iterations yield a satisfactory approximation to the solution and due to exploitation of the parallel computation, a noticeable speedup over just using the fine integrator is achieved. As similar results were obtained for different scenarios utilizing for example a different reactor sequence (first DR, then RWGS) or having recycles switched off, they are omitted here.

One reason for the observed quick convergence surely is the use of an implicit RKM as the coarse integrator, as the accuracy of the coarse integrator directly influences the convergence of the PARAREAL approximation, see, e.g., [17]. At the same time, this also puts a limit on the achievable speed-up. Considering perfect parallelism, i.e., no time is lost due to processor communication or transfer of data to processors, and letting t_G and t_F denote the time required by the coarse and fine integrator for one coarse or fine time step and letting $R = \delta_{t,c}/\delta_{t,f}$ (resulting in $R = 1$ and $R = 4$ in the first and second setup considered above), we obtain for K PARAREAL iterations ($T_{parareal}$) and the fine solution (T_{fine}) the

$$\text{speedup} = \frac{T_{fine}}{T_{parareal}} = \frac{N_t R t_F}{N_t t_G + K(R t_F + N_t t_G)}.$$

Clearly, an expensive coarse integrator (larger t_G) results in a smaller speedup. Therefore, even more speed-up can be expected, if a faster coarse solver is utilized and we

Table 7.2 Results of the second PARAREAL experiment using different Newton tolerances

| # Iter | $\delta_{t,f} = \delta_{t,c}$ | | | $\delta_{t,f} = \delta_{t,c}/4$ | | |
	Speedup	err_fine	err_ref	Speedup	err_fine	err_ref
1	7.556	1.114e-02	1.201e-04	9.922	1.412e-02	1.204e-04
2	3.681	4.788e-03	8.937e-05	4.991	6.241e-03	9.007e-05
3	2.463	2.112e-03	2.853e-05	3.343	3.151e-03	2.888e-05
4	1.847	1.029e-03	1.855e-05	2.515	1.814e-03	1.875e-05
5	1.476	5.112e-04	4.671e-06	2.017	1.016e-03	4.383e-06
6	1.228	2.487e-04	6.559e-06	1.669	5.654e-04	6.736e-06
7	1.052	1.207e-04	6.656e-06	1.434	3.156e-04	6.508e-06
8	0.913	5.893e-05	6.262e-06	1.257	1.736e-04	5.137e-06
9	0.812	2.879e-05	6.596e-06	1.119	9.508e-05	9.807e-06
10	n/a	n/a	n/a	1.007	5.282e-05	4.252e-06

will discuss this in Sect. 7.4. To mimic this scenario already, a rerun of the above experiment is carried out, where the tolerance for the Newton-solver inside the RKMs is kept at 10^{-6} for the fine integrator, but is reduced to 10^{-4} for the coarse integrator. As a result, the coarse integrator should be faster than in the first experiment, and more speed-up is expected. The results can be seen in Table 7.2 and are already quite promising.

7.3.4 Model Order Reduction

While PARAREAL aims at the temporal dimension in order to speedup the forward simulation, model order reduction (MOR) aims at reducing the number of state variables of the system in question. Instead of applying MOR to the full DAE system (7.18), various MOR approaches will be applied directly to the model equations of both the RWGS-FBR and the DR-FBR reactors (7.5) as well as one PSA adsorption unit (7.10). First, the well-known method of Proper Orthogonal Decomposition (POD), see, e.g., [18, 19], is introduced.

7.3.4.1 Proper Orthogonal Decomposition (POD)

In order to describe the POD approach [6, 20], a generic autonomous system

$$\dot{y}(t) = \tilde{f}(y(t)) \tag{7.20}$$

with state variable $y \in \mathbb{R}^n$ is considered and one could think of a spatially discretized DR-FBR equation according to Sect. 7.3.1. The aim of POD then is a Galerkin projection onto a suitable reduced space, represented by a projection matrix $V_r \in \mathbb{R}^{n \times r}$, where the reduced dimension r should be significantly smaller than the state dimension n. Introducing the reduced state $y_r(t) = V_r^\mathsf{T} y(t) \in \mathbb{R}^r$, the reduced system reads

$$\dot{y}_r(t) = V_r^\mathsf{T} \tilde{f}(V_r y_r(t)) \tag{7.21}$$

and it remains to describe the construction of the projection matrix V_r. In a first step, so-called *snapshots* are collected, where a snapshot is the solution of the state equation (7.20) at a certain time point $t_i \in [0, T]$ and these snapshots are then collected in the *snapshot matrix*

$$\mathbf{S} = [y(t_0), y(t_1), \ldots, y(t_d)] \in \mathbb{R}^{n \times d},$$

where the snapshots should capture all important dynamics of the model. Afterwards, a truncated singular value decomposition (SVD) of the snapshot matrix is performed, such that

$$\mathbf{S} = V \Theta Y^\mathsf{T},$$

where $V \in \mathbb{R}^{n \times d'}$ contains the singular vectors associated to the singular values $\theta_1 \geq \cdots \geq \theta_{d'} > 0$ collected on the diagonal of $\Theta \in \mathbb{R}^{d' \times d'}$ and $d' \leq d$. The POD then takes the first $r \leq d'$ singular vectors to form V_r such that r controls the amount of information taken from the SVD. Furthermore, an Offline/Online decomposition is employed which is a common technique in projection based model order reduction, see, e.g., [21], which enables an efficient implementation of the approach. It is important to note that $\tilde{f}$, the right hand side of the state equation (7.20), is nonlinear. Therefore, evaluating the reduced state equation (7.21) still requires operations depending on the full state dimension n. This problem will be tackled in the next subsection.

7.3.4.2 Hyperreduction

The reduced model (7.21) still depends on high dimensional quantities as $V_r y_r(t) \in \mathbb{R}^n$. To further reduce the computational time required, so-called *hyperreduction techniques* are available and as a prominent method from this class, the orthogonal discrete empirical interpolation method (QDEIM) is shortly reviewed in the following.

The aim of QDEIM, see, [22], is the selection of $m \geq r$ interpolation vectors at which the nonlinear function $\tilde{f}$ is to be evaluated and the algorithm, based on the QR-decomposition of the modes matrix V, aims at choosing these vectors such that the resulting interpolant is a good approximation of $\tilde{f}$. As a result, it is then sufficient to evaluate the *cheap* interpolant at the m interpolation vectors instead of evaluating the *expensive* nonlinear function $\tilde{f}$ everywhere and the interested reader is referred to [23] for further details of the algorithm. Of course, this increased reduction in computational time comes at the cost of accuracy and this among other things will be investigated in the upcoming numerical discussion.

7.3.4.3 MOR: Numerical Results

The POD (without hyperreduction) is applied to the DR-FBR model such that the state $y(t)$ collects the spatial discretization of the various chemical species that are present in the unit. In contrast to the previous sections, each species is spatially discretized using 500 points here (instead of only 50 points before). The reasoning for this is that MOR can only be reasonable, if the state dimension is large such that the solution of the forward model is expensive, otherwise MOR is not required. Therefore, the discretization is refined in this section and such a refinement is expected to be necessary if the model complexity is increased, see Sect. 7.4.

As four species are involved in the reactions of the DR-FBR units, this results in $n = 2000$ and the remaining model parameters applying to the various units are chosen as in Sect. 7.3.2. Furthermore, the MOR experiments were done in Python using the Scipy solver. For the given time horizon [0, 1], the solver chooses an adaptive time grid and the resulting snapshot matrix is then utilized to generate the projection matrix V_r. As there is one PDE per chemical species, the projection matrix V_r is of block-diagonal form where each block corresponds to a POD-projection matrix for the corresponding species.

Having V_r at hand, the reduced solution is then compared to the full one with respect to solution quality and solution time. Regarding the quality, the relative error between both the full and the reduced solution at the last time point is investigated. Regarding the solution time, the relative solution time, i.e., the solution time of the POD-reduced solution divided by the time of the full solution is calculated. Both quantities are investigated for increasing r and the results can be seen in Fig. 7.4.

It can be observed, that the error gradually decays for all species as the amount of singular vectors, and thus the approximation quality, is increased. Regarding the

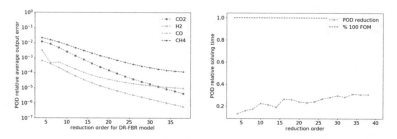

Fig. 7.4 DR-FBR: POD relative average error (left) and POD relative solution time (right) over the size r of the reduced state

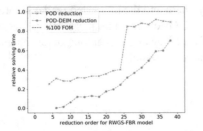 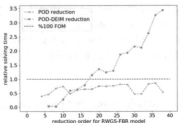

Fig. 7.5 Relative solution time of POD only versus POD-DEIM for the RWGS-FBR (left) and PSA(CO_2) (right) over r (POD) and m (QDEIM)

relative solution time, it can be seen that the reduced solution requires between 90% and 60% less time than the full order model solution depending on the size of the projection matrix, while it is clear that the reduced solution time grows with r. Exemplarily, for $r = 10$, a time reduction of around 75% is achieved due to the reduced system being only of dimension $4 \times 10 = 40$ compared to 2000, the dimension of the full system, while the error for different species is between 10^{-2} and 10^{-4}.

As discussed after Eq. (7.7), the velocity in the RWGS-FBR is constant, such that hyperreduction via QDEIM can be applied to further reduce the computational time. For the DR-FBR and the PSA/TSA units, the velocity is modelled via an ODE and there is further coupling of the PDEs for the different species via the reaction rate term, see Eqs. (7.5) and (7.13). Therefore, it is not clear how hyperreduction can be applied in the case of DR-FBR and this will be further discussed in Sect. 7.4, while for PSA(CO_2) a comparison for regular POD and hyperreduction is presented in this section.

The experiment described above is repeated using the same model parameters once using only the POD and once using the POD with hyperreduction via QDEIM for both RWGS-FBR and PSA(CO_2). For the number m of interpolation points in QDEIM, the fixed difference of $m = r + 5$ is chosen. It is now interesting how much additional speed-up can be achieved by using hyperreduction and how this additional layer of approximation affects the error. The results regarding the computational time can be seen in Fig. 7.5 and the results regarding the approximation quality can be seen in Figs. 7.6 and 7.7.

It can clearly be observed that hyperreduction can further decrease the computational time for RWGS-FBR, while the increase in the approximation error is only about one order of magnitude. Meanwhile, hyperreduction for PSA(CO_2) does not pay off for $m \geq 12$ while at $m = 10$, sufficient time reduction with reasonable accuracy can be reached. Note that this behaviour was expected due to the non-constant velocity in the PSA(CO_2) and future research has to investigate how hyperreduction can be improved in this case. Furthermore, it is worth mentioning that a better effect of hyperreduction on the solution time is expected when the model is parameterized.

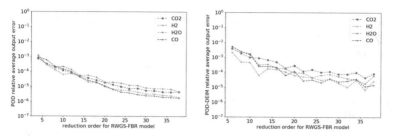

Fig. 7.6 RWGS-FBR: POD relative average error (left) and POD-DEIM relative average error (right) over r (POD) and m (QDEIM)

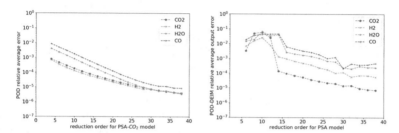

Fig. 7.7 PSA(CO_2): POD relative average error (left) and POD-DEIM relative average error (left) over the size r of the reduced state

7.4 Outlook and Challenges

For a more realistic representation of the original problem, the PSA/TSA model must be refined as it is a rough approximation at the moment. This will entail at least two separation units operating simultaneously (one in separation mode and one in regeneration mode) for each separation task, which gives rise to an inherently dynamic operation of the whole process. Furthermore, the assumption of perfect separation must be relaxed in the future and impurities in the resources must be considered, which will increase the necessary separation steps before a suitable syngas can be obtained. Separation may also be necessary ahead of the reactors to ensure the stability of the utilized catalyst materials.

Another important aspect that needs to be considered in the future are energy balances for all process units to obtain a reasonable estimate for the energy consumption. Moreover, the isobaric and isothermal model assumptions made in Sect. 7.2.2 are hard to meet in practice. Thus, a more accurate description of the model relaxing these assumptions and allowing non-zero temperature and pressure gradients in the gas flows needs to be derived. This will also be the foundation for a detailed analysis of the operating costs, which may be used for an economic optimization of the process. The operating temperature and pressure for the reactors and separators may also be considered as an optimization variable in the future. Finally, the velocity equation in tubular units should be derived from the continuity equation for the conservation of the total mass, where a quasi-steady state assumption should be employed as suggested in [11].

The DAE system described in Sect. 7.2 and discretized in Sect. 7.3.1 can be subjected to a sensitivity analysis to obtain both directional derivatives and gradients of real-valued functions that depend on the solution of the DAE, e.g., suitable objective functions. Due to the fact that the DAE system encompasses spatially discretized PDEs, adjoint sensitivity analysis is of particular interest because it allows the efficient calculation of gradients of a relatively low number of objective and constraint functions in a DAE system with a high number of state variables. The interested reader may refer to [24] for methodological notes on this topic. This opens the DAE model up to being used in reduced-space optimization with gradient-based and quasi-Newton methods.

If the superstructure is assumed to be fixed, the dimensions and, to a limited extent, operating pressures and temperatures of FBRs, PSAs, and TSAs, the flow rates of the inlet streams (see streams 1, 2, and 3 of Fig. 7.1), as well as the split ratios of some splitters can be optimized. In particular, the inlet stream flow rates and the amount of H_2 integrated into the product stream from splitter 1 can, in principle, be regulated dynamically over time to transition between operating states while maintaining the desired syngas ratio under most circumstances. Using these controls, the production cost of syngas can then be optimized utilizing a suitable objective function. Such an optimization problem would be continuous in nature and would require constraints to ensure that an adequate amount of sufficiently pure syngas at the correct ratio is produced. Optimizing a dynamic process rather than a steady state opens up the possibility of exploring the optimal plant operation subject to changing resource and energy availability and cost.

While the local optimization of a DAE-constrained coupled design and control problem with state-dependent constraints can already be challenging, further difficulties arise if discrete choices are added. Such discrete choices may influence the splitting factors and thus the current operation outline inside the superstructure, with some configurations allowing savings in installation cost by completely removing units from the superstructure. Optimization problems involving such discrete decisions are difficult and must generally be solved using heuristics or enumerative methods such as variants of the Branch-and-Bound algorithm [25]. Enumerative methods generally use continuous optimization methods in sub-problems and may require their rigorous solution to global optimality using techniques such as spatial Branch-and-Bound [26, 27]. Generally, Branch-and-Bound type algorithms require the solution of many hundreds or thousands of continuous optimization problems, thus making acceleration methods such as MOR or PARAREAL extremely appealing, even if the speedup achieved by them may not be deemed necessary in the context of a single continuous optimization problem.

Besides integrating PARAREAL in the continuous optimization, further improvements of PARAREAL in the context of possibly large-scale DAEs are of considerable interest. On the one hand, sophisticated linear algebra could be used to handle the (large-scale) Newton systems inside the RKMs, when the state dimension becomes large, and on the other hand a faster coarse solver is desireable in the context of

DAEs. Regarding the coarse solver, the use of an implicit explicit (IMEX) numerical scheme, see, e.g., [28] for IMEX methods in PARAREAL for DAEs, should be investigated, and one could examine the application of the coarse solver to a model with easier/reduced dynamics. Combining PARAREAL and MOR, where MOR tackles the large state dimension and PARAREAL then covers the temporal dimension, is another interesting topic for future research.

Finally, MOR should be integrated into the solution of the DAE system by replacing the full order models of reactor and absorption units in the superstructure with the corresponding reduced models. This MOR-based DAE solution can then be integrated in the continuous optimization process (possibly together with PARAREAL) to speed up the overall Branch-and-Bound algorithm that requires many solves of such optimization problems. Furthermore, identifying suitable model parameters (e.g. the unit length) for which the reduced model can be trained for, see, e.g., [7], should lead to a more efficient offline/online decomposition and thus more speed up for such parametrized problems. Lastly, applying MOR not only to single units of the superstructure, but to the whole DAE system is of certain interest, see, e.g., [19, 29] for MOR for DAEs.

Acknowledgements D. Garmatter, M. Hahn, S. Sager and M. Stoll acknowledge the financial support by the Federal Ministry of Education and Research of Germany with in the project P2Chem (support code 05M18OCB). P. Benner and S. Monem acknowledge funding by the German Research Foundation (DFG) research training group 2297 "MathCoRe", Magdeburg.

Appendix

Table 7.3 Collection of parameters and constants introduced throughout the chapter

Symbol	Description	Value	Unit
A	Connectivity matrix	–	–
A_{cross}	Cross-sectional area	1.06E-03	m^2
b	Kinetic parameter	2.75E-04	$mol\,s^{-1}\,Pa^{-1}$
$c_{tot,DR}$	Concentration DR	6.10E+01	$mol\,m^{-3}$
$c_{tot,PSA/TSA}$	Concentration PSA/TSA	4.09E+01	$mol\,m^{-3}$
$c_{tot,RWGS}$	Concentration RWGS	4.57E+01	$mol\,m^{-3}$
D	Dispersion coefficient	1.00E-04	$m^2\,s^{-1}$
D_{eff}	Effective diffusion coefficient	1.00E-07	$m^2\,s^{-1}$
k	Kinetic parameter	4.33E-10	$mol\,kg^{-1}\,s^{-1}\,Pa^{-2}$
k_H	Henry constant	1.63E+01	Pa

(continued)

Table 7.3 (continued)

Symbol	Description	Value	Unit
K_1	Kinetic parameter	5.13E-06	Pa^{-1}
K_2	Kinetic parameter	9.87E-05	Pa^{-1}
K_3	Kinetic parameter	2.67E-04	Pa^{-1}
$K_{eq,DR}$	Equilibrium constant	1.79E+01	–
$K_{eq,WGS}$	Equilibrium constant	4.38	–
$L_{PSA/TSA}$	Separator length	5.00	m
$L_{RWGS/DR}$	Reactor length	7.00	m
N	Molar flow	–	$mol\,s^{-1}$
p_{DR}	Pressure DR	5.07E+05	Pa
$p_{PSA/TSA}$	Pressure PSA/TSA	1.01E+05	Pa
p_{ref}	Normal pressure	1.01E+05	Pa
p_{RWGS}	Pressure RWGS	3.03E+05	Pa
q	Adsorbent loading	–	$mol\,m^{-3}$
q^*	Equilibrium adsorbent loading	0	$mol\,m^{-3}$
R	Universal gas constant	8.31	$J\,mol^{-1}\,K^{-1}$
R_p	Adsorbent particle radius	1.00E-03	m
t	Time	–	s
v	Interstitial velocity	–	$m\,s^{-1}$
$x, \tilde{x}$	Mole fraction	–	–
z	Axial coordinate	–	m
α	Chemical component	–	–
γ_1	Kinetic parameter	1.26E+01	–
γ_2	Kinetic parameter	5.90E-03	–
γ_3	Kinetic parameter	1.34E+01	–
ε_{FBR}	Void fraction FBR	4.00E-01	–
$\varepsilon_{PSA/TSA}$	Void fraction PSA/TSA	3.00E-01	–
$\nu_{RWGS/DR}$	Stoichiometric coefficients	–	–
ρ_{cat}, ρ_{ads}	Solid density	2.00E+03	$kg\,m^{-1}$
σ	Molar reaction source term	–	$mol\,kg^{-1}\,s^{-1}$

References

1. Dry, M.E.: High quality diesel via the Fischer–Tropsch process–a review. J. Chem. Technol. Biotechnol. Int. Res. Process Environ. Clean Technol. **77**(1), 43–50 (2002)
2. Daza, Y.A., Kuhn, J.N.: CO2 conversion by reverse water gas shift catalysis: comparison of catalysts, mechanisms and their consequences for CO2 conversion to liquid fuels. RSC Adv. **6**(55), 49675–49691 (2016)
3. Usman, M., Daud, W.M.A.W., Abbas, H.F.: Dry reforming of methane: influence of process parameters-a review. Renew. Sustain. Energy Rev. **45**, 710–744 (2015)
4. Lions, J.: Résolution d'EDP par un schéma en temps "pararéel" A "parareal" in time discretization of PDE's. Academie des Sciences Paris Comptes Rendus Serie Sciences Mathematiques **332**, 661–668 (2001)
5. Baffico, L., Bernard, S., Maday, Y., Turinici, G., Zérah, G.: Parallel-in-time molecular-dynamics simulations. Phys. Rev. E **66**(5), 057701 (2002)
6. Benner, P., Sachs, E., Volkwein, S.: Model order reduction for PDE constrained optimization. In: Trends in PDE Constrained Optimization, pp. 303–326. Springer (2014)
7. Dihlmann, M., Haasdonk, B.: Certified nonlinear parameter optimization with reduced basis surrogate models. PAMM **13**(1), 3–6 (2013)
8. Bird, R.B., Lightfoot, E.N., Stewart, W.E.: Transport Phenomena. Wiley (2002). ISBN 9780471364740
9. Bremer, J., Rätze, K.H.G., Sundmacher, K.: CO2 methanation: optimal start-up control of a fixed-bed reactor for power-to-gas applications. AIChE J. **63**(1), 23–31 (2017)
10. Olsbye, U., Wurzel, T., Mleczko, L.: Kinetic and reaction engineering studies of dry reforming of methane over a ni/la/al2o3 catalyst. Ind. Eng. Chem. Res. **36**(12), 5180–5188 (1997)
11. Bremer, J., Sundmacher, K.: Operation range extension via hot-spot control for catalytic CO2 methanation reactors. React. Chem. Eng. **4**(6), 1019–1037 (2019)
12. Ko, D., Siriwardane, R., Biegler, L.T.: Optimization of a pressure-swing adsorption process using zeolite 13x for CO2 sequestration. Ind. Eng. Chem. Res. **42**(2), 339–348 (2003). https://doi.org/10.1021/ie0204540
13. Bentley, J., Huang, Q., Kawajiri, Y., Eic, M., Seidel-Morgenstern, A.: Optimizing the separation of gaseous enantiomers by simulated moving bed and pressure swing adsorption. Adsorption **17**(1), 159–170 (2011). https://doi.org/10.1007/s10450-010-9299-x
14. Kunkel, P., Mehrmann, V.: Differential-Algebraic Equations: Analysis and Numerical Solution, vol. 2. European Mathematical Society (2006)
15. Wanner, G., Hairer, E.: Solving Ordinary Differential Equations II. Springer, Berlin, Heidelberg (1996)
16. Bischof, C.H., Bücker, H.M., Lang, B., Rasch, A., Vehreschild, A.: Combining source transformation and operator overloading techniques to compute derivatives for MATLAB programs. In: Proceedings of the Second IEEE International Workshop on Source Code Analysis and Manipulation (SCAM 2002), pp. 65–72, Los Alamitos, CA, USA (2002). IEEE Computer Society. https://doi.org/10.1109/SCAM.2002.1134106
17. Gander, M.J., Vandewalle, S.: Analysis of the parareal time-parallel time-integration method. SIAM J. Sci. Comput. **29**(2), 556–578 (2007)
18. Benner, P., Ohlberger, M., Patera, A., Rozza, G., Urban, K.: Model Reduction of Parametrized Systems. Springer (2017)
19. Benner, P., Stykel, T.: Model order reduction for differential-algebraic equations: a survey. In: Surveys in Differential-Algebraic Equations IV, pp. 107–160. Springer (2017)
20. Gräßle, C., Hinze, M., Lang, J., Ullmann, S.: POD model order reduction with space-adapted snapshots for incompressible flows. https://link.springer.com/article/10.1007/s10444-019-09716-7 (2018)
21. Haasdonk, B.: Reduced basis methods for parametrized PDEs–a tutorial introduction for stationary and instationary problems. Model Reduct. Approx. Theory Algorithms **15**, 65 (2017)
22. Chaturantabut, S., Sorensen, D.C.: Nonlinear model reduction via discrete empirical interpolation. SIAM J. Sci. Comput. **32**(5), 2737–2764 (2010)

23. Drmač, Z., Gugercin, S.: A new selection operator for the discrete empirical interpolation method—improved a priori error bound and extensions. SIAM J. Sci. Comput. **38**(2), A631–A648 (2016)
24. Burger, M., Gerdts, M.: A survey on numerical methods for the simulation of initial value problems with sDAEs. In: Surveys in Differential-Algebraic Equations IV, pp. 221–300. Springer International Publishing (2017). https://doi.org/10.1007/978-3-319-46618-7_5
25. Land, A.H., Doig, A.G.: An automatic method of solving discrete programming problems. Econometrica **28**(3), 497–520 (1960). ISSN 00129682, 14680262
26. McCormick, Garth P.: Computability of global solutions to factorable nonconvex programs: part I - convex underestimating problems. Math. Program. **10**(1), 147–175 (1976). https://doi.org/10.1007/BF01580665
27. Falk, J.E., Soland, R.M.: An algorithm for separable nonconvex programming problems. Manag. Sci. **15**(9), 550–569 (1969)
28. Steiner, J., Ruprecht, D., Speck, R., Krause, R.: Convergence of parareal for the Navier-Stokes equations depending on the Reynolds number. In: Numerical Mathematics and Advanced Applications-ENUMATH 2013, pp. 195–202. Springer (2015)
29. Ahmad, M.I., Benner, P., Goyal, P.: Krylov subspace-based model reduction for a class of bilinear descriptor systems. J. Comput. Appl. Math. **315**, 303–318 (2017)

Part III
Energy Networks

Chapter 8
Optimization and Stabilization of Hierarchical Electrical Networks

Tim Aschenbruck, Manuel Baumann, Willem Esterhuizen, Bartosz Filipecki, Sara Grundel, Christoph Helmberg, Tobias K. S. Ritschel, Philipp Sauerteig, Stefan Streif, and Karl Worthmann

Abstract Triggered by the increasing number of renewable energy sources, the German electricity grid is undergoing a fundamental change from mono to bidirectional power flow. This paradigm shift confronts grid operators with new problems but also

T. Aschenbruck · W. Esterhuizen · S. Streif
Automatic Control and System Dynamics Laboratory, Technische Universität Chemnitz, Chemnitz, Germany
e-mail: tim.aschenbruck@etit.tu-chemnitz.de

W. Esterhuizen
e-mail: willem.esterhuizen@etit.tu-chemnitz.de

S. Streif
e-mail: stefan.streif@etit.tu-chemnitz.de

B. Filipecki (✉) · C. Helmberg
Faculty of Mathematics, Technische Universität Chemnitz, Chemnitz, Germany
e-mail: bartosz.filipecki@mathematik.tu-chemnitz.de

C. Helmberg
e-mail: helmberg@mathematik.tu-chemnitz.de

M. Baumann · S. Grundel · T. K. S. Ritschel
Max Planck Institute for Dynamics of Complex Technical Systems, Magdeburg, Germany
e-mail: baumann@mpi-magdeburg.mpg.de

S. Grundel
e-mail: grundel@mpi-magdeburg.mpg.de

T. K. S. Ritschel
e-mail: ritschel@mpi-magdeburg.mpg.de

P. Sauerteig · K. Worthmann
Faculty of Mathematics and Natural Sciences, Technische Universität Ilmenau, Ilmenau, Germany
e-mail: philipp.sauerteig@tu-ilmenau.de

K. Worthmann
e-mail: karl.worthmann@tu-ilmenau.de

© Springer Nature Switzerland AG 2021
S. Göttlich et al. (eds.), *Mathematical Modeling, Simulation and Optimization for Power Engineering and Management*, Mathematics in Industry 34, https://doi.org/10.1007/978-3-030-62732-4_8

new opportunities. In this chapter we point out some of these problems arising on different layers of the grid hierarchy and sketch mathematical methods to handle them. While the transmission system operator's main concern is stability and security of the system in case of contingencies, the distribution system operator aims to exploit inherent flexibilities. We identify possible interconnections among the layers to make the flexibility from the distribution grid available within the whole network. Our presented approaches include: the distributed control of energy storage devices on a residential level; transient stability analysis via a new set-based approach; a new clustering-based model-order reduction technique; and a modeling framework for the power flow problem on the transmission level which incorporates new grid technologies.

Keywords Smart grid · Distributed optimization · Safety sets · Transient stability · Model order reduction · Optimal power flow

8.1 Introduction

The energy transition is causing a paradigm shift towards decentralized energy supply. It is accompanied by an increasing share of volatile producers (renewable energy sources), flexible customers (e-mobility) and introduction of novel types of energy storage (large batteries, heat storage, power-to-gas). The power networks are continuously expanded, together with addition of new elements such as Flexible AC Transmission Systems (FACTS) devices and High-Voltage DC (HVDC) transmission lines. The result is increased controllability, flexibility and efficiency potential that is incompatible with traditional, semi-automated system management of network operators. A fully automated control system that can use these potentials requires new mathematical procedures and concepts to understand the complex interactions of the high-voltage grid of Transmission System Operators (TSOs) and the medium to low voltage grid of Distribution System Operators (DSOs). This leads to new challenges and opportunities in modelling, simulation and control of power grids.

In our work, we attempt to identify and address some of these challenges. In particular, we present a framework of mathematical methods to exploit flexibility and control uncertainty within and between individual power network layers (Fig. 8.1). These levels include a transmission system and a distribution system, operated by TSOs and DSOs, respectively. The transmission system covers high, ultra-high and extra-high voltages that are used to transfer electricity over long distances according to the demand of the energy markets. The primary concern here is stable and secure operation. The former is addressed as a dynamic problem with requirements such that the system does not change its prescribed set-point inadvertently. The latter is a static problem and follows the notion of $n - 1$ security, which means that the network should maintain operation even if a single element fails. This is often further extended to $n - 2$ security, where two elements can fail without causing widespread problems. The above operational requirements can be achieved by using new power system technologies such as Thyristor Controlled Series Capacitors (TCSCs) and Phase Shifting Transformers (PSTs), which are becoming popular solutions for improving transmission network operation in an environment that is becoming increasingly strained. The distribution network, on the other hand, consists of medium and low voltages that are used to transmit energy over short distances, from substations to residential and industrial consumers. The main interest here is to exploit new small-scale generation and storage technologies (batteries, solar and wind generation, electric vehicles and "prosumers") which provide flexibility to improve operations.

We address a number of problems in this hierarchy that, overall, aim to increase the efficiency of electrical transmission while ensuring stable operation in the current infrastructure. As a result, this will reduce losses at all grid levels and serve as a foundation for the increase of the proportion of renewable energy and reduction of fossil fuel consumption. To this end, we consider dynamic models at microgrid level (residential energy systems), which include flexibilities and uncertainties from local energy sources and storage. In particular, facilitating the use of renewable energies, energy storage and energy-saving measures in the network through single-family homes supports sustainable energy management and thus the environment and the quality of life. The flexibility of local energy sources and storage, which can be used in the wider area, increases the attractiveness of energy measures and the possibilities of contributing to the energy transition on a broad basis. The key to control a large number of energy storage devices on a residential level is distributed optimization yielding both scalability [21] and plug-and-play capability [38]. At the second stage, the resulting flexibility data is included in substitute medium and high voltage networks obtained by model reduction. The main idea here is to cluster nodes of the network in a way that approximates the system's behavior This allows us to further work on mathematically manageable models of multiscale systems and focus on the essential network elements. In the third stage, these replacement models are used to study various stability-related problems, as well as safe operation of the grid due to large faults. From this stage, auxiliary constraints may be communicated upwards in the hierarchy, to be included in the power flow optimization with discrete decisions on new network components and semidefinite programming techniques. Consistent interaction between these four mathematical sub-steps: the power flow optimization, stability and safety, reduced-order models and microgrid flexibilities make up the mathematical research core of our framework.

The outline of this chapter is as follows. In Sect. 8.2 we briefly summarize our modeling paradigm, emphasizing the link between power-flow (which involves static mathematical programming) and the dynamics that describe the evolution of physical machines in the grid. Then in Sect. 8.3, and starting at the bottom of the hierarchy, we present a distributed model predictive control (MPC) approach for microgrids to shave the peaks in power demanded from the large power grid, and report on the advantages of using a new algorithm, called Augmented Lagrangian based Alternating Direction Inexact Newton method (ALADIN), in the MPC controller. We also investigate the use of surrogate models, obtained using machine learning techniques, in a new optimization scheme where power is exchanged between microgrids. In Sect. 8.4, and moving up the hierarchy, we consider the power grid modeled as a

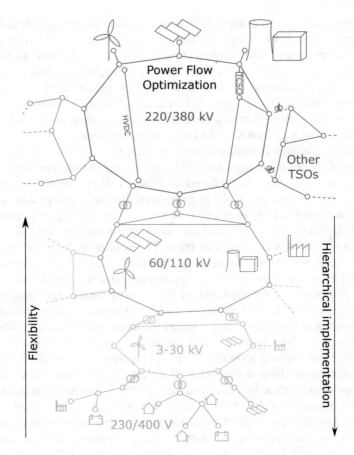

Fig. 8.1 Hierarchical structure of the German electricity grid

graph, with the dynamics at the nodes given by the well-known swing equation. We describe how we can obtain *safety sets* for use in transient stability analysis. Then, in Sect. 8.5, we present a new model-order reduction technique, where the nodes of this model are clustered. Section 8.6 is dedicated to a new power flow modeling framework that incorporates advanced grid technologies, such as TCSCs and PSTs. We also present a semidefinite programming relaxation of the model. We conclude the chapter with Sect. 8.7, and point out the focus of future research.

8.2 Modeling of the Electrical Power Grid

The simplest model of electrical power transmission/distribution network consists of buses (nodes) $b \in \mathcal{B}$ and lines (edges) $\ell \in \mathcal{L}$, and a graph structure identifying the topology of the network. There are two complex variables for each node in the network. One is the voltage $V = |V|e^{i\phi}$ with magnitude $|V|$ and angle ϕ and the other is the power generated at that node $S = P + iQ$, consisting of real power P and reactive power Q. Finally, each node has a known power demand $S^D = P^D + iQ^D$, which could be zero. Each element in the network, buses as well as lines are characterized by complex admittance $Y = G + iB$, which consists of conductance G and susceptance B. This parameter represents how simple it is to transfer electricity through a given piece of equipment. In addition, lines also have a shunt admittance parameter $Y^{SH} = iB^{SH}$ consisting only of susceptance.

Given the above notation, following the more detailed information contained in [4, 31, 49], we write down the equality constraints, which describe the operation of a static network. Each line connects two nodes and therefore we can write $\ell = (b_s, b_e)$. Similarly, S^S and S^E denote the complex power injections at 'start' and 'end' nodes. These are defined as follows:

$$
\begin{aligned}
S_\ell^S &= \left(I_\ell^S\right)^* V_{b_s} \\
S_\ell^E &= \left(I_\ell^E\right)^* V_{b_e}
\end{aligned}
\quad \forall \ell = (b_s, b_e) \in \mathcal{L},
\tag{8.1}
$$

where I_ℓ^S and I_ℓ^E are complex current injections:

$$
\begin{aligned}
I_\ell^S &= (Y_\ell + Y_\ell^{SH})V_{b_s} - Y_\ell V_{b_e} \\
I_\ell^E &= (Y_\ell + Y_\ell^{SH})V_{b_e} - Y_\ell V_{b_s}
\end{aligned}
\quad \forall \ell = (b_s, b_e) \in \mathcal{L}.
\tag{8.2}
$$

Finally, we write down the nodal power balance constraint, which uses the above power injections:

$$
S_b - S_b^D = \sum_{\ell:b_s=b} S_\ell^S + \sum_{\ell:b_e=b} S_\ell^E + (Y_b)^* |V_b|^2 \quad \forall b \in \mathcal{B},
\tag{8.3}
$$

where the left hand side is the difference between complex power generation and demand at the node, and the right hand side is a sum of all complex power injections to/from edges and losses at the bus. This is one complex equation at each node for two complex variables, namely S and V. The power demand is a given quantity and the admittances are parameters describing the network. We will refer to a solution that satisfies these equations as a *power-flow set-point*, and we will label the tuples of this solution as $(\hat{P}_b, \hat{Q}_b, |\hat{V}_b|, \hat{\phi}_b)$, with $b \in \mathcal{B}$.

We now describe the link between a power-flow set-point and the differential equations that govern the time evolution of synchronous machines and loads, as described by various models, see for example [3, 13, 33, 35, 43]. The reader may consult Fig. 8.2 for clarification. As detailed in the paper [35], a set-point $(\hat{P}_b, \hat{Q}_b, |\hat{V}_b|, \hat{\phi}_b)$ determines the parameters of the differential equations, as well as their initial conditions. Let $\mathcal{B}_G$ and $\mathcal{B}_L$ denote the indices of generator and load terminal buses, respectively, and let $\mathcal{B} = \mathcal{B}_G \cup \mathcal{B}_L$. Each generator has an inertia J_b and a damping coefficient D_b, and is represented by a voltage source which has constant magnitude $|E_b|$, and whose time-varying voltage angle, $\delta_b(t)$, is assumed to be equal

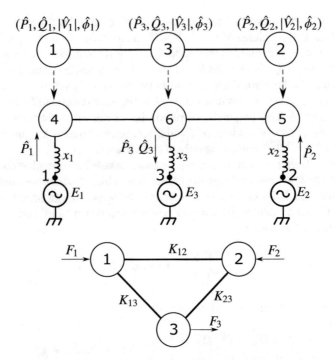

Fig. 8.2 Figure adapted from [35] showing how the parameters of the dynamics in (8.4) are obtained from a power-flow set-point, $(\hat{P}_b, \hat{Q}_b, |\hat{V}_b|, \hat{\phi}_b), b \in \mathcal{B}$. Generators and loads are modeled as voltage sources, $E_b = |E_b| \angle \delta_b(t)$, connected via reactances to their terminal nodes, which are re-labeled (indicated by dashed arrows). New nodes are introduced, called the internal nodes, that are located between the voltage sources and reactances. A new graph is formed, with the internal nodes connected via the "effective admittances", K_{bh}. The constants in the dynamics are then obtained from the power-flow set-point. In the example, nodes 1 and 2 are generator nodes, and node 3 is a load

to the machine's rotor angle. This voltage source is connected to its terminal bus via a reactance $x_b > 0$ (this is the so-called "classic model"). For every generator a new node is introduced that is located between the generator's transient reactance x_b and the voltage source, thus making the number of nodes $\mathcal{B} + \mathcal{B}_G$. We will refer to these new nodes as *generator internal nodes* (versus the *generator terminal nodes*). Loads may be modeled in various ways. In this chapter, we focus on the case where the loads are assumed to be synchronous motors, also with inertia and damping J_b and D_b, respectively. Their modeling is identical to the classic model of the generators, except that the power term appearing in the differential equations has opposite sign. Thus, with the *load internal nodes*, the total number of nodes increases to $2\mathcal{B}$. Let $n_G = |\mathcal{B}_G|, n_L = |\mathcal{B}_L|$, and $N = n_G + n_L$, and let the index set $\{1, 2, \ldots, n_G, n_G + 1, \ldots, n_G + n_L\}$, refer to the generator and load internal nodes. It can then be shown that the time evolution of the generators and loads (where the state is $(\delta_b, \dot{\delta}_b)$) is described by:

$$\frac{2J_b}{\omega_R}\ddot{\delta}_b + \frac{D_b}{\omega_R}\dot{\delta}_b = F_b - \sum_{h=1,h\neq b}^{N} K_{bh}\sin(\delta_b - \delta_h - \gamma_{bh}), \quad b \in \{1, \ldots, N\}. \quad (8.4)$$

with

$$F_b = \hat{P}_b - |\hat{E}_b|^2 \text{Re}(Y_{bb}) \qquad b = 1, \ldots, N, \qquad (8.5a)$$

$$K_{bh} = |\hat{E}_b \hat{E}_h Y_{bh}|, \qquad (8.5b)$$

$$\gamma_{bh} = \theta_{bh} - \frac{\pi}{2}, \qquad (8.5c)$$

where $\hat{P}_b$ is the constant real power supplied (for generators) or consumed (for loads), $Y_{bh} = |Y_{bh}|\angle\theta_{bh}$ is the "effective" admittance between node b and h (refer to [35]), and ω_R is the system's reference frequency. The constants F_b and K_{bh} are determined from the power-flow set-point (which we again emphasize is represented with hats). The values for $|\hat{E}_b|$ and the initial conditions for the dynamics (8.4) may be determined from the power-flow set-point and the fact that the bus voltage, V_b, and the voltage source, E_b, are related via the reactance x_b.

8.3 Distributed Optimization in Low-Voltage Smart Grids

This section is dedicated to the lowest level of the electricity network - the residential microgrid (green part in Fig. 8.1). Nowadays, due to the use of renewable energy sources households are able to generate power and feed it into the grid in addition to simply withdrawing it. Hence, one can observe a bidirectional power flow [22]. This paradigm shift comes along with several problems such as (uncertain) volatile infeed [34], the need for dynamic dispatch methods [47], and degradation of power quality, and overvoltage yielding outages [45]. In this section, we will focus on the aggravation of fluctuations in the power demand. One idea to tackle this problem is to use locally distributed energy storage devices (e.g., residential batteries) in order to compensate these fluctuations. A straightforward strategy to control the batteries is to charge them whenever the local generation exceeds the local demand and discharge otherwise. When the state of charge reaches the limits of the battery the grid operator has to deliver the required or withdraw the superfluous energy. The main advantage of this approach is that no additional communication between the households and the grid operator is needed. However, as elaborated for example in [45, 46], the performance is far from being optimal. One direction for improvement is to additionally incorporate predicted values of the future net consumption (load minus generation) [36].

With respect to information exchange we distinguish three optimization techniques: decentralized, centralized, and distributed optimization. Using a decentralized approach each agent solves its own optimization problem based on its local

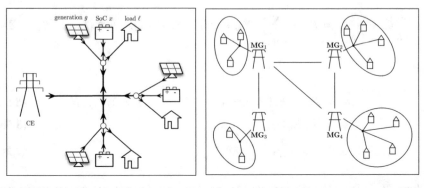

Fig. 8.3 Microgrid consisting of residential energy systems each equipped with its load and some generation and storage device (left) and a network of (partially) coupled microgrids (right)

information only. Thus, achieving any optimality criterion which is defined over the entire smart grid is unlikely. Centralized methods assume the existence of some Central Entity (CE) gathering the information of all agents yielding a better overall performance. In practice, however, agents are not willing to share their complete data with some CE, e.g., the grid operator. Therefore, distributed optimization techniques have been developed combining the advantages of both centralized and decentralized approaches, i.e., a good overall performance and privacy as well as flexibility in form of scalability and plug-and-play capability, see e.g., [5, 8, 24, 50].

Note that in this paper we only consider optimization problems at a fixed time. Typically, these problems are solved online within an MPC scheme in order to compute an optimal control sequence, see e.g., [20, 37] for an introduction to MPC.

8.3.1 Modelling Microgrids

We consider a grid of $B = |\mathcal{B}| \in \mathbb{N}$ residential energy systems (buses) each equipped with some energy generation device (e.g., rooftop photovoltaic panels) and some energy storage device (e.g., battery) as visualized in Fig. 8.3 (left), see also [7] and the references therein for details.

The dynamics of the b-th system are given by

$$x_b(n+1) = \alpha_b x_b(n) + T\beta_b^T u_b(n) \tag{8.6}$$

$$P_b(n) = \gamma_b^T u_b(n) + w_b(n), \tag{8.7}$$

where the state $x_b(n) \in \mathbb{R}_{\geq 0}$, output $P_b(n) \in \mathbb{R}$, exogenous input $w_b(n) \in \mathbb{R}$, and control input $u_b(n) = (u_b^+(n), u_b^-(n))^T \in \mathbb{R}^2$ denote the state of charge, active power demand, net consumption (load minus generation), and charging and discharging rate of system $b \in \mathcal{B}$ at time instant $n \in \mathbb{N}_0$. The parameter $T > 0$ encodes the length of a time step while $\alpha_b \in (0, 1]$ and $\beta_b, \gamma_b \in (0, 1]^2$ model the system dynamics. The current state of charge of the energy storage device b is denoted by $x_b(k) = \hat{x}_b, b \in \mathcal{B}$, where $k \in \mathbb{N}_0$ is the current time instant. Note that the future net consumption $w_b(n)$, $n \geq k$, is not known a priori. However, it is assumed to be forecasted over a prediction horizon of $N \in \mathbb{N}_{\geq 2}$ time steps. State and control input are subject to constraints such as battery capacity and (dis-)charging rate at each time step $n \in \mathbb{N}_0$, i.e.,

$$0 \leq x_b(n) \leq C_b \tag{8.8a}$$

$$\underline{u}_b \leq u_b^-(n) \leq 0 \tag{8.8b}$$

$$0 \leq u_b^+(n) \leq \overline{u}_b \tag{8.8c}$$

$$0 \leq \frac{u_b^-(n)}{\underline{u}_b} + \frac{u_b^+(n)}{\overline{u}_b} \leq 1 \tag{8.8d}$$

for some $-\underline{u}_b, \overline{u}_b, C_b \geq 0, b \in \mathcal{B}$. Constraints (8.8d) ensure that the bounds on charging (8.8c) and discharging (8.8b) hold even if both charging and discharging occur during one time step. Initial conditions, system dynamics, and constraints for the next N time steps yield the feasible set depending on the current state of charge $\hat{x} = (\hat{x}_1, \ldots, \hat{x}_B)^T \in \mathbb{R}^B_{\geq 0}$,

$$\mathbb{U} = \mathbb{U}(\hat{x}) = \left\{ u \in \mathbb{R}^{2BN} \mid \begin{array}{l} x_b(k) = \hat{x}_b \, \forall \, b \in \mathcal{B} \\ (6) \text{ and } (8) \text{ hold } \forall \, n \in [k : k + N - 1] \end{array} \right\}.$$

Here, the control inputs $u_b = (u_b(k)^T, \ldots, u_b(k + N - 1)^T)^T \in \mathbb{R}^{2N}$ over the prediction horizon of all agents $b \in \mathcal{B}$ are stacked in $u \in \mathbb{R}^{2BN}$. Furthermore, we used the notation $[p : q] = \{p, p + 1, \ldots, q\}$ for some integers $p \leq q$.

One goal from a grid operator's perspective is peak shaving, i.e., flattening the aggregated power demand in order to have an almost constant control energy. These considerations motivate the Quadratic Program (QP)

$$\min_{u \in \mathbb{U}} \bar{f}(\bar{P}) \quad \text{s.t.} \quad \bar{P} = \sum_{b=1}^B A_b u_b + w_b, \tag{8.9}$$

where $\bar{f} \colon \mathbb{R}^N \to \mathbb{R}, \bar{P} \mapsto \left\| \bar{P} - \bar{\zeta} \right\|_2^2$, penalizes the deviation of the aggregated future active power demand $\bar{P} = (\bar{P}(k), \ldots, \bar{P}(k + N - 1))^T \in \mathbb{R}^N$ from some reference trajectory $\bar{\zeta} = (\bar{\zeta}(k), \ldots, \bar{\zeta}(k + N - 1))^T \in \mathbb{R}^N$ and $A_b \in \mathbb{R}^{N \times 2N}$ are chosen according to (8.7), see also [26] for details. Numerical simulations show the potential of controlling distributed energy storage devices, see Fig. 8.4 (left) for details. Here, the goal is to trace the overall average net consumption (dashed red line) by controlling local batteries (solid blue line). Note that different objectives can be considered, see e.g., [7]. Optimization with respect to several objectives at the same time yields a multiobjective optimization problem as e.g., studied in [41]. In [7, 26, 40] local costs are introduced to spare the lifetime of the batteries. Therefore,

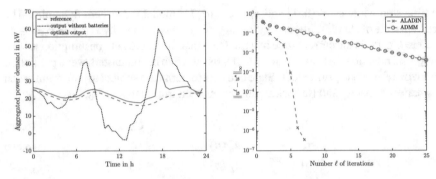

Fig. 8.4 Impact of controlling local batteries on aggregated active power demand of 50 households (left) and convergence comparison: ALADIN vs. ADMM (right), where u^* denotes the optimal solution

the problem becomes

$$\min_{(\bar{P},u)\in\mathbb{R}^N\times\mathbb{U}} \bar{f}(\bar{P}) + \sum_{b=1}^{B} f_b(u_b) \quad \text{s.t.} \quad \bar{P} = \sum_{b=1}^{B} A_b u_b + w_b \qquad (8.10)$$

for (strictly convex) local cost functions $f_b\colon \mathbb{R}^{2N} \to \mathbb{R}$ given by

$$f_b(u_b) \quad = \quad \frac{1}{2}\, \|u_b\|_{F_b}^2$$

with some positive definite matrices $F_b \in \mathbb{R}^{2N\times 2N}$, $b \in \mathcal{B}$. From a mathematical point of view such strictly convex cost functions have the nice side effect that they ensure uniqueness of the optimal solution. Note that, for some positive definite matrix $H \in \mathbb{R}^{n\times n}$ we denote $\|\cdot\|_H = \langle H\cdot, \cdot\rangle^{1/2}$.

8.3.2 Distributed Optimization via ALADIN

Typically, one is interested in solving (8.9) or (8.10) in a distributed manner, i.e., with as few communication as possible while approximately attaining the performance of a centralized solver. To this end, several optimization techniques have been established [5] and the references therein. In [26] the authors apply the recently published Augmented Lagrangian based Alternating Direction Inexact Newton (ALADIN) method [24] to solve the optimization problem (8.10). The main advantage of ALADIN compared to the state-of-the-art Alternating Direction Method of Multipliers (ADMM) is its locally quadratic convergence. We briefly recap ALADIN and refer to [24–26] for details. Consider the structured optimization problem (8.10). Since (8.6) and (8.8) are linear, problem (8.10) can be written as

$$\min_{(\bar{P},u)\in\mathbb{R}^N\times\mathbb{R}^{2BN}} \bar{f}(\bar{P}) + \sum_{b=1}^{B} f_b(u_b)$$

$$\text{s.t.} \quad \bar{P} = \sum_{b=1}^{B} A_b u_b + w_b$$

$$D_b u_b \leq d_b \quad \forall b \in \mathcal{B}$$

with matrices $D_b \in \mathbb{R}^{m\times 2N}$, and vector $d_b \in \mathbb{R}^m$ for some $m \in \mathbb{N}$. ALADIN consists of two main steps. First, the decoupled problem

$$\min_{v_b} \quad f_b(v_b) - \lambda^T A_b v_b + \frac{1}{2} \|v_b - u_b\|_{H_b}^2 \quad \text{s.t.} \quad D_b v_b \leq d_b$$

with some positive definite scaling matrix $H_b \in \mathbb{R}^{2N\times 2N}$ is solved for each agent. Note that this can be done in parallel. Then, a coupled QP of the form

$$\min_{(\bar{P},u,s)\in\mathbb{R}^N\times\mathbb{R}^{2BN}\times\mathbb{R}^{m^{\text{act}}}} \bar{f}(\bar{P}) + \sum_{b=1}^{B}\left(\frac{1}{2}\|u_b - v_b\|_{F_b}^2 + (\nabla f_b(v_b))^T u_b + \frac{\mu}{2}\|s_b\|_2^2 \right)$$

$$\text{s.t.} \quad \bar{P} = \sum_{b=1}^{B} A_b u_b + w_b$$

$$D_b^{\text{act}}(u_b - v_b) = s_b, \quad b \in \mathcal{B}$$

is solved to update the iterate u_b, $b \in \mathcal{B}$, until some stopping criterion is met. Here, we used the shorthand $D_b^{\text{act}} \in \mathbb{R}^{m_b^{\text{act}} \times 2N}$ to denote the active constraints at plant $b \in \mathcal{B}$ at the current iterate, i.e., $D_b^{\text{act}} u_b = d_b^{\text{act}}$. Updating the approximate Hessian

$$H_b \quad = \quad F_b + \mu \cdot D_b^{\text{act}^T} D_b^{\text{act}}$$

with properly chosen scaling parameter $\mu \geq 0$, yields the locally quadratic convergence of ALADIN similar to Newton-type methods as depicted in Fig. 8.4 (right).

8.3.3 Surrogate Models in Optimization of Coupled Microgrids

Exploiting local energy storage devices might not suffice to compensate the fluctuations of the power demand in one MicroGrid (MG). Therefore, additional energy exchange among several connected MGs might be beneficial. To this end, the concept of (partially) coupled MGs was introduced in [9]; as depicted in Fig. 8.3 (right). Thus, the optimization problem incorporates an additional (upper) level yielding a hierarchical structure.

Consider $M \in \mathbb{N}_{\geq 2}$ coupled MGs each consisting of B_κ residential energy systems, $B_\kappa \in \mathbb{N}, \kappa \in [1 : M]$. The basic idea is to minimize the overall objective function $J : \mathbb{R}^{MN} \times \mathbb{R}^{M \times M \times N} \to \mathbb{R}$,

$$(\bar{P}, r) \mapsto \sum_{n=k}^{k+N-1} \sum_{\kappa=1}^{M} \left(\bar{\zeta}_\kappa(n) B_\kappa - \sum_{\nu=1}^{M} r_{\nu\kappa}(n) \eta_{\nu\kappa} \bar{P}_\kappa(n) \right)^2, \tag{8.11}$$

where $\bar{P} = (\bar{P}(k)^T, \ldots, \bar{P}(k+N-1)^T)^T \in \mathbb{R}^{MN}$ stacks the aggregated power demand of all MGs over the prediction horizon and $r \in [0, 1]^{M \times M \times N}$ denotes the energy exchange rates among the MGs. We use $\bar{\zeta}_\kappa(n)$ to describe the reference trajectory of MG_κ at time n. Moreover, $\eta \in [0, 1]^{M \times M}$ models losses along the transmission lines.

The constraints on $\bar{P}$ are implicitly given by the sets $\mathbb{U}(\hat{x}_1, \ldots, \hat{x}_{B_\kappa}), \kappa \in [1 : M]$, yielding a polyhedral feasible set $\mathbb{D}_{\bar{P}}$. The energy exchange rate on the other hand is subject to

$$r \in \mathbb{D}_r := \left\{ r \in [0, 1]^{M \times M \times N} \mid \begin{array}{l} \sum_\kappa r_{\nu\kappa}(n) = 1 \\ r_{\nu\kappa}(n) \cdot r_{\kappa\nu}(n) \leq 0 (\kappa \neq \nu) \end{array} \right\}$$

in order to ensure that the hole amount of energy at each MG is scheduled and energy only flows in one direction. In [2] a bidirectional optimization scheme is proposed to solve

$$\min_{(\bar{P}, r) \in \mathbb{D}_{\bar{P}} \times \mathbb{D}_r} J(\bar{P}, r) \tag{8.12}$$

in an alternating fashion: First, problem (8.12) is minimized with respect to $\bar{P}$ and then with respect to r. To this end, optimization problem (8.9) has to be solved several times in each MG. In [19] the complex optimization step is replaced by an approximation using Radial Basis Functions (RBFs) in order to speed up the computation and avoid communication. As a follow-up a bidirectional optimization scheme incorporating surrogate models in form of RBFs and Artificial Neural Nets (ANNs) has been proposed in [2].

The main results are illustrated in Fig. 8.5. The iterative improvement via the bidirectional optimization scheme is visualized on the left side. In each iteration the blue crosses denote the overall costs based on battery control while the red circles depict the costs after additional energy exchange. Note that the first exchange yields a huge improvement whereas after five iterations the values before and after energy exchange (almost) coincide indicating that additional exchange does not further improve the performance. Figure 8.5 (right) illustrates the impact of surrogate models on the costs. Both surrogate models are able to reduce communication effort while attaining good overall performance. In particular, the approximation via RBFs yields almost the same costs as the optimal solution. Note that training an ANN requires huge amounts of data. In our case this means that the (complex) optimiza-

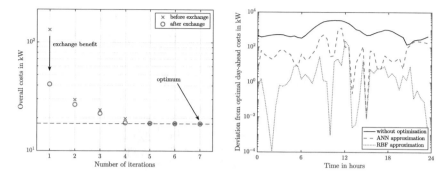

Fig. 8.5 Performance of bidirectional optimization scheme (left) and impact of surrogate models on the day-ahead costs (8.11) of the bidirectional optimization of coupled MGs (right)

tion routine has to be run many thousands/millions times. However, the ANN used in Fig. 8.5 (right) was trained on a dataset with less than 1000 samples. Nevertheless, it is able to reduce the costs significantly compared to the uncontrolled system.

8.3.4 Outlook

So far in this section we ignored the physical topology of the grid. Typically, a microgrid can be viewed as a tree-structured graph, where the nodes represent the local households. In [39] the authors study an Optimal Power Flow (OPF) problem, where the active power P_b at the bus b, $b \in \mathcal{B}$, is assumed to be given, while the reactive power Q_b can be controlled. A future goal is to combine the battery model with this OPF problem to be able to control both active and reactive power. The resulting flexibility can then be used on the higher grid levels, see Fig. 8.1, e.g., in order to enlarge the safety sets introduced in the subsequent Sect. 8.4.

8.4 Safety Sets for Transient Stability in Power Networks

Section 8.3 was concerned with optimizing energy management on the distribution grid level. These decisions affect the power drawn from higher up in the hierarchy, which affects the power-flow in the transmission grid. We are interested in studying how decisions in the distribution grid may affect the transmission grid's robustness against large faults. We focus on the synchronous machine model, Eq. (8.4), and first show how this high-dimensional model may be decomposed into many coupled low dimensional systems. We then show how the *theory of barriers* may be applied to construct *safety sets* and explain how they may be used to address the so-called transient stability problem. Finally, we outline the focus of future research: how these

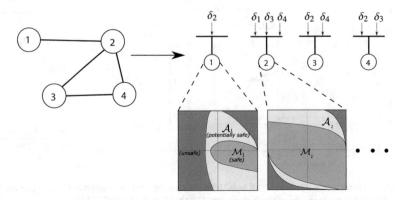

Fig. 8.6 Decomposition of the high dimensional system: we consider each node by itself, treating its neighbors' state variables, δ_b's, as arbitrary input functions, bounded within the state constraints of the b-th machine's neighbors. For example, the state variables δ_1, δ_3 and δ_4 appear on the right-hand side of the dynamics of node 2

sets might be used to arrive at decisions on the distribution grid that are a compromise between energy efficiency and robustness against faults.

8.4.1 Decomposition of Model for Synchronization Problems

The state dimension of the model in (8.4) could be extremely large, but it may be observed that the coupling between nodes is through the state variables δ_b. We propose focusing on each node on its own, and treating its neighbors' δ_b's as time-varying *input functions*, which may behave arbitrarily within some bounded set. (Below we impose the assumption that these input functions are Lebesgue-measurable, as is required to apply the theory of barriers.) Thus, the idea is to abstract away the effects of the large network on individual machines. We then find the safety sets for each node, which are only two dimensional, see Fig. 8.6 for an example.

8.4.2 Safety Sets

Consider the following constrained nonlinear system:

$$\dot{x}(t) = f(x(t), d(t)), \quad x(0) = x_0, \tag{8.13}$$
$$g_i(x(t)) \leq 0, \quad d(t) \in D, \quad \forall t \in [0, \infty), \quad \forall i \in \{1, \ldots, p\},$$

where $x \in \mathbb{R}^n$ is the state, $d \in \mathbb{R}^m$ is an input, x_0 is the initial condition, the g_i's are state constraints, and $D \subset \mathbb{R}^m$ is compact and convex. We impose a number of assumptions, as in [14, 17]. Briefly, they are that the input function $d\colon [0, \infty) \to D$ is Lebesgue-measurable; f is C^2, and for every bounded x_0 and every d the solution does not blow up in finite time; and g_i is C^2 for all i. We let $\mathcal{D}$ refer to the set of all Lebesgue-measurable functions that map $[0, \infty)$ to D, and let $x^{(d, x_0)}(t)$ refer to the solution of (8.13) at time $t \in [0, \infty)$ with initial condition x_0 and input $d \in \mathcal{D}$. The focus of our study is two sets. The admissible set, which is the set of initial conditions for which there exists an input such that the constraints are always satisfied:

$$\mathcal{A} \triangleq \{x_0 : \exists d \in \mathcal{D} | g_i(x^{(d, x_0)}(t)) \leq 0, \ \forall t \in [0, \infty), \ \forall i \in \{1, \ldots, p\}\},$$

and the *maximal robust positively invariant set* (MRPI), which is the set of initial conditions for which every solution satisfies the state constraints, regardless of the input function:

$$\mathcal{M} \triangleq \{x_0 : g_i(x^{(d, x_0)}(t)) \leq 0, \ \forall t \in [0, \infty), \ \forall i \in \{1, \ldots, p\}, \ \forall d \in \mathcal{D}\}.$$

Note that $\mathcal{M} \subseteq \mathcal{A}$. If "safety" refers to satisfying the state constraints for all future time, then the sets $\mathcal{M}$ and $\mathcal{A}^C$ are very useful in that they indicate whether the system's state is *safe or unsafe*, respectively. The set $\mathcal{A}$ gives *potentially safe* states, and so encapsulates a type of conservatism in classifying the system's state.

The main observation in the papers [14, 17] is that parts of the boundaries of these two sets may be constructed by integrating the system dynamics backwards in time using a special input according to a maximum/minimum-like principle, from points on the state constraint boundary that satisfy an "ultimate tangentiality" condition. We refer the reader to these references for details on how the sets may be constructed.

8.4.2.1 Use of Safety Sets in Transient Stability Analysis

Transient stability analysis is concerned with the grid's behavior after a large fault occurs, see [12, 13, 15, 27, 43]. As detailed in [43], one can model the phenomenon as a system governed by the pre-fault dynamics, which then switches to the fault-on dynamics, and then to the post-fault dynamics after the fault is cleared. The problem is to classify the post-fault state (which we label $(\tilde{\delta}, \tilde{\omega}) \subset \mathbb{R}^{2N}$, with $\tilde{\delta} := (\tilde{\delta}_1, \tilde{\delta}_2, \ldots, \tilde{\delta}_N)^T$ and $\tilde{\omega} := (\tilde{\omega}_1, \tilde{\omega}_2, \ldots, \tilde{\omega}_N)^T$) as needing corrective action from the system operator or not. In particular, a safe post-fault state settles to an equilibrium with all the rotor angles not diverging too far from each other. As detailed in [1], the above-introduced safety-sets may be used to address the problem as follows:

1. Model the post-fault dynamics according to (8.4).
2. Impose state constraints on each node.
3. Decompose the system, and consider each node by itself. Let the new bounded inputs to each node be contained in the neighboring node's state constraints.
4. Find the sets $\mathcal{A}_b$ and $\mathcal{M}_b$ for each node.
5. If $(\tilde{\delta}_b, \tilde{\omega}_b) \in \mathcal{M}_b$ for all $b \in \{1, \ldots, N\}$, then the post-fault state is *safe*.

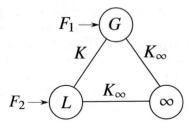

Fig. 8.7 Three-node example

6. If $(\tilde{\delta}_b, \tilde{\omega}_b) \in \mathcal{A}_b^c$ for at least one $b \in \{1, \dots, N\}$, then the post-fault state is *unsafe*.
7. If $(\tilde{\delta}_b, \tilde{\omega}_b) \in \mathcal{A}_b$ for all $b \in \{1, \dots, N\}$, and $(\tilde{\delta}_b, \tilde{\omega}_b) \in \mathcal{M}_b^c$ for at least one $b \in \{1, \dots, N\}$, then the post-fault state is *potentially safe*.

Thus, a post-fault state may be classified according to its location in the state space, as opposed to relying on intensive simulation-based techniques that are often used in industry. It is also possible to identify individual nodes that will experience a constraint violation, which can then inform how corrective action should be focused.

8.4.2.2 Example

To demonstrate the idea, consider the three node network as in Fig. 8.7, with one load and one generator connected to the "infinite bus" (this is a node with constant voltage and bus angle, and models the larger power grid) via lossless lines. The dynamics are fourth order, which we can rewrite as four first order equations, which fits to the dynamics (8.13):

$$\dot{\delta}_1 = \omega_1,$$
$$\dot{\omega}_1 = (F_1 - K \sin(\delta_1 - \delta_2) - K_\infty \sin(\delta_1) - D\omega_1)/H,$$
$$\dot{\delta}_2 = \omega_2,$$
$$\dot{\omega}_2 = (F_2 - K \sin(\delta_2 - \delta_1) - K_\infty \sin(\delta_2) - D\omega_2)/H,$$

where we have assumed identical normalized parameters for the generator and load: $H = 1$, $D = 1$, $K = 0.8$, $K_\infty = 2$, and $F_1 = 0.1 = -F_2$. We impose constraints on the bus angles, $\delta_b(t) \in [-0.1, 0.1]$, $b = 1, 2$, and find the safety sets, see Fig. 8.8, where we also show the solution of the full-order system from a number of post-fault states.

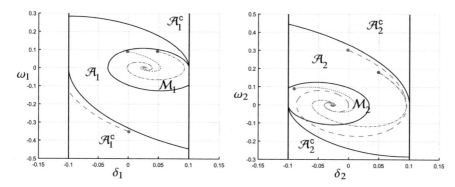

Fig. 8.8 The admissible set and MRPI for the generator and load node (left and right figure respectively) in the three-node example, with simulations of the full order system from three different post-fault states. The dashed magenta trajectory corresponds to a post-fault state that is unsafe, the finely-dashed red trajectory to one that is safe, and the dash-dotted blue trajectory to one that is potentially safe (and indeed turns out to be)

8.4.3 Future Research: Interaction Between the Distribution and Transmission Levels

Power-flow set points on the distribution grid level may be optimal in terms of energy efficiency, but may not be the safest (in terms of robustness to faults) for the transmission grid above. We now briefly outline an idea (our current focus of research) of how to arrive at compromising decisions. The tree networks of the distribution grid connect to the higher level via a reference node, which is seen as a single load bus from the transmission grid's perspective. The distribution grid could provide a number of possible power-flow set-points along with their respective costs (which reflect energy loss). For each low level set-point, which affects the power-flow on the higher level and results in new model parameters in (8.4), the safety sets could be computed and a measure of "safety" (possibly a measure involving the average area of the sets) could be determined for each set-point. Then, a final compromising power-flow set-point could be chosen that Pareto dominates the remaining ones. Otherwise, the Pareto front could be estimated by using various popular methods, such as evolutionary algorithms, see for example [16], and used as a guide to how the power-flow set-point could be changed depending on the system's operating conditions. For example, if bad weather is approaching it may make sense to change the power-flow set-point to one that is "safer", but still Pareto efficient.

8.5   Clustering-Based Model Order Reduction for the Synchronous Machine Model

As explained in Sect. 8.4, the synchronous machine model is used to evaluate stability among other things. The size of such a model depends on the number of nodes in the network. We assume here that the set of nodes or buses $\mathcal{B}$ is given by $\{1, \ldots, N\}$. Still considering the model as detailed in (8.4), with $x = [\delta_1, \ldots, \delta_N]$ the overall system is given as

$$H\ddot{x} + D\dot{x} = F - f(x), \tag{8.14}$$

with $H = \operatorname{diag}(\frac{2J_1}{\omega_r}, \ldots, \frac{2J_N}{\omega_r})$ and $D = \operatorname{diag}(\frac{D_1}{\omega_r}, \ldots, \frac{D_N}{\omega_r})$, $F = [F_1, \ldots, F_N]^T$ and $f(x)_b = \sum_{h \neq b} K_{bh} \sin(\delta_b - \delta_h - \gamma_{bh})$. This is a system of Ordinary Differential Equations (ODEs) and if the power grid is large, one may be able to profit from finding a reduced order model approximating this. A reduced model is a model of similar structure for a vector $\hat{x} \in \mathbb{R}^n$, namely

$$\hat{H}\ddot{\hat{x}} + \hat{D}\dot{\hat{x}} = \hat{A} - \hat{f}(\hat{x})$$

such that for a matrix $V \in \mathbb{R}^{N \times n}$ we can bound $\|V\hat{x} - x\|$ in an appropriate norm and $n << N$.

One standard tool to create reduced models is to do so-called projection based model order reduction. This is done either by Galerkin or Petrov-Galerkin Projektion of the ODE. Given the projection matrices $V, W \in \mathbb{R}^{N \times n}$, which are identical for Galerkin projection the reduced ODE is given by

$$W^T H V\ddot{\hat{x}} + W^T D V\dot{\hat{x}} = W^T F V\hat{x} - W^T f(V\hat{x}).$$

For generic projection matrices V and W the structure of the reduced model will be possibly very different from the structure of the original model. For example, even if H and D are sparse matrices the projected matrices $\hat{H}$ and $\hat{D}$ will not necessarily be sparse. Furthermore, the evaluation of the function $W^T f(V\hat{x})$ in the reduced system is an evaluation that scales with large N, as we take the vector $\hat{x}$ and lift it into the large space then compute f and then project back. One way to avoid these fallbacks is the use of a Petrov-Galerkin projection as in Paper [32] with

$$\begin{aligned} V_P &:= P, \\ W_P &:= P, \end{aligned} \tag{8.15}$$

where $P \in \mathbb{R}^{N \times n}$ is the characteristic matrix of a partition of the set $\mathcal{B}$ of natural numbers less or equal to N into n subsets B_j, $j = 1, \ldots, n$:

$$\mathcal{B} = \{1, \ldots, N\} = \bigcup_{j=1}^{n} B_j.$$

This means

$$P_{bj} = \begin{cases} 1 & \text{if } b \in B_j \\ 0 & else \end{cases}$$

Using such a projection will lead to a clustering of the nodes in the network. We will first show that a Petrov-Galerkin projection coming from such a clustering will result in a reduced model with the same structure as the original model and will then present an algorithm on how to find a clustering, which keeps the dynamical behaviour of the large system based on Proper Orthogonal Decomposition (POD), and present some results for a small network system.

8.5.1 Structure-Preserving

In the following we show that if we project with matrices V_P, W_P as in (8.15) the reduced ODE can still be interpreted as a system of swing equations on a reduced (clustered) network. Let $\mathbf{1}_N := [1, \ldots, 1]^T \in \mathbb{R}^N$. Then (8.14) reads,

$$H\ddot{x} + D\dot{x} = F - \left(K \odot \sin(x\mathbf{1}_N^T - \mathbf{1}_N x^T - \Gamma)\right)\mathbf{1}_N, \tag{8.16}$$

where the sine function acts elementwise on a matrix, and $\odot$ denotes the hadamard product, which is just componentwise multiplication. The diagonal matrices H and D and the vector F are are as in (8.14). The matrix K is given by (8.5b), but with the diagonal elements $K_{bb} \equiv 0$ in order to mimic the summation in (8.4) and $\Gamma_{bh} = \gamma_{bh}$ for $b \neq h$ and $\Gamma_{bb} = 0$ on the diagonal. Note that $P\mathbf{1}_n = \mathbf{1}_N$ because P resembles a clustering, but $PP^T \neq I_N$, $P^T P \neq I_n$.

Consider the projection for the non-linear term and the case when $\Gamma = 0$:

$$W_P^T \left(K \odot \sin(V_P\hat{x}\mathbf{1}_N^T - \mathbf{1}_N(V_P\hat{x})^T)\right)\mathbf{1}_N$$
$$= P^T \left(K \odot \sin(P\hat{x}\mathbf{1}_n^T P^T - P\mathbf{1}_n\hat{x}^T P^T)\right)P\mathbf{1}_n$$
$$= P^T \left(K \odot P \sin(\hat{x}\mathbf{1}_n^T - \mathbf{1}_n\hat{x}^T)P^T\right)P\mathbf{1}_n$$
$$=: \left(\hat{K} \odot \sin(\hat{x}\mathbf{1}_n^T - \mathbf{1}_n\hat{x}^T)\right)\mathbf{1}_n,$$

with $\hat{K} = P^T K P$. A clustering-based Reduced-Order Model (ROM) of (8.16) is, thus, given by,

$$\hat{H}\ddot{\hat{x}} + \hat{D}\dot{\hat{x}} = \hat{A} - \left(\widehat{K^c} \odot \sin(\hat{x}\mathbf{1}_n^T - \mathbf{1}_n\hat{x}^T) - \widehat{K^s} \odot \cos(\hat{x}\mathbf{1}_n^T - \mathbf{1}_n\hat{x}^T)\right)\mathbf{1}_n, \tag{8.17}$$

where the linear parts are $\hat{H} = P^T H P$, $\hat{D} = P^T D P$, $\hat{A} = P^T A$, and

$$\widehat{K^c} = P^T(K \odot \cos(\Gamma))P, \quad \widehat{K^s} = P^T(K \odot \sin(\Gamma))P.$$

Note that $\hat{H}$, $\hat{D}$ are again diagonal matrices, and (8.17) can be re-interpreted as a network of clustered nodes.

8.5.2 POD-Based Clustering

A classical tool to find a good projection V and W for a dynamical system is by Proper Orthogonal Decomposition (POD). In POD, solution snapshots $x(t_i)$ of the trajectory x at certain time instances are collected into a snapshot matrix $X = [x(t_0), x(t_1), \ldots]$ and by a singular value decomposition of this matrix, the dominant subspaces in this solution are identified. Since we also have a dynamical system, we can do exactly that and identify a Galerkin projection which gives a very good reduced order model. Let V be said projection matrix. The goal would now be to find a P which is a matrix coming from a clustering such that the image of P and the image of V are as close as possible.

In [30], the authors consider this problem in detail and present an algorithm based on the QR decomposition with column pivoting [48]. In this work, however, we use a heuristic K-means algorithm to find a clustering based on the projection matrix V.

8.5.3 Numerical Results

We use a test network with 10 nodes presented in Fig. 8.9.

In each node we have the swing equation and simulate the full system to create a snapshot matrix. From that matrix we create a projection matrix through POD, which we then use to find a good clustering. In Fig. 8.10 we see the time simulation for the 10 nodes as well as for the five clusters. It is clear from that picture that some

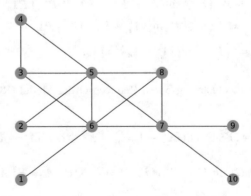

Fig. 8.9 Network structure of the test network

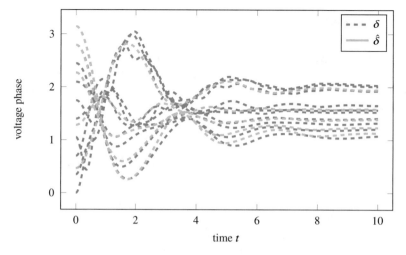

Fig. 8.10 Simulation of the original swing equation (dashed lines) and its clustering-based ROM (solid lines)

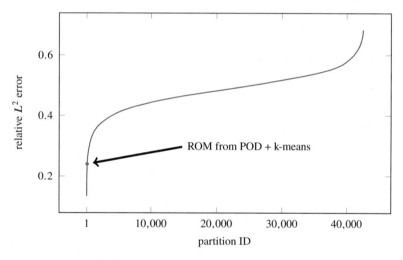

Fig. 8.11 Relative error of all possible five cluster partitions

nodes behave very similar in their time series and it is therefore possible to cluster them into one node.

The clustering that is presented in this picture is done by the POD-based clustering algorithm presented in this section. Since this problem is so simple we can test all possible five cluster partitions and compute the relative L^2 error. Here we see (Fig. 8.11) that our algorithm doesn't give the best one, but a very good one.

8.5.4 Summary

These reduced network will in general give a good representation of the dynamical behaviour of the network and are structurally completely identical to the original one. They can therefore easily be used in the stability analysis of a large system and future work will be to classify error bounds.

8.6 Optimal Power Flow for Future Power Networks

The optimal power flow problem [10, 11] is used in electrical network operation to determine a set-point, at which the market demand is satisfied subject to physical constraints and other operational requirements. This includes identifying voltages at each network node and real and reactive power produced by the generators or supplied by external sources. In the current setting, we specifically consider application of the problem to high voltage grids. We extend the standard formulation with new permissible control measures. These include circuit deactivation, phase-shifting transformers, thyristor-controlled series capacitors and HVDC transmission lines. Furthermore, flexibility (Sect. 8.3), stability (Sect. 8.4) and subnetwork (Sect. 8.5) information can be used to provide additional constraints.

Current solution methods for OPF in the industry involve semi-automatic planning with expert knowledge based on DC load flow models, sensitivity analyses and linearizations. In recent years, a popular approach in research has been to use semidefinite programming (SDP) relaxations based on the work of Lasserre et al. [28] and Lavaei and Low [29]. SDP extends linear programming by allowing, in addition to variables in the cone of nonnegative vectors, variables in the cone of positive semidefinite matrices, see Boyd and Vandenberghe [6] for gentle introduction and Wolkowicz et al. [44] for detailed explanations. Semidefinite matrices describe convex quadratic forms. In the current context, they admit the exact representation of sums of squares relaxations of optimization problems over higher order polynomials. This, however, comes at a considerable cost, as matrix sizes grow rapidly and the actual number of variables within the matrices grows quadratically in this order. By exploiting structural properties or employing decomposition techniques it may still be possible to solve the corresponding relaxations reasonably efficiently with interior point methods [6] or the spectral bundle method [23]. Due to the improved quality of the model this may even be worth the effort if only essential parts of the problem at hand are represented that way. Semidefinite programming relaxations for optimal power flow have an advantage of producing better results than linear approximations. In our work, we consider a formulation with added binary decisions in topology and controllable transformers and TCSCs. On top of that, the final goal is inclusion of network security in the form of the $n - 1$ principle. This means that a failure of a single network element will not have long-lasting adverse effects on network operation. Such results can be achieved by identifying possible contingencies within the network and dynamically adding critical failure scenarios. Due to the power balances this requires case distinctions with extensive starting points for polyhedral investigations.

Given the electrical power network model described in Sect. 8.1, we take into consideration additional network elements, namely, transformers (conventional and phase-shifting) $t \in \mathcal{T}$ and thyristor-controlled series capacitors (TCSCs) $c \in C$. These elements are modeled similarly to the lines in a sense that each is characterized by complex admittance $Y = G + \mathbf{i}B$ and has predefined start and end points from the set of buses $b \in \mathcal{B}$. The main difference between these two elements is that while the value of admittance is fixed for transformers, it is modifiable for TCSCs and is used to control power flow in the network. Since there is a discrete number of set points for each device, we introduce binary variables z_j associated with different settings Y_j, each corresponding to a single admittance value for $j \in \{-m, \ldots, 0, \ldots, m\}$, where m depends on the number of settings available for a device. The shunt admittance Y^{SH} remains constant for both types of devices. The transformers are additionally characterized by their complex influence on voltage, consisting of the tap ratio $|\tau|$ and phase shift angle ξ. Based on these properties, we can distinguish between standard transformers that do not change the voltage angle and the phase-shifting transformers that do. For some transformers, it is also possible for their setting to be modifiable in discrete steps, similarly to the TCSCs. Therefore, we define binary variables y_j associated with settings $\tau_j = |\tau_j|e^{\mathbf{i}\xi_j}$ for $j \in \{-n, \ldots, 0, \ldots, n\}$, where n is, once again, the number of settings available for a device.

We introduce one more set of binary variables x associated with line switching decisions, meaning that they take value 1 if the line is in operation and 0 otherwise. Moreover, we assign lower and upper limits V^-, V^+ to voltage magnitude and S^-, S^+ to generated power. The latter are applied elementwise to real and imaginary parts. Finally, each line, transformer and TCSC has thermal limit T^+ on the current that it can carry. This limit is observed at both ends of each device and represents allowed congestion before it is switched off for safety reasons. Most of the parameters mentioned above are clearly defined in UCTE [42] and CGMES [18] data exchange formats for power systems, which are currently used for load flow calculations in Europe. An exception here is the description of TCSCs, which is unavailable at the moment. This is subject to change as the CGMES format can be extended in the future to include new data.

Given the above notation, we write down the non-linear non-convex formulation of the optimal power flow problem with new variables. We can distinguish between two categories of constraints. The first is the set of equality constraints related to line current/power injections and power balance preservation at all nodes in the network. The second is category of inequality constraints, which cover the operation bounds such as voltage magnitude limits, thermal limits and power generation limits. In order to present the constraints in a concise manner, we use the so-called uniform line model, which is used to represent lines, transformers and TCSCs together. The current injections into the lines can be written as follows:

$$
\begin{aligned}
I_\ell^S &= x_\ell \left(\tau_\ell (Y_\ell + Y_\ell^{SH}) V_{b_s} - Y_\ell V_{b_e} \right) \\
I_\ell^E &= x_\ell \left((Y_\ell + Y_\ell^{SH}) V_{b_e} - \tau_\ell Y_\ell V_{b_s} \right)
\end{aligned}
\qquad \forall \ell = (b_s, b_e) \in \mathcal{L} \cup \mathcal{T} \cup \mathcal{C}, \qquad (8.18)
$$

where $x_\ell = 1$ for transformers and TCSCs and $\tau_\ell = 1$ for lines and TCSCs. The values of τ for transformers and Y for TCSCs are chosen from the set of available settings for each such device. This is modeled by use of binary variables, each corresponding to a single setting. Only one setting can be chosen at a time. Therefore, we obtain the following sets of constraints:

$$
\begin{aligned}
\sum_{j=-n_t}^{n_t} y_{t,j} \tau_{t,j} &= \tau_t \\
\sum_{j=-n_t}^{n_t} y_{t,j} &= 1 \\
y_{t,j} &\in \{0, 1\} \quad \forall j = -n_t, \ldots, 0, \ldots, n_t
\end{aligned}
\qquad \forall t \in \mathcal{T} \qquad (8.19)
$$

$$
\begin{aligned}
\sum_{j=-m_c}^{m_c} z_{c,j} Y_{c,j} &= Y_c \\
\sum_{j=-m_c}^{m_c} z_{c,j} &= 1 \\
z_{c,j} &\in \{0, 1\} \quad \forall j = -m_c, \ldots, 0, \ldots, m_c
\end{aligned}
\qquad \forall c \in \mathcal{C} \qquad (8.20)
$$

We can then define complex power injections into the lines, transformers and TCSCs, based on the current injections:

$$
\begin{aligned}
S_\ell^S &= \left(I_\ell^S \right)^* \tau_\ell V_{b_s} \\
S_\ell^E &= \left(I_\ell^E \right)^* V_{b_e}
\end{aligned}
\qquad \forall \ell = (b_s, b_e) \in \mathcal{L} \cup \mathcal{T} \cup \mathcal{C}, \qquad (8.21)
$$

Finally, we write down the nodal power balance constraint, which uses the above power injections:

$$
S_b - S_b^D = \sum_{\ell : b_s = b} S_\ell^S + \sum_{\ell : b_e = b} S_\ell^E + (Y_b)^* |V_b|^2 \qquad \forall b \in \mathcal{B}, \qquad (8.22)
$$

where the left hand side is the difference between complex power generation and demand, and the right hand side is a sum of the losses at bus and all complex power injections to/from lines, transformers and TCSC.

Moreover, we define several inequality constraints. Most importantly, flow along the lines, transformers and TCSCs is limited by thermal capacity of the equipment. This is defined as follows:

$$
\begin{aligned}
|I_\ell^S| &\leq T_k^+ \\
|I_\ell^E| &\leq T_k^+
\end{aligned}
\qquad \forall \ell = (b_s, b_e) \in \mathcal{L} \cup \mathcal{T} \cup \mathcal{C}, \qquad (8.23)
$$

The remaining inequality constraints are straightforward limits on voltage magnitude and elementwise bounds on complex power generation:

$$
V_b^- \leq |V_b| \leq V_b^+ \qquad \forall b \in \mathcal{B}, \qquad (8.24)
$$

$$
S_b^- \leq S_b \leq S_b^+ \qquad \forall b \in \mathcal{B}. \qquad (8.25)
$$

From this point on, we build a semidefinite programming relaxation. We construct a vector consisting of an entry of value 1 followed by voltage variables, line binary variables and bilinear products of voltage and line binary variables, voltage and transformer setting variables and voltage and TCSC setting variables. This can be illustrated as follows:

$$w = [1, V_b, x_l, x_l V_b, \tau_t V_b, Y_c^{TCSC} V_b]^T,$$

where $x_l V_b$, $\tau_t V_b$ and $Y^{TCSC} V_b$ are only defined for lines/transformers/TCSCs and buses adjacent to each other (for example, for each line l we only need $x_l V_n$ and $x_l V_m$, where n and m are beginning and end of the line respectively). SDP formulations of the standard OPF problem depend on the observation that, for a column vector of voltages V, $V V^H$ is a positive semidefinite rank one matrix whose entries can replace the products $V_n V_m^*$. Afterwards, the rank one constraint is dropped in order to obtain a relaxation. We follow a similar chain of reasoning, but we apply it to our vector w instead and consider matrix $W = w w^H$. We let W^R and W^I denote real and imaginary parts of W respectively. After relaxing the rank constraint, we obtain a set of equalities. First, the positive semidefinite constraint on the variable matrix

$$\begin{bmatrix} W^R, -W^I \\ W^I, W^R \end{bmatrix} \succeq 0 \text{ or } W \succeq 0,$$

where the real part W^R is symmetric and the imaginary part W^I is skew symmetric. Moreover, $W_{1,1}^R = 1$ and $\text{diag}(W^I) = 0$.

We replace all variables in constraints with their corresponding entries of the new W matrix. This covers the current and power injections, bus balance equations, thermal limits and voltage limits. We keep generation limit constraints since power generation variables are not included in the W matrix. Finally, we adapt the constraints for the setting points of transformers and TCSCs. We use a standard approach for binary variables, in which for a binary variable r the equality $r^2 = r$ holds. For each transformer/TCSC we define a positive semidefinite matrix that is formed by taking a vector of its corresponding binary variables $y_t = [y_{t,-n}, \ldots, y_{t,0}, \ldots, y_{t,n}]^T$ and forming a product $y_t y_t^T \succeq 0$. Then, we express the constraints (4) and (5) in terms of elements of this matrix.

The above process gives us a complete semidefinite programming relaxation of the extended optimal power flow problem. Additional constraints based on flexibility and stability inputs can be included to provide more complex control of network operation. However, there are still major challenges to be overcome, which include computational efficiency and scalability of this approach.

8.7 Conclusion

We have considered a number of challenges appearing in the modern power grid due to the proliferation of renewable energy sources and flexible consumers. Furthermore, we have explained how the grid may be analyzed in a hierarchical fashion, with power-flow set-points at the top (the transmission grid level) and bottom (the distribution grid level) of the hierarchy affecting the dynamics of coupled machines. We have presented an overview of the application of new mathematical methods to improve energy management at the bottom of the hierarchy; to classify large faults in the grid as safe or unsafe; to obtain structure-preserving reduced-order models; and to model new grid technologies and solve the associated power-flow problems via semi-definite relaxations.

All of these tasks fit a framework that aims to make electrical network operation more efficient through cooperation between different levels in the hierarchy. This includes bringing together information about network structure, flexibility, stability and controllability. Model order reduction provides us with smaller networks, which are easier to use in other tasks. Flexibility information contributes useful parameters for network study and operation planning. Transient stability analysis gives us information on whether a set-point is safe and can be used at other levels. Finally, solutions of the optimal power flow can be fed back to the above and further analysed in terms of possible flexibilities and stability.

Acknowledgements The authors collaborate within the project *KONSENS: Konsistente Optimierung uNd Stabilisierung Elektrischer NetzwerkSysteme* funded by the German Federal Ministry of Education and Research (BMBF, grants 05M18OCA, 05M18SIA, and 05M18EVA).

References

1. Aschenbruck, T., Esterhuizen, W., Streif, S.: Transient stability analysis of power grids with admissible and maximal robust positively invariant sets. Automatisierungstechnik (2020). In Press
2. Baumann, M., Grundel, S., Sauerteig, P., Worthmann, K.: Surrogate models in bidirectional optimization of coupled microgrids. Automatisierungstechnik **67**(12), 1035–1046 (2019)
3. Bergen, A.R., Hill, D.J.: A structure preserving model for power system stability analysis. IEEE Trans. Power Appar. Syst. **100**(1), 25–35 (1981)
4. Bienstock, D.: Electrical transmission system cascades and vulnerability. SIAM (2015)
5. Boyd, S., Parikh, N., Chu, E., Peleato, B., Eckstein, J.: Distributed optimization and statistical learning via the alternating direction method of multipliers. Found. Trends Mach. Learn. **3**(1), 1–122 (2011)
6. Boyd, S., Vandenberghe, L.: Convex Optimization. Cambridge University Press, Cambridge (2004)
7. Braun, P., Faulwasser, T., Grüne, L., Kellett, C.M., Weller, S.R., Worthmann, K.: Hierarchical distributed ADMM for predictive control with applications in power networks. IFAC J. Syst. Control **3**, 10–22 (2018)

8. Braun, P., Grüne, L., Kellett, C.M., Weller, S.R., Worthmann, K.: A distributed optimization algorithm for the predictive control of smart grids. IEEE Trans. Autom. Control **61**(12), 3898–3911 (2016)
9. Braun, P., Sauerteig, P., Worthmann, K.: Distributed optimization based control on the example of microgrids. In: Blondin, M.J., Pardalos, P.M., Saéz, J. (eds.), *Computational Intelligence and Optimization Methods for Control Engineering,* vol. 150, Springer Optmization and Its Applications, pp. 173–200. Springer, Berlin (2019)
10. Cain, M.B., O'Neill, R.P., Castillo, A.: History of optimal power flow and formulations, optimal power flow paper 1. Federal Energy Regulatory Comission (2012)
11. Carpentier, J.: Contribution to the economic dispatch problem. Bulletin de la Societe Francoise des Electriciens **3**(8), 431–447 (1962)
12. Chan, K.W., Cheung, C.H., Su, H.T.: Time domain simulation based transient stability assessment and control. Proc. Internat. Conf. Power Syst. Technol. **3**, 1578–1582 (2002)
13. Chiang, H.D.: Direct Methods for Stability Analysis of Electric Power Systems: Theoretical Foundation, BCU Methodologies, and Applications. Wiley, New york (2011)
14. De Dona, J., Lévine, J.: On barriers in state and input constrained nonlinear systems. SIAM J. Control Optim. **51**(4), 3208–3234 (2013)
15. El-Guindy, A., Chen, Y.C., Althoff, M.: Compositional transient stability analysis of power systems via the computation of reachable sets. In: American Control Conference (ACC), pp. 2536–2543 (2017)
16. Emmerich, M.T., Deutz, A.H.: A tutorial on multiobjective optimization: fundamentals and evolutionary methods. Nat. Comput. **17**, 585–609 (2018)
17. Esterhuizen, W., Aschenbruck, T., Streif, S.: On maximal robust positively invariant sets in constrained nonlinear systems. Automatica **119**, 109044 (2020)
18. European Network of Transmission System Operators for Electricity (ENTSO-E): Common Grid Model Exchange Specification (CGMES) (2017)
19. Grundel, S., Sauerteig, P., Worthmann, K.: Surrogate models for coupled microgrids. In: Faragó, I., Izsák, F., Simon, P. (eds.) Progress in Industrial Mathematics at ECMI 2018, vol. 30, pp. 477–483 (2019)
20. Grüne, L., Pannek, J.: Nonlinear Model Predictive Control. Theory and Algorithms, 2nd edn. Springer, Berlin (2017)
21. Guggilam, S.S., Dall'Anese, E., Chen, Y.C., Dhople, S.V., Giannakis, G.B.: Scalable optimization methods for distributed networks with high PV integration. IEEE Trans. Smart Grid **7**(4), 2061–2070 (2016)
22. Hans, C.A., Braun, P., Raisch, J., Grüne, L., Reincke-Collon, C.: Hierarchical distributed model predictive control of interconnected microgrids. IEEE Trans. Sustain. Energy **10**(1), 407–416 (2019)
23. Helmberg, C., Rendl, F.: A spectral bundle method for semidefinite programming. SIAM J. Optim. **10**(3), 673–696 (2000)
24. Houska, B., Frasch, J.V., Diehl, M.: An augmented Lagrangian based algorithm for distributed nonconvex optimization. SIAM J. Optim. **26**(2), 1101–1127 (2016)
25. Houska, B., Jiang, Y.: Distributed Optimization and Control with ALADIN, p. (provisionally accepted). Springer. http://faculty.sist.shanghaitech.edu.cn/faculty/boris/AladinChapter.html
26. Jiang, Y., Sauerteig, P., Houska, B., Worthmann, K.: Distributed Optimization using ALADIN for Model Predictive Control in Smart Grids. IEEE Trans. Control Syst. Technol. https://doi.org/10.1109/TCST.2020.3033010
27. Kundur, P., Paserba, J., Ajjarapu, V., Andersson, G., Bose, A., Canizares, C., Hatziargyriou, N., Hill, D., Stankovic, A., Taylor, C., Van Cutsem, T., Vittal, V.: Definition and classification of power system stability. IEEE Trans. Power Syst. **19**(2), 1387–1401 (2004)
28. Lasserre, J.B.: Global optimization with polynomials and the problem of moments. SIAM J. Optim. **11**(3), 796–817 (2001)
29. Lavaei, J., Low, S.H.: Zero duality gap in optimal power flow problem. IEEE Trans. Power Syst. **27**(1), 92–107 (2012)

30. Mlinarić, P., Grundel, S., Benner, P.: Efficient model order reduction for multi-agent systems using QR decomposition-based clustering. In: 54th IEEE Conference on Decision and Control (CDC), Osaka, Japan, pp. 4794–4799 (2015)
31. Momoh, J.: Electric Power System Applications of Optimization. Power Engineering (Willis). CRC Press (2017)
32. Monshizadeh, N., Trentelman, H.L., Çamlibel, M.K.: Projection-based model reduction of multi-agent systems using graph partitions. IEEE Trans. Control Netw. Syst. **1**(2), 145–154 (2014)
33. Motter, A.E., Myers, S.A., Anghel, M., Nishikawa, T.: Spontaneous synchrony in power-grid networks. Nat. Phys. **9**(3), 191–197 (2013)
34. Mühlpfordt, T., Hagenmeyer, V., Faulwasser, T.: Uncertainty quantification for optimal power flow problems. Proc. Appl. Math. Mech. **19**(SI) (2019)
35. Nishikawa, T., Motter, A.E.: Comparative analysis of existing models for power-grid synchronization. New J. Phys. **17**(1), 015012 (2015)
36. Ratnam, E.L., Weller, S.R., Kellett, C.M., Murray, A.T.: Residential load and rooftop PV generation: an Australian distribution network dataset. In: International Journal of Sustainable Energy (2015)
37. Rawlings, J.B., Mayne, D.Q., Diehl, M.: Model Predictive Control: Theory, Computation, and Design. Nob Hill Publishing (2017)
38. Riverso, S., Boem, F., Ferrari-Trecate, G., Parisini, T.: Plug-and-play voltage and frequency control of islanded microgrids with meshed topology. IEEE Trans. Smart Grid **6**(3), 1176–1184 (2015)
39. Sauerteig, P., Baumann, M., Dickert, J., Grundel, S., Worthmann, K.: Reducing Transmission Losses via Reactive Power Control. In: Göttlich, S., Herty,M., Milde, A. (eds.) Mathematical MSO for Power Engineering and Management, Mathematics in Industry. Springer (2021)
40. Sauerteig, P., Jiang, Y., Houska, B., Worthmann, K.: Distributed Control Enforcing Group Sparsity in Smart Grids. In: Proceedings of the 21st IFAC World Congress. Berlin, Germany (2020)
41. Sauerteig, P., Worthmann, K.: Towards multiobjective optimization and control of smart grids. Optim. Control Appl. Methods (OCAM) **41**, 128–145 (2020)
42. Union for the Co-ordination of Transmission of Electricity (UCTE): UCTE data exchange format for load flow and three phase short circuit studies (2007)
43. Varaiya, P., Wu, F.F., Chen, R.L.: Direct methods for transient stability analysis of power systems: recent results. Proc. IEEE **73**(12), 1703–1715 (1985)
44. Wolkowicz, H., Saigal, R., Vandenberghe, L.: Handbook of Semidefinite Programming. Theory, Algorithms, and Applications. Springer US (2000)
45. Worthmann, K., Kellett, C.M., Braun, P., Grüne, L., Weller, S.R.: Distributed and decentralized control of residential energy systems incorporating battery storage. IEEE Trans. Smart Grid **6**(4), 1914–1923 (2015)
46. Worthmann, K., Kellett, C.M., Grüne, L., Weller, S.R.: Distributed control of residential energy systems using a market maker. IFAC Proc. Vol. **47**(3), 11641–11646 (2014)
47. Xiaoping, L., Ming, D., Jianghong, H., Pingping, H., Yali, P.: Dynamic economic dispatch for microgrids including battery energy storage. In: The 2nd International Symposium on Power Electronics for Distributed Generation Systems, pp. 914–917 (2010)
48. Zha, H., He, X., Ding, C., Simon, H., Gu, M.: Spectral relaxation for k-means clustering. In: NIPS, pp. 1057–1064 (2001)
49. Zhu, J.: Otimization of Power System Operation. IEEE Press Series on Power Engineering. Wiley, New York (2009)
50. Zimmermann, J., Tatarenko, T.: Optimales Energie-Management über verteilte, beschränkte Gradientenverfahren. Automatisierungstechnik **67**(11), 922–935 (2019)

Chapter 9
New Time Step Strategy for Multi-period Optimal Power Flow Problems

Nils Schween, Philipp Gerstner, Nico Meyer-Hübner, and Vincent Heuveline

Abstract The computation of the optimum of a dynamical or multi-period Optimal Power Flow problem assuming an Interior Point Method (IPM) leads to linear systems of equations whose size is proportional to the number of considered time steps. In this paper we propose a new approach to reduce the amount of time steps needed. Assuming that the power grid's dynamic is mainly determined by changes of the residual demand, we drop time steps in case it does not change substantially. Hence, the size of the linear systems can be reduced. We tested this method for the German Power Grid of the year 2023 and a synthetic 960 h profile. We were able to reduce the amount of time steps by 40% without changing the quality objective function's value significantly.

9.1 Introduction

A dynamical or multi-period Optimal Power Flow problem addresses the question how to improve the operation of a power grid with respect to a given objective in a given period of time. This question is answered by computing the dispatch of power generation among the available sources.

The state of a given power grid can be described by a set of continuous variables. This set, among others, contains the active power generated by each power plant

N. Schween (✉)
Max Planck Institute for Nuclear Physics, Saupfercheckweg 1, 69117 Heidelberg, Germany
e-mail: nils.schween@mpi-hd.mpg.de

P. Gerstner · V. Heuveline
Heidelberg University, Im Neuenheimer Feld 205, 69120 Heidelberg, Germany
e-mail: philipp.gerstner@uni-heidelberg.de

V. Heuveline
e-mail: vincent.heuveline@uni-heidelberg.de

N. Meyer-Hübner
Karlsruhe Institute of Technology, Engesserstraße 11, 76131 Karlsruhe, Germany
e-mail: nico.meyer-huebner@kit.edu

© Springer Nature Switzerland AG 2021
S. Göttlich et al. (eds.), *Mathematical Modeling, Simulation and Optimization for Power Engineering and Management*, Mathematics in Industry 34,
https://doi.org/10.1007/978-3-030-62732-4_9

connected to the grid. The task to find the optimal operating state of a given power grid is known as *Optimal Power Flow* (OPF) and is formalised as the minimisation of a cost function with respect to this set of variables. The cost function takes as input the current state of the power grid and its output is, for example, the amount of green house gases or the actual operating costs. At a specified moment in time, a particular amount of power is demanded by the consumers (e.g. households, industry, etc.), which are connected to the grid. This demand can be satisfied by different states of the power grid, e.g. on a sunny day it is possible to substitute a part of the power provided by coal plants with the power provided by photovoltaic panels. Optimising the cost function means to find the state which satisfies the power demand, while keeping the operation costs (or the green house gas emissions) as low as possible.

It is possible to approximately predict the power demand. These predictions enable power grid operators to determine the optimal operation state at times in the future. To this end, *a dynamic or multi-period Optimal Power Flow* simultaneously solves multiple OPF problems each corresponding to one moment in time. Unfortunately, the individual OPFs are not independent of each other, for example, a conventional power plant is not able to ramp up its power supply arbitrarily. If, for example, the optimal state of operation required a coal plant to increase its power input by 50% from one moment in time to the next, it might very well be that this is not possible due to technical restrictions [2].

From a mathematical point of view, an OPF problem is a constrained optimisation problem. A multi-period OPF is a series of single OPFs, one associated to each discrete instance in time, which is supplemented with additional constraints modeling, for example, the ramping restrictions of conventional power plants and, thus, the coupling between the different times, see e.g. [5] and references therein. In the next view paragraphs a sketch of the mathematical formulation of the multi-period OPF is presented.

The period of time during which the operation of the power grid should be optimised is denoted H and called optimisation period. It is normally given in hours and is splitted into a set of time steps of equal length. The distance between time steps is called the temporal resolution and denoted by Δt_0 . The time steps are addressed with the help of the index set $\mathcal{T}_0$. Hence, the cardinality of $\mathcal{T}_0$ equals the number of time steps considered, which we denote N_T.

The power grid consists of $N_\mathbf{B}$ nodes denoted by $\mathbf{B} := \{1, \ldots, N_\mathbf{B}\}$ and $N_\mathbf{E}$ transmission lines denoted by $\mathbf{E} \subset \mathbf{B} \times \mathbf{B}$. These lines are modeled by a symmetric admittance matrix

$$Y = G + jB \in \mathbb{C}^{N_\mathbf{B} \times N_\mathbf{B}}$$

$$\text{with } G, B \in \mathbb{R}^{N_\mathbf{B} \times N_\mathbf{B}} \text{ and } j = \sqrt{-1}. \tag{9.1}$$

$Y_{lk} = Y_{kl} \neq 0$ if and only if there is a branch connecting node k and l, i.e. $(k, l) \in \mathbf{E}$ and $(l, k) \in \mathbf{E}$. Therefore, Y is a sparse matrix for most real world power grids.

The complex voltage at node $k \in \mathbf{B}$ for time step t_i, with $i \in \mathcal{T}_0$, is given as the complex number

$$V_k^{t_i} = E_k^{t_i} + jF_k^{t_i}. \tag{9.2}$$

We define $E^{t_i} := [E_k^{t_i}]_{k \in \mathbf{B}}$ and $F^{t_i} := [F_k^{t_i}]_{k \in \mathbf{B}}$.

An example set of variables describing the state of the power grid at time t_i, $i \in \mathcal{T}_0$ may look like

$$
x^{t_i} = \begin{pmatrix} [P_g^{t_i}] \\ [P_r^{t_i}] \\ [SOC^{t_i}] \\ [P_{ex}^{t_i}] \\ [P_{im}^{t_i}] \end{pmatrix} \begin{array}{l} \text{(injected conventional power)} \\ \text{(injected renewable power)} \\ \text{(stored energy)} \\ \text{(power export)} \\ \text{(power import)} \end{array} , \tag{9.3}
$$

where every square bracket [ ] denotes a vector with as much elements as there are sources, e.g. conventional power plants. The complete vector of optimisation variables is then given by $x = [x^{t_i}]_{i \in \mathcal{T}_0}$.

The cost (or objective) function f_0 may for example account for the expenditure and income and can be defined as

$$
f_0(x) = \sum_{i=1}^{N_T} \Delta t_0 f_t(x^{t_i}) \tag{9.4}
$$

where f_t denotes an objective function, which models the costs per unit of time at any time. It includes, for example, the costs for a specific generator to provide a given amount of power for a specified amount of time. f_t can be defined as follows [5]

$$
\begin{aligned}
f_t(x^{t_i}) &= \sum (a_{g,2}(P_g^{t_i})^2 + a_{g,1} P_g^{t_i}) \\
&+ \sum (a_{r,2}(P_r^{t_i})^2 + a_{r,1} P_r^{t_i}) \\
&+ \sum (a_{ex,2}(P_{ex}^{t_i})^2 + a_{ex,1} P_{ex}^{t_i}) \\
&+ \sum (a_{im,2}(P_{im}^{t_i})^2 + a_{im,1} P_{im}^{t_i})
\end{aligned} \tag{9.5}
$$

and coefficients $a_{*,*} \in \mathbb{R}$. Note that the sums are taken over all available sources, for example, the first sum is taken over all available conventional power plants.

Eventually, we have to include the technical limits of the power grid in the formulation of the optimisation problem. First of all, Ohm's and Kirchhoff's law have to be fulfilled. In the case of a power grid, the two laws are transformed into the requirement that the power available at a node k must make up the difference between the power supplied and the power demanded at this node. If we denote the active power demand (or load) for a time step t^i by $L_p^{t_i}$ and assign the loads to the nodes with the help of the matrix $C_L \in \{0, 1\}^{N_B \times N_L}$, where N_L is the number of loads, the above requirement can be formulated in the form of the following balance equation:

$$
C_P P^{t_i} - C_L L_p^{t_i} - Pf_p(E^{t_i}, F^{t_i}) = 0 \in \mathbb{R}^{N_B} \tag{9.6}
$$

where the k-th component of Pf_p,

$$Pf_{p,k}(E, F) = \sum_{l=1}^{N_B} E_k(G_{kl}E_l - B_{kl}F_l) + F_k(G_{kl}F_l + B_{kl}E_l), \qquad (9.7)$$

describes the sum of power flows at node k. Note that we dropped the index t^i to improve readability. This equation is called the AC Power Flow Equation and is, for example, derived in [10]. Another example for a technical constraint of the power grid is that the change in the amount of stored energy must equal the difference between the newly stored and supplied energy during one time step. These constraints are given by

$$[SOC]^{t_i} = [SOC]^{t_{i-1}} + \Delta t_0(\nu_l[P_{sl}]^{t_i} - \nu_g^{-1}[P_{sd}]^{t_i}) \qquad (9.8)$$

with factors $\nu_l, \nu_g \in (0, 1)$ describing the efficiency of the loading process $[P_{sl}]^{t_i}$ and the discharging process $[P_{sd}]^{t_i}$, respectively. Note that in this constraint the variables $[SOC]^{t_{i-1}}$, which belong to the previous time step, are used. This induces the coupling between the individual OPFs. Due to technical restrictions arising from power plants, renewable energy sources, storage facilities, it is necessary to impose lower and upper bounds for active power injections:

$$P_{min}^{t_i} \le P^{t_i} \le P_{max}^{t_i}. \qquad (9.9)$$

Finally, we impose ramping constraints for conventional power plants to bound the rate of change for injected power between consecutive time steps by

$$|P_g^{t_i} - P_g^{t_{i-1}}| \le \tau P_{max}^{t_i}\alpha \ \text{ for all } i \in T_0 \setminus \{1\}, \qquad (9.10)$$

where α is the ramping factor in percentage per unit time.

For a more complete account of the constrained optimisation problem we refer the reader to [4, 8] and references there in.

We summarise this in the following equations describing a general constrained optimisation problem:

$$\begin{aligned}
\min_x \ & f(x) \text{ s.t.} \\
& g^{t_i}(x^{t_i}) = 0, && \text{for all } i \in T_0, \\
& g_{td}^{t_i}(x^{t_{i-1}}, x^{t_i}) = 0, && \text{for all } i \in T_0 \setminus \{1\}, \qquad (9.11) \\
& h^{t_i}(x^{t_i}) \le 0, && \text{for all } i \in T_0, \\
& h_{td}^{t_i}(x^{t_{i-1}}, x^{t_i}) \le 0, && \text{for all } i \in T_0 \setminus \{1\}.
\end{aligned}$$

The dimension of the vector of optimisation variables x is n^x and, furthermore, $x^{t_i} \in \mathbb{R}^{n^{x,t}}$. Thus, $n^x = N_T n^{x,t}$. The number of optimisation variables and the number of constraints are proportional to the number of time steps N_T (see e.g. [3, 8]). Since the cost function is quadratic and the AC Power Flow equation is non-convex, the described optimisation problem is a non-linear and non-convex optimisation problem.

Depending on the explicit form of the constraints and the objective function, there are many different algorithms to solve such an optimisation problem. We consider an approach based on the Primal Dual Interior Point Method (*PDIPM*) (see e.g. [7]). This method allows for solving a general class of optimisation problems, i.e. non-convex and non-linear optimisation problems. Its computational most intensive part is the solution of a linear system of equations, whose size is proportional to the number of considered time steps as we will show in the next paragraphs. The following description is in line with the implementation found in the MATLAB package MATPOWER [11].

As in the well-known case of unconstrained optimisation, there are necessary and sufficient conditions for a solution to be the optimum of a constrained optimisation problem. The necessary conditions are the Karush-Kuhn-Tucker conditions (*KKT* conditions), which we summarised in the following equation [3, 8][1]:

$$\mathbf{F}(x, s, \lambda, \mu) := \begin{pmatrix} \nabla_x \mathcal{L}_0(x, \lambda, \mu) \\ g(x) \\ h(x) + s \\ S\mu \end{pmatrix} = 0, \; s, \mu \geq 0. \tag{9.12}$$

$\mathcal{L}_0(x, \lambda, \mu)$ denotes the Lagrangian function, which is given by

$$\begin{aligned} \mathcal{L}_0 \colon \mathbb{R}^{n^x} \times \mathbb{R}^{n^\lambda} \times \mathbb{R}^{n^\mu} &\to \mathbb{R} \\ (x, \lambda, \mu) &\mapsto f(x) + \lambda^T g(x) + \mu^T h(x), \end{aligned} \tag{9.13}$$

and $s \in \mathbb{R}^{n^\mu}$ is a vector containing slack variables and $S \in \mathbb{R}^{n^\mu \times n^\mu}$ denotes a diagonal matrix whose diagonal elements are $S_{kk} = s_k$. λ and μ are called Lagrange multipliers. Their dimensions (n^λ and n^μ) equal the number of equality and inequality constraints respectively. Hence, they are proportional to the number of time steps considered, i.e. $n^\lambda = n^{\lambda,t} N_T$ and $n^\mu = n^{\mu,t} N_T$.

The PDIPM aims at solving the Eq. (9.12) and thus to find a point (x^*, λ^*, μ^*) which complies with the necessary conditions to be a local minimum of the multi-period Optimal Power Flow problem. Mathematically it solves a slightly modified version of this equation by applying Newton's method. The modification consists in considering an additional term, which decreases with iteration index k. This term keeps the slack variables s away from zero and, thus, the inequality constraints inactive. This means that solutions $(x^{(k)}, s^{(k)}, \lambda^{(k)}, \mu^{(k)})$ to this modified version of Eq. (9.12) are inside the feasible region and *not* on its surface, hence, the name Interior Point Method. More rigorously formulated, we find

---

[1]For the sake of correctness, it is also necessary to assume an additional constraint qualification, e.g. the linear independence constraint qualification (*LICQ*) [7].

$$\mathbf{F}_\gamma(x, s, \lambda, \mu) := \mathbf{F}(x, s, \lambda, \mu) - \begin{pmatrix} 0 & 0 & 0 & \gamma e \end{pmatrix}^T = 0, \tag{9.14}$$
$$s, \mu \geq 0,$$

for a sequence of *barrier parameters*

$$\gamma_k := \sigma \frac{(\mu^{(k-1)})^T s^{(k-1)}}{n^\mu} \tag{9.15}$$

where $\sigma = 0.1$ is the *centering parameter* and $(\mu^{(k-1)})^T s^{(k-1)}$ is called *complementary gap*. Additionally, $e = (1, \ldots, 1) \in \mathbb{R}^{n^\mu}$. For details we refer the inclined reader to [7, 11].

The application of Newton's method to Eq. (9.14) yields iterations in which the following linear systems of equations must be solved:

$$\nabla \mathbf{F}_{\gamma_k}(x^{(k)}, s^{(k)}, \lambda^{(k)}, \mu^{(k)}) \Delta^{(k)} = -\mathbf{F}_{\gamma_k}(x^{(k)}, s^{(k)}, \lambda^{(k)}, \mu^{(k)}). \tag{9.16}$$

Henceforth we omit the iteration index k. Assuming that $s, \mu > 0$, the Newton system (9.16) can be transformed into a reduced, symmetric saddle point form

$$\underbrace{\begin{pmatrix} L(x, \lambda, \mu) & (\nabla_x g(x))^T \\ \nabla_x g(x) & 0 \end{pmatrix}}_{=:A(x,s,\lambda,\mu)} \underbrace{\begin{pmatrix} \Delta x \\ \Delta \lambda \end{pmatrix}}_{=:\Delta_R} = -\underbrace{\begin{pmatrix} r_x \\ r_\lambda \end{pmatrix}}_{=:b} \tag{9.17}$$

with

$$L(x, \lambda, \mu) = \nabla^2_{xx} \mathcal{L}_0(x, \lambda, \mu) + (\nabla_x h(x))^T \Sigma (\nabla_x h(x)),$$
$$r_x = \nabla_x \mathcal{L}_0(x, \lambda, \mu) + (\nabla_x h(x))^T Z^{-1} (\gamma e + \text{diag}(\mu) h(x)), \tag{9.18}$$
$$r_\lambda = g(x),$$

and diagonal matrix $\Sigma = \text{diag}\left(\frac{\mu_1}{s_1}, \ldots, \frac{\mu_{n^\mu}}{s_{n^\mu}}\right)$.

Please note the different meanings of the ∇-operator. If it is applied to a scalar function, it is the gradient of this function. Applied to a vector function, it denotes the Jacobian of that function. Furthermore, if there are no variables specified in its subscript, the derivative of the (vector) function has to be computed with respect to all its arguments. $\nabla^2_{xx} \mathcal{L}_0$ denotes the Hessian matrix of the scalar function $\mathcal{L}_0$.

A look at the *KKT matrix* A, defined in Eq. (9.17), shows that its size is proportional to the number of time steps: First of all, the size of the Hessian matrix is by definition proportional to the number of optimisation variables x, which in turn are proportional to the number of time steps. Secondly, the number of rows of the Jacobian $\nabla_x g(x)$ equals the number of equality constraints and, consequently, it is also proportional to the number of time steps. Finally, $A \in \mathbb{R}^{n \times n}$ and $n = n^x + n^\lambda = (n^{x,t} + n^{\lambda,t}) N_T$. Since the computationally most intensive part of the PDIPM algorithm is the solution of Eq. (9.17), the reduction of its size by means of a time step reduction method may strongly improve its time to solution.

Before we start describing the proposed time step reduction method, we introduce the concept of a load profile [9]. As described above, the optimisation of the operation of a power grid heavily depends on the power demanded by different kinds of consumers. A load profile depicts the evolution of this demand with time. More importantly, it is an integral part of the formulation of the constraints of the multi-period Optimal Power Flow problem as can be seen in the AC Power Flow Eq. (9.6). In Eq. (9.11) they are represented by the function $g(x^{t_i})$. Note that we use the terminology "load" and "demand" interchangeably. Of course, the prediction of power demand is inherently uncertain and a stochastical load profile which mirros this uncertainity is needed for realistic use cases. In this paper, however, we restricted ourselves to a deterministic load profile, since we solely like to demonstrate our time step reduction method without further complicating its description.

9.2 Methodology

The proposed time step reduction method is based on the assumption that if the load's change is small, the change in the optimisation variables is almost linear and, thus, it should be possible to drop time steps, i.e. to decrease the temporal resolution, and to compute the optimisation variables' values at the left out time steps by linear interpolation. In contrast, if the load changes rapidly, the non-linear character of the optimisation problem (9.11) becomes dominant and forbids linear interpolation and therefore a high temporal resolution is needed to map the power grid's dynamic.

First of all, we approximately compute the load's rate of change at node k of the power grid at the time steps t_i, $i \in \mathcal{T}_0 = \{1, \ldots, N_T\} \setminus \{N_T\}$, with the help of the difference quotient:

$$\dot{P}^{L,k}(t_i) = \left. \frac{dP^{L,k}}{dt} \right|_{t_i} \approx \frac{P_{t_{i+1}}^{L,k} - P_{t_i}^{L,k}}{t_{i+1} - t_i}. \tag{9.19}$$

If the rate of change is small, the load's change will be small, and, accordingly, we propose to drop some time steps. For this purpose, we segment the optimisation period H accordingly and choose different temporal resolutions Δt for the individual sections.

9.2.1 Measuring the Change in Residual Load

In order to adjust the temporal resolution Δt based on the load's rate of change, it is necessary to differentiate small rate changes apart from large ones. On a mathematical level, this can be done by defining a measure, i.e. a function which maps its arguments onto the interval [0, 1]. In a first step towards this definition, we compute the mean value of the rate of change at time step t_i:

$$\dot{P}^{L,MV}(t_i) = \frac{1}{N_B} \sum_{k=1}^{N_B} \dot{P}^{L,k}(t_i).$$
(9.20)

N_B denotes the number of nodes, which belong to the investigated power grid. In the second step, we divide these mean values by their maximum $\dot{P}_{max}^{L,MV}$ and, thus, obtain the definition of a measure of the rate of change at time step t_i:

$$\phi: \mathcal{T}_0 \to [0, 1]$$
$$i \mapsto \frac{\dot{P}^{L,MV}(t_i)}{\dot{P}_{max}^{L,MV}}.$$
(9.21)

With this definition at hand, we may say that the load's rate of change is big if the value of ϕ is close to one.

9.2.2 Different Δt for Different Changes

Moreover, we need to relate the rate of change's measure to different temporal resolutions. To this end, we define an additional step function Φ, which maps the rate of change's size of every time step t_i to a temporal resolution Δt. The possible temporal resolutions are obtained by deciding on the maximal temporal resolution Δt_{max}. Since we are only dropping time steps, it is clear from the outset, that possible temporal resolutions can only be integer multiples of the load profile's original temporal resolution Δt_0. Hence, the number of possible temporal resolutions $N_{\Delta t}$ is $\Delta t_{max}/\Delta t_0$. Furthermore, the definition of Φ requires, that we split the interval $(0, 1]$ into $N_{\Delta t}$ parts:

$$(0, 1] = \bigcup_{l=0}^{N_{\Delta t}-1} (l\frac{1}{N_{\Delta t}}, (l+1)\frac{1}{N_{\Delta t}}].$$

We define the function Φ as

$$\Phi: [0, 1] \to \mathbb{R}$$
$$x \mapsto (l+1)\Delta t_0 \text{ if } x \in (\frac{l}{N_{\Delta t}}, \frac{l+1}{N_{\Delta t}}].$$
(9.22)

In Fig. 9.1 we plotted one example Φ function to illustrate the way it relates different rates of change to different temporal resolutions Δt.

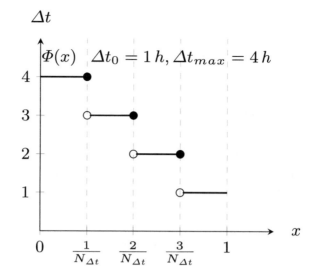

Fig. 9.1 Step function $\Phi(x)$. Smaller changes $\phi(i) \lesssim 0.5$ are related to bigger temporal resolutions Δt

9.2.3 Segmenting the Optimisation Period

Due to the measure ϕ and the step function Φ we are able to assign a Δt to every time step t_i of the optimisation period H. Since the aggregated load does not change abruptly, i.e. the load profile is assumed to be continuously differentiable, it is likely that a time step t_i is surrounded by a set of time steps whose values of ϕ do not differ much and to whom the step function Φ consequently assigns the same Δt. Thus, it is possible to segment the optimisation period H into N_A sections by grouping the time steps with an equal Δt. We write $H = \sum_{k=1}^{N_A} H_k$ and denote the temporal resolution, which corresponds to section H_k, Δt_k.

The section's different temporal resolutions Δt_k reduce the original set of indices $\mathcal{T}_0$, used to denote the time steps and to formulate the original optimisation problem, to the subset

$$\mathcal{T} = \bigcup_{k=1}^{N_A} \mathcal{T}_k \subset \mathcal{T}_0 .$$

Where $\mathcal{T}_k = \{i_k + m(\Delta t_k / \Delta t_0) \mid m \in \{0, \ldots, (H_k/\Delta t_k) - 1\}\}$ with

$$i_k = \sum_{l=1}^{k-1} \frac{H_l}{\Delta t_0} + \frac{\Delta t_k}{\Delta t_0} \quad \text{for } k > 1 .$$

For the sake of clarity, we depicted in Fig. 9.2 a section H_k and a subset $\mathcal{T}_k$ of the optimisation period H.

It is important to note, that the length of an optimisation section H_k is not always an integral multiple of its temporal resolution Δt_k. In this case we add the "non-

Fig. 9.2 Segmentation of the optimisation period H into sections H_k

fitting" time steps to the next section. If there are time steps left at the end of the optimisation period, we replenish it with an additional section of temporal resolution Δt_0.

9.2.4 Temporal Resolution Profile

We would like to demonstrate the explained time step reduction method by considering a concrete example. We take a look at the power grid and the synthetic 960 hour load profile referred to in Sect. 9.3. In Fig. 9.3a we plotted the first ninety-six hours of the arithmetic mean of the load. The load profile repeats itself after forty-eight hours. In Fig. 9.3b the load's rate of change is plotted. A brief check of the minima shows that the difference quotient used to compute it is a good approximation. Figure 9.3c shows the result of measuring the load's rate of change with the help of ϕ and plugging its values into the step function Φ: If the load does not change drastically in an optimisation section H_k, the maximum Δt_{max} is chosen. The values of the step function Φ form the basis of the segmentation of the optimisation period, by grouping its values as described above we obtain the *temporal resolution profile* shown in Fig. 9.3d.[2]

9.2.5 Interpolation Operator

After having applied the time step reduction method, we have the reduced set of time step indices $\mathcal{T}$ to formulate the multi-period Optimal Power Flow problem. It is solved using an Interior Point Method. Thus, after an unknown number of iterations the computed solution is an accord with the Karush-Kuhn-Tucker conditions. This solution is used to determine the values of the optimisation variables at the time steps we have dropped. This is done via linear interpolation. In principle, it is possible to use higher-order interpolation schemes, for example Spline Interpolation. But since

---

[2]The observant reader may recognise that the first section H_1 should have temporal resolution $\Delta t_{max} = 4$. The reason it does not is that we have not defined an extrapolation operator, which would be needed to compute the values of the first time steps. We just set the temporal resolution of H_1 to Δt_0 to avoid extrapolation.

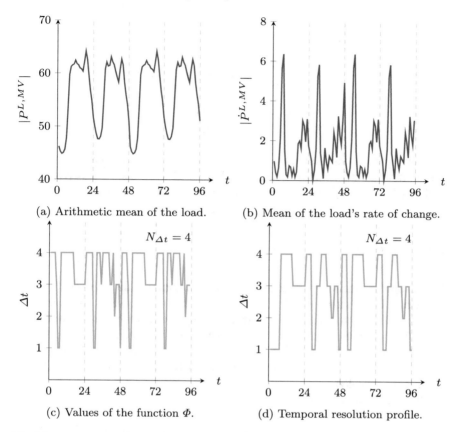

(a) Arithmetic mean of the load.

(b) Mean of the load's rate of change.

(c) Values of the function Φ.

(d) Temporal resolution profile.

Fig. 9.3 Time step adaption based on load profiles

this is a proof-of-concept for the proposed time step reduction method, we restricted ourselves to a linear interpolation scheme. The definition of an according linear interpolation operator $\mathcal{I}_{\mathcal{T}}$ consists of different parts for different sections of the optimisation period, because it depends on the temporal resolution of the respective section. Before defining one part of it for a specific section of the optimisation period, we remind the reader that the optimisation variables are sorted by time steps: $\bar{x} = [\bar{x}^{t_i}]_{i \in \mathcal{T}}$. In the section H_k the corresponding part of the interpolation operator is defined as:

$$\tilde{x}^{t_{a_k}+mw_k+l} = l\frac{\Delta t_0}{\Delta t_k}\left(\bar{x}^{t_{a_k}+(m+1)w_k} - \bar{x}^{t_{a_k}+mw_k}\right) + \bar{x}^{t_{a_k}+mw_k} . \qquad (9.23)$$

The following definitions and values are used:

$$w_k = \frac{\Delta t_k}{\Delta t_0},$$

$$a_k = \sum_{i=1}^{k-1} \frac{H_i}{\Delta t_0} \quad (\text{if } k = 0 \text{, then } a_0 = 0),$$

$$l \in \{1, \ldots, \Delta t_k / \Delta t_0 - 1\},$$

$$m \in \{0, \ldots, (H_k / \Delta t_k) - 1\}.$$

The complete interpolation operator $\mathcal{I}_T$ is defined as the linear interpolation of all optimisation variables via Eq. (9.23). We abbreviate this operation to $\tilde{x} = \mathcal{I}_T(\bar{x})$. The dual variables $\bar{\lambda}$ and $\tilde{\mu}$ are computed analogously, i.e. $\tilde{\lambda} = \mathcal{I}_T(\bar{\lambda})$ and $\tilde{\mu} = \mathcal{I}_T(\bar{\mu})$.

9.2.6   Estimating the Error

Since dropping time steps will cause an error in the optimisation problem's objective function, it is important to estimate this error in the worst case. To find out about this, we construct an error estimator, which is supposed to present an approximation of the absolute value of the error. We take advantage of the interpolated optimisation variables $\tilde{x}$ and introduce x^* to denote the solution of the multi-period Optimal Power Flow problem without time step reduction, which, henceforth, will be referred to as the reference problem.

The error of the objective function is given by:

$$f(\bar{x}) - f_0(x^*) = \underbrace{f(\bar{x}) - f_0(\tilde{x})}_{=:\Delta_1} + \underbrace{f_0(\tilde{x}) - f_0(x^*)}_{=:\Delta_2}. \tag{9.24}$$

Since $\bar{x}$ and $\tilde{x}$ are known after the IPM finished, Δ_1 can be computed directly. In the following we will derive a heuristic expression for the computation of the error Δ_2:

$$\Delta_2 = f_0(\tilde{x}) - f_0(x^*)$$

$$= \mathcal{L}_0(\tilde{x}, \tilde{\lambda}, \tilde{\mu}) - \tilde{\lambda}^T g(\tilde{x}) - \tilde{\mu}^T h(\tilde{x}) - \mathcal{L}_0(x^*, \lambda^*, \mu^*)$$

$$= \int_0^1 l'(s)\, ds - \tilde{\lambda}^T g(\tilde{x}) - \tilde{\mu}^T h(\tilde{x}) \tag{9.25}$$

$$= \frac{1}{2}\left(l'(1) + l'(0)\right) + \mathcal{O}(\|e\|^3) - \tilde{\lambda}^T g(\tilde{x}) - \tilde{\mu}^T h(\tilde{x}) \tag{9.26}$$

where $l(s) := \mathcal{L}_0(x^* + s e_x, \lambda^* + s e_\lambda, \mu^* + s e_\mu)$. Moreover,

$$e_x := \tilde{x} - x^*, \quad e_\lambda := \tilde{\lambda} - \lambda^*, \quad e_\mu := \tilde{\mu} - \mu^* \text{ and } e := \left(e_x \ e_\lambda \ e_\mu\right)^T.$$

The integral in Eq. (9.25) was approximated by applying the trapezoidal rule to it. Computing the derivative $l'(s)$ yields

$$
\begin{aligned}
l'(s) &= \nabla_x \mathcal{L}_0(\dots)e_x + \nabla_\lambda \mathcal{L}_0(\dots)e_\lambda + \nabla_\mu \mathcal{L}_0(\dots)e_\mu \\
&= \nabla_x \mathcal{L}_0(\dots)e_x + g^T(\dots)e_\lambda + h^T(\dots)e_\mu,
\end{aligned}
$$

and plugging into it the limits of the integral results in

$$
l'(1) = \nabla_x \mathcal{L}_0(\tilde{x}, \tilde{\lambda}, \tilde{\mu})e_x + g^T(\tilde{x})e_\lambda + h^T(\tilde{x})e_\mu,
$$
$$
l'(0) = \nabla_x \mathcal{L}_0(x^*, \lambda^*, \mu^*)e_x + g^T(x^*)e_\lambda + h^T(x^*)e_\mu = h^T(x^*)e_\mu = h^T(x^*)\tilde{\mu}.
$$

With the evaluated derivatives at hand Eq. (9.26) becomes

$$
\begin{aligned}
2\Delta_2 = \nabla_x \mathcal{L}_0(\tilde{x}, \tilde{\lambda}, \tilde{\mu}) \cdot (\tilde{x} - x^*) - g(\tilde{x})^T(\tilde{\lambda} + \lambda^*) \\
- h(\tilde{x})^T(\tilde{\mu} + \mu^*) + h(x^*)^T\tilde{\mu} + \mathcal{O}(\|e\|^3).
\end{aligned} \tag{9.27}
$$

Since the exact solution of the reference problem (x^*, λ^*, μ^*) is needed and because of the error term $\mathcal{O}(\|e\|^3)$, it is not possible to actual compute Δ_2. Further modifications are needed:

$$
\Delta_2 \approx \frac{1}{2}\nabla_x \mathcal{L}_0(\tilde{x}, \tilde{\lambda}, \tilde{\mu}) \cdot (\tilde{x} - \tilde{x}_{Spline}) - g(\tilde{x})^T\tilde{\lambda} - \frac{1}{2}h(\tilde{x})^T\tilde{\mu} \tag{9.28}
$$

We dropped the error term $\mathcal{O}(\|e\|^3)$ and replaced the solution of the reference problem x^*, λ^*, μ^* with its interpolated counterpart $\tilde{x}, \tilde{\lambda}, \tilde{\mu}$. But note, to avoid that the first term of Eq. (9.27) vanishes, we could not replace the first appearance of x^* with $\tilde{x}$. Instead, we need a interpolation scheme of higher order and thus chose $\tilde{x}_{Spline}$. The choice of the Spline interpolation was arbitrary, any other higher order interpolation scheme would also have been possible. These modifications are motivated by an analogous construction of an error estimator for Partial Differential Equations, which, for example, can be found in [1]. Numerical experiments showed, that Δ_2 is usually much bigger than $f(\tilde{x})$ and, consequently, not suited to provide a good error estimation. Thus, we attempted to drop the last term of Eq. (9.28), which yielded reasonable estimations. Hence, we settled on the following formula for Δ_2:

$$
\Delta_2 \approx \frac{1}{2}\nabla_x \mathcal{L}_0(\tilde{x}, \tilde{\lambda}, \tilde{\mu}) \cdot (\tilde{x} - \tilde{x}_{Spline}) - g(\tilde{x})^T\tilde{\lambda}. \tag{9.29}
$$

We would like to emphasise that the above construction of an error estimator is based on a heuristic which is investigated in the upcoming numerical experiments.

9.3 Results

We implemented the proposed time step reduction method in an expanded version of the free and open-source software MATPOWER. MATPOWER already contains the Primal Dual IPM, but it needed to be expanded by a mechanism, which is able to work with multiple time steps and different temporal resolutions to set up the KKT matrix for a multi-period Optimal Power Flow problem. We developed an Octave function, which creates temporal resolution profiles for different Δt_{max}. Furthermore, we implemented the interpolation operator by means of Octave's interp1 function and provided an additional function to compute an estimation of the objective function's error.

We tested the time step reduction method on a power grid, which consists of 1215 nodes, 2247 AC lines, 8 DC lines, 181 conventional power plants, 355 renewable energy sources and 67 storage facilities. This model is based on scenario B for the German Transmission Grid in the year 2023 by the German Network Regulatory Body. For more details, we refer to [6].

At the outset we had a synthetic $48h$ load profile, which was concatenated to $N_T = 96h$, see Fig. 9.3a. Then, we modified the load profiles by either smoothing or adding white noise. In this way, we obtained a family of profiles of varying temporal roughness. To be precise, temporal roughness of load profiles $P^L := \{P^{L,k}(t_i)\}_{i=1,k=1}^{N_T,N_B}$ is measured in terms of the following semi-norm

$$|P^L|_* := \frac{1}{N_T} \sum_{i=2}^{N_T} \sum_{k=1}^{N_B} |P^{L,k}(t_i) - P^{L,k}(t_{i-1})|. \tag{9.30}$$

The smoothed profiles are obtained by subsequently averaging the original profile P^L over a number of q time steps, followed by a linear interpolation over the next q steps. Rougher profiles are constructed by adding white noise in the following way:

$$\tilde{P}^{L,k}(t_i) := c_i P^{L,k}(t_i) \text{ with } c_i \sim \mathcal{U}([1 - \delta, 1 + \delta]) \text{ for given } \delta \in (0, 1). \tag{9.31}$$

The plots in Fig. 9.4 illustrate the number of time steps of the reduced problem and the number of IPM iterations, required for reaching the specified convergence criterion. These quantities are plotted for six different load profiles: smoothed profiles with $q \in \{8, 4, 2\}$, corresponding to $|P^L|_* \in \{4155, 4718, 5325\}$, original profile, corresponding to $|P^L|_* = 5629$ and noisy profiles obtained by choosing $\delta \in \{0.05, 0.1\}$ which corresponds to $|P^L|_* \in \{6908, 8611\}$. The smoothest profile ($q = 8$) thereby leads to the strongest reduction of time steps by employing the algorithm presented in Sect. 9.2.3, while the roughest profile ($\delta = 0.1$) only allows a significantly lower reduction. Between these extrema, however, the time step reduction appears to be non-monoton w.r.t. $|P^L|_*$. All considered profiles have in common that the number of IPM iterations significantly increases for $\Delta t_{max} \geq 3$. A reason for this phenomenon could be the fact, that a reduced number of time steps also reduces the number of control variables and thereby shrinking the feasible space of the optimization prob-

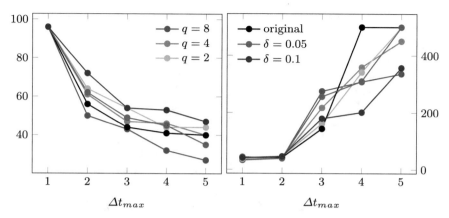

(a) Number of time steps N_T of OPF.   (b) Number of IPM iterations.

Fig. 9.4 Dependence of problem quantities w.r.t. Δt_{max} for load profiles of increasing roughness

lem. In fact, we observed that the computed optimal cost value $f(\bar{x})$ monotonically increases with increasing Δt_{max}.

The plots in Fig. 9.5 depict how the optimal cost value depends on the number of reduced time steps, with reference costs $f_0(x^*)$ denoting the optimal value of the respective unreduced problems, i.e. $\Delta t_{max} = 1$. As expected, the proposed method yields optimal values, which stronger deviate from the reference value, the more time steps are dropped. However, these deviations stay on a moderate level for all considered profiles. In particular, the post-processed optimal cost value $f_0(\tilde{x})$ differs from the reference value for at most 3% in all considered cases. Up to a time step size reduction of 50%, the relative difference is even below 1%. Moreover, this deviation appears to be rather robust w.r.t. $|P^L|_*$.

We further investigate the efficiency of the error estimator derived in Sect. 9.2.6. According to Fig. 9.6a, the estimator $\Delta_1 + \Delta_2$ generally underestimates the raw error $|f(\bar{x}) - f_0(x^*)|$ when increasing the number of dropped time steps. However, for up to 50% reduction, the estimated error is still of the same order of magnitude as the real error. Keeping in mind that the actual error is of rather moderate magnitude, see Fig. 9.5a, one may expect that $[f(\bar{x}) - |\Delta_1 + \Delta_2|, f(\bar{x}) + |\Delta_1 + \Delta_2|]$ yields a fairly accurate range for the unknown, reference costs $f_0(x^*)$.

Concerning the estimator Δ_2 for the post-processed error $|f_0(\tilde{x}) - f_0(x^*)|$, no clear trend can be observed and the ratio of estimator and real error varies of several orders of magnitude, see Fig. 9.6b. Thus, further research is required here to improve the accuracy in estimating the error of the post-processed optimal cost value.

We finally investigate how our method performs when applied to a large-scale OPF problem. To this end, we concatenated the original, $48h$ profile to obtain a test case comprising 48 days with a temporal resolution of $\Delta t_0 = 1h$, i.e. $N_T = 960$ time steps. This causes the linear system (9.17), which has to be solved in every PDIPM iteration, to be of size $n = n^x + n^\lambda = 6.168 \cdot 10^6$.

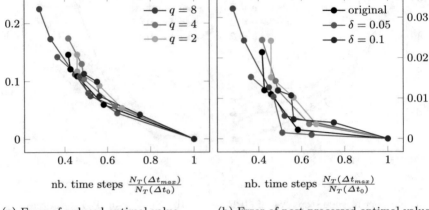

(a) Error of reduced optimal value
$|f(\bar{x}) - f_0(x^*)||f_0(x^*)|^{-1}$.

(b) Error of post-processed optimal value
$|f_0(\bar{x}) - f_0(x^*)||f_0(x^*)|^{-1}$.

Fig. 9.5 Errors of optimal cost value w.r.t. reference optimal cost values obtained for $\Delta t_{max} = 1$

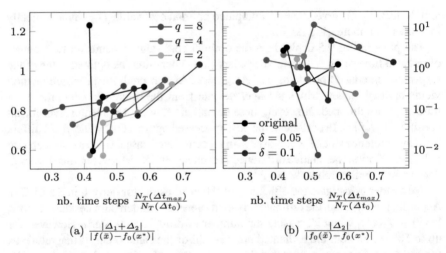

(a) $\dfrac{|\Delta_1 + \Delta_2|}{|f(\bar{x}) - f_0(x^*)|}$

(b) $\dfrac{|\Delta_2|}{|f_0(\bar{x}) - f_0(x^*)|}$

Fig. 9.6 Effectiveness of error estimators, plotted over problem size for load profiles of increasing roughness

As in the previous tests, reference optimal costs $f_0(x^*)$ are obtained by solving the unreduced problem, i.e. $H = 960h$ without dropping any time steps.[3] In a next step, we tested how many time steps can be dropped without causing too big errors in the objective function. Due to the construction of the proposed time step reduction

---

[3] Since the size of the KKT matrix A was very big (being proportional to $N_T \cdot N_B$), we solved the optimisation problems on a cluster taking advantage of a parallelization method, which we had developed previously. For more details, we refer the reader to [3, 8].

Fig. 9.7 Deviation of the objective function in %

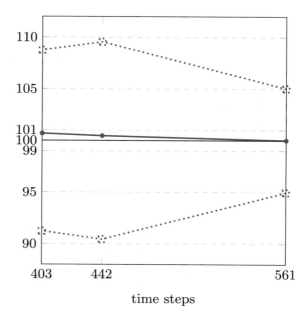

method, the more time steps will be ignored, the bigger Δt_{max} is chosen. We tested the following values for Δt_{max}: $2h$, $3h$ and $4h$.

Figure 9.7 shows on the x-axis the number of time steps, which had to be taken into account after we had applied the time step reduction method. For example, $\Delta t_{max} = 3\Delta t_0 = 3h$ resulted in 442 time steps. The dark blue line represents the deviation of the objective function from the reference value in percentage. The dashed grey lines depict the estimated error of the objective function, which we computed with the help of Eq. (9.29). The numerical experiments showed, that the choice $\Delta t_{max} = 2\Delta t_0$ yields, from a practical point of view, the most promising results. Although we were able to further decrease the number of time steps by setting $\Delta t_{max} = 3\Delta t_0$ or $\Delta t_{max} = 4\Delta t_0$ without deviating too much from the reference value of the objective function, we observed a destabilisation of the Interior Point Method. Similar to the previous numerical tests, increasing Δt_{max} leads to a dramatic increase in IPM iterations in comparison to the reference. Moreover, convergence of the PDIPM algorithm could only be obtained by easing the convergence criteria. It's tempting to assume that the reduced size of the KKT matrix A and the according decrease in the solution time compensates for the increase in iterations, but that's not the case for choices $\Delta t_{max} > 2\Delta t_0$. In contrast, in case Δt_{max} is set to $2\Delta t_0$, we observed only a slight increase in IPM iterations and a negligible deviation from the objective function's reference value. At the same time the number of time steps could be reduced to 561, i.e. by around 40 percent. This led to a large decrease in the solution time: in the reference problem an iteration took on average $441.05s$, which is much more than the $200.36s$ needed in the $2\Delta t_0$ case.

In an attempt to further benchmark the proposed time step reduction method, we dropped every second time step and solved the multi-period Optimal Power Flow problem again. Dropping every second time step reduced the number of time steps to 486.[4] The IPM needed 58 iterations to converge and every iteration took on average 190.61s. Hence, it took longer to solve the optimisation problem, even though less time steps were considered. Furthermore, the deviation from the reference value of the objective function was 25 times bigger in comparison with the $2\Delta t_0$ case.

9.4   Conclusion and Outlook

In case the parameter Δt_{max} was set to $2\Delta t_0$, the proposed time step reduction method yielded good results. The amount of time steps could be reduced by $\sim40\%$ while the objective function's deviation from its reference value was only 0.000176%. Moreover, the PDIPM algorithm was almost twice as fast. And in comparison with a naïve reduction of time steps, i.e. dropping every second time step, it performed well.

The proposed approach shows up very good properties in the practical setup of the German Power Grid. The proposed heuristical error estimator however over estimates the objective function's error. Further investigations including a theoretical analysis will be conducted in order to improve the reliability of the proposed method.

Acknowledgements This work was carried out with the financial support of the German Research Foundation (DFG) within the project 242471205. Moreover, the authors acknowledge support by the state of Baden-Württemberg through bwHPC and the German Research Foundation (DFG) through grant INST 35/1134-1 FUGG.

References

1. Becker, R., Braack, M., Rannacher, R.:. Numerical simulation of laminar ames at low Mach number by adaptive finite elements. Combust. Theory Model. **3**(3), 503–534 (1999). https://doi.org/10.1088/1364-7830/3/3/305
2. Frank, S., Rebennack, S.: An introduction to optimal power flow: Theory, formulation, and examples. IIE Trans. **48**(12), 1172–1197 (2016). https://doi.org/10.1080/0740817X.2016.1189626
3. Gerstner, P., Schick, M., Heuveline, V., Meyer- Hübner, N., Suriyah, M., Leibfried, T., Slednev, V., Fichtner, W., Bertsch, V.: A Domain Decomposition Approach for Solving Dynamic Optimal Power Flow Problems in Parallel with Application to the German Transmission Grid. Tech. rep. 01. 37.06.01; LK 01. Universität Heidelberg, Heidelberg, 21 pp (2016). https://journals.ub.uni-heidelberg.de/index.php/emcl-pp/article/download/33784/27475

---

[4]This value deviates slightly from the expected 480 time steps. Because of a missing definition of an extrapolation operator, the first time steps of the optimisation period could not be included in time step reduction method, c.f. with footnote 2.

4. Meyer-Hübner, N., Gielnik, F., Suriyah, M., Leibfried, T.: Dynamic optimal power flow in AC networks with multiterminal HVDC and energy storage. In: 2016 IEEE Innovative Smart Grid Technologies - Asia (ISGT-Asia) (2016). https://doi.org/10.1109/isgt-asia.2016.7796402
5. Meyer-Hübner, N., Suriyah, M., Leibfried, T.: On efficient computation of time constrained optimal power flow in rectangular form. In: PowerTech, 2015 IEEE Eindhoven, pp. 1–6 (2015). https://doi.org/10.1109/PTC.2015.7232378
6. Meyer-Hübner, N., Suriyah, M.R., Leibfried, T., Slednev, V., Bertsch, V., Fichtner, W., Gerstner, P., Schick, M., Heuveline, V.: Time constrained optimal power flow calculations on the German power grid. In: International ETG Congress 2015; Die Energiewende - Blueprints for the new energy age, pp. 1–7 (2015)
7. Nocedal, J., Wright, S.J.: Numerical Optimization, 2nd edn. Springer, New York, NY, pp. xxii + 664 (2006). ISBN: 0-387-30303-0/hbk
8. Schween, N., Gerstner, P., Meyer-Hübner, N., Slednev, V., Leibfried, T., Fichtner, W., Bertsch, V., Heuveline, V.: A domain decomposition approach to solve dynamic optimal power flow problems in parallel. In: Proceedings of the second International Symposium on Energy System Optimization (2020)
9. Von Meier, A.: Electric Power Systems: A Conceptual Introduction. Wiley (2006)
10. Zhu, J.: Optimization of Power System Operation, vol. 47. Wiley (2015)
11. Zimmerman, R.D., Murillo-Sánchez, C.E.: Matpower 6.0 User's Manual (2016)

Chapter 10
Reducing Transmission Losses via Reactive Power Control

Philipp Sauerteig, Manuel Baumann, Jörg Dickert, Sara Grundel, and Karl Worthmann

Abstract Modern smart grids are required to transport electricity along transmission lines from the renewable energy sources to the customer's demand in an efficient manner. It is inevitable that power is lost along these lines due to active as well as reactive power flows. However, the losses caused by reactive power flows can be reduced by optimizing the power factor. Therefore, we propose a power flow optimization problem aiming to reduce losses by controlling the power factors within the low-voltage electricity grid online. Furthermore, we show the potential of the proposed scheme in a numerical case study for two scenarios based on real-world data provided by a German distribution system operator.

Keywords Reactive power control · Loss reduction · Smart grid · Optimal power flow

10.1 Introduction

Electrical grids typically possess a hierarchical structure [2]. On the top there is the extra-high voltage grid operated by the Transmission System Operator (TSO). Below that from the high over the medium-voltage to the low-voltage grid the Dis-

P. Sauerteig (✉) · K. Worthmann
Faculty of Mathematics and Natural Sciences, Technische Universität Ilmenau, Ilmenau, Germany
e-mail: philipp.sauerteig@tu-ilmenau.de

K. Worthmann
e-mail: karl.worthmann@tu-ilmenau.de

M. Baumann · S. Grundel
Max Planck Institute for Dynamics of Complex Technical Systems, Magdeburg, Germany
e-mail: baumann@mpi-magdeburg.mpg.de

S. Grundel
e-mail: grundel@mpi-magdeburg.mpg.de

J. Dickert
ENSO NETZ GmbH, Dresden, Germany
e-mail: joerg.dickert@enso.de

© Springer Nature Switzerland AG 2021
S. Göttlich et al. (eds.), *Mathematical Modeling, Simulation and Optimization for Power Engineering and Management*, Mathematics in Industry 34,
https://doi.org/10.1007/978-3-030-62732-4_10

tribution System Operator (DSO) provides energy to the consumer. Grid operators have to ensure that this supply chain works at any time. To this end, researchers are investigating various optimization problems in modern smart grids [5, 11, 12, 18, 24].

One class of optimization problems arising in smart grids concerns Optimal Power Flow (OPF) [1, 3, 8]. Typically, the objective in OPF problems is to minimize generation costs, see e.g. [9] and the references therein. However, different objectives have been considered in the literature. Our focus lies on reducing transmission losses within the grid. In [6] the impact of distributed generation on the line losses is studied. For simplicity voltage drop is neglected. An algorithm for loss minimization in radial distribution grids is presented in [14]. In [4] a special case of the capacitor placing problem is tackled, which aims to minimize transmission losses by optimally placing and sizing capacitors within the grid. Due to the problem structure incorporating several local agents nature-inspired algorithms have been applied to grid optimization problems. In [10, 17] particle swarm optimization algorithms are used to minimize transmission losses in power grids. Based on a bacteria foraging optimization algorithm the authors in [23] minimize such losses by optimizing on-load tap changers using Flexible AC Transmission Systems (FACTS). In some energy markets the generation and consumption is scheduled via hourly day-ahead auctions neglecting the grid topology. Therefore, the resulting transmission losses have to be allocated to the buyer or seller of energy, respectively. In [7] four algorithms for ex post transmission loss allocation are compared, i.e. after the losses have occured. The power flow equation incorporates several variables at each node: the complex voltage and the complex power. The ratio between imaginary and real part of the latter is referred to as power factor. See [13, 20] for power factor corrections techniques in a single-phase or three-phase scenario, respectively.

In this paper, we aim to minimize current-dependent transmission losses within the low-voltage DSO grid subject to the power flow equation for online network operation in contrast to ex-post loss allocation. Transmission losses depending on the voltage are not in focus. The novelty of the approach compared to the above-mentioned works lies within the problem formulation. We assume the active power at each node to be given in advance. However, the reactive power will be used as a control parameter, but subject to constraints depending on the active power. More precisely, we assume the power factor to be controllable within certain box constraints being possible with the advent of inverters. With more inverters connected to the grid and the increase of communication DSO's may use this potential of reducing losses. We investigate robustness of the solution of the optimization problem with respect to disturbances in the active power numerically. Note that this is not a rigorous robustness analysis of this optimization problem, which could hopefully be done in the future. However, those disturbances can be interpreted as additional generators or loads within the grid, and therefore, they have practical meaning.

The remainder is structured as follows. First, we provide some basic notion and important relationships with respect to modelling electrical grids in Sect. 10.2. Section 10.3 introduces the considered optimization problem. The main contribution of this paper is a feasibility study based on real-world data of a German DSO presented in Sect. 10.4.

Notation: Throughout this paper we use $\mathbb{N} = \{1, 2, \ldots\}$ and $\mathbb{N}_0 = \mathbb{N} \cup \{0\}$ to denote the natural numbers including and excluding zero, respectively. The set of integer numbers from m to n is given by $[m : n] = \{m, m + 1, \ldots, n\}$ for some integers $m \le n$. Following the notation typically used in electronics we denote the imaginary unit by $j = \sqrt{-1}$.

10.2   Power Flow Equation

In this section we briefly recap some fundamental aspects in modelling power grids. The low-voltage grid consists of several tree-structured graphs each connected to the medium-voltage grid via so-called slack nodes. Trees are undirected connected graphs without circles or loops. In this paper we consider one of these trees $\mathcal{G} = (\mathcal{V}, \mathcal{E})$ with a set of nodes $\mathcal{V}$ and a set of edges $\mathcal{E} \subseteq \mathcal{V} \times \mathcal{V}$. In electrical grids the nodes are referred to as buses, while the edges represent transmission lines. The low-voltage subnet is rooted at a transformer station the slack node, which can be seen as the connection to the medium-voltage grid. We highlight the particular role of the slack node by denoting it by 0. The remainders are enumerated as $[1 : n - 1]$ with $n = |\mathcal{V}| \in \mathbb{N}_{\ge 2}$, i.e. $\mathcal{V} = [0 : n - 1]$ and therefore, $|\mathcal{E}| = n - 1 \in \mathbb{N}$. If we want to highlight that line $E_l \in \mathcal{E}$ connects buses i and $j, i, j \in \mathcal{V}$, we use the notation $E_l = (i, j)$. Figure 10.1 shows a tree with 93 buses in total. The slack node is highlighted via a red square. The triangles are referred to in the numerical simulations in Sect. 10.4, where the impact of minor (green) and major disturbances (red) on the active power demand at a single bus is investigated. In this context we call a node j *following node* of node $i, i, j \in \mathcal{V}$, if the shortest path[1] (and hence, every path) from slack 0 to j includes i. For example, node 92 is a following node of 63.

Each bus $i, i \in \mathcal{V}$, incorporates an active power demand P_i in kW, a reactive power demand Q_i in kvar, and a complex voltage $V_i = |V_i| e^{j\delta_i}$, where $|V_i|$ and δ_i denote the voltage magnitude and angle, respectively. Typically, the voltage $V_0 = |V_0| e^{j\delta_0}$ at the slack node is given by $V_0 = 1$ [p.u.] (per unit), i.e. $|V_0| = 1$ and $\delta_0 = 0$. Note that in our setting, 1 p.u. at the slack node corresponds to 10 kV while the other buses are normalized to $V_i = 1$ p.u. $= 400$ V, $i > 0$. For $i > 0$, we assume that the active power demand is known in advance. Given the power factor $\tan(\varphi_i)$ the reactive power demand is determined by the relation

$$Q_i = P_i \cdot \tan(\varphi_i), \tag{10.1}$$

where φ_i denotes the argument of the complex power $S_i = P_i + jQ_i = |S_i| e^{j\varphi_i}$.

---

[1] A shortest path from i to j in a graph $\mathcal{G} = (\mathcal{V}, \mathcal{E})$ is a set of nodes $i_0 = i, i_1, \ldots, i_k = j$ in $\mathcal{V}$ such that $(i_m, i_{m+1}) \in \mathcal{E}$ for all $m \in [0 : k - 1]$ and k is minimal.

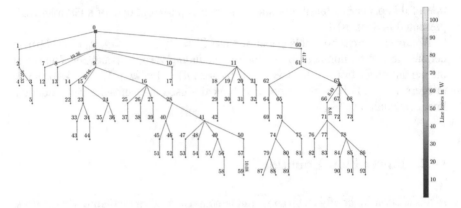

Fig. 10.1 Tree-structured graph with slack node 0 (red square) and 92 (following) nodes

In order to describe how one bus affects another the links between those buses needs to be specified. In graph theory typically, the adjacency matrix $A = (A_{ij}) \in \mathbb{R}^{n \times n}$, given by

$$A_{ij} = \begin{cases} 1, & \text{if } (i, j) \in \mathcal{E} \\ 0, & \text{else,} \end{cases}$$

is used to encode the graph's structure. In electrical engineering a similar technique is used. However, additionally to the topology the (complex) admittance $y_{ij} \in \mathbb{C}$ along the transmission line $(i, j) \in \mathcal{E}$ is encoded via the bus admittance matrix $Y \in \mathbb{R}^{n \times n}$ given by

$$Y_{ij} = \begin{cases} y_i + \sum_{k \in \mathcal{V} \setminus \{i\}} y_{ik}, & \text{if } i = j \\ -y_{ij}, & \text{else.} \end{cases}$$

Here, y_i collects the so-called shunt admittances of node i, i.e. the admittances of linear loads connected to the bus as well as its admittance-to-ground [15]. In [9] the authors neglect so-called shunts. Hence, y_i is omitted. The bus admittance matrix is decomposed into its real and imaginary part, i.e. $Y = G + jB$ with matrices $G = (G_{ij})$, $B = (B_{ij}) \in \mathbb{R}^{n \times n}$, which are parameters of the system.

The Power Flow Equation (PFE) reads as

$$\begin{bmatrix} P_i - |V_i| \sum_{j=1}^{n} |V_j|(G_{ij} \cos(\delta_{ij}) + B_{ij} \sin(\delta_{ij})) \\ Q_i - |V_i| \sum_{j=1}^{n} |V_j|(G_{ij} \sin(\delta_{ij}) - B_{ij} \cos(\delta_{ij})) \end{bmatrix} = 0 \quad \forall i \in \mathcal{V}, \qquad (10.2)$$

see also [8]. Note that (10.2) consists of $2n$ equations with $4n$ variables namely $(|V_i|, \delta_i, P_i, Q_i)^\top \in \mathbb{R}^4$ and $\delta_{ij} = \delta_i - \delta_j$ denotes the angle difference for $i, j \in \mathcal{V}$.

10.3   Optimization Problem

The transportation of energy along transmission lines comes along with losses depending on the length and material of the line and the amount of the current flow. Our goal is to minimize these losses. To this end, we consider the objective function $P_L : \mathbb{C}^{n-1} \to \mathbb{R}$,

$$P_L(I) = 3 \cdot \sum_{l=1}^{n-1} R'_l \ell_l \, |I_l|^2 . \tag{10.3}$$

Here, R'_l and ℓ_l denote the specific resistance in Ω/km and the length in m while $|I_l|$ represents the magnitude of the complex current along line E_l, $l \in [1 : n-1]$, in A. Therefore, P_L represents the aggregated transmission losses in W within $\mathcal{G}$. The factor 3 reflects the fact that the lines consist of three phases. However, it does not affect the optimal solution. According to *Ohm's law* the input and output currents I_i^{in} and I_i^{out} at each bus i, $i \in \mathcal{V}$, can be determined via $\check{I} = YV$. These currents again yield the current I_l along each transmission line l, $l \in [1 : n-1]$, i.e. $I = (I_1, \dots, I_{n-1})^\top \in \mathbb{C}^{n-1}$. Alternatively, one can use the *from* and *to* admittance matrices Y^f, $Y^t \in \mathbb{C}^{(n-1)\times n}$ to compute the current via

$$I \;=\; \frac{1}{\sqrt{3}} \, \max\{|Y^f \cdot V|, |Y^t \cdot V|\},$$

where $|\cdot|$ and max are to be understood component-wise, see also [15, 22].

In the DSO grid the active power P_i describes the active power demand of a household located at node i, $i > 0$. In the following we consider P_i, $i > 0$, to be given. Furthermore, we assume the buses $i > 0$ to be able to control the reactive power by setting the power factor $\mu_i = \cos(\varphi_i)$ in a range

$$\underline{\mu}_i \;\le\; \mu_i \;\le\; \overline{\mu}_i$$

for some $\underline{\mu}_i \le \overline{\mu}_i$, $i > 0$. Note that the power factors can only be set indirectly by manipulating inverters. This work aims to check whether such manipulations are worth the effort for generators. Loads connected via inverters to the grid may also be governed in this way.

Typically, the grid operator has to make sure that the voltages stay within some corridors, which can be enforced by introducing the constraints

$$\underline{V}_i \;\le\; |V_i| \;\le\; \overline{V}_i$$

for some constants $\underline{V}_i \le \overline{V}_i$ for all $i > 0$. Having the interface to the upper grid level in mind, we assume some bounds on Q_0 to be given by the DSO, i.e.

$$\underline{Q}_0 \leq Q_0 \leq \overline{Q}_0$$

for some constants $\underline{Q}_0 \leq \overline{Q}_0$.

Taking these considerations into account yields the optimization problem

$$\min_{\mu_i \in [\underline{\mu}_i, \overline{\mu}_i]} \quad P_L(I) = 3 \cdot \sum_{l=1}^{n-1} R'_l \ell_l \, |I_l|^2 \tag{10.4a}$$

subject to $\forall i \in \mathcal{V}$:

$$\begin{bmatrix} P_i - |V_i| \sum_{j=1}^{n} |V_j| (G_{ij} \cos(\delta_{ij}) + B_{ij} \sin(\delta_{ij})) \\ P_i \cdot \tan(\varphi_i) - |V_i| \sum_{j=1}^{n} |V_j| (G_{ij} \sin(\delta_{ij}) - B_{ij} \cos(\delta_{ij})) \end{bmatrix} = 0 \tag{10.4b}$$

$$I = \frac{1}{\sqrt{3}} \max\{|Y^{\mathrm{f}} \cdot V|, |Y^{\mathrm{t}} \cdot V|\} \tag{10.4c}$$

$$\underline{V}_i \leq |V_i| \leq \overline{V}_i \quad \forall i > 0 \tag{10.4d}$$

$$\mu_i = \cos \varphi_i \quad \forall i \in \mathcal{V} \tag{10.4e}$$

$$|V_0| = 1, \quad \delta_0 = 0 \tag{10.4f}$$

$$\underline{Q}_0 \leq P_0 \tan(\varphi_0) \leq \overline{Q}_0 \tag{10.4g}$$

for given local active power $P_i, i > 0$. Here, we directly replaced Q_i via (10.1), $i \in \mathcal{V}$. Note that we omit constraints on the voltage angles $\delta_i, i \in \mathcal{V}$. Due to constraint (10.4b) the problem becomes non-convex.

Remark 1 One might be interested in additionally reducing the voltage losses. This, however, is another objective and, hence, would yield a multiobjective optimization problem, see e.g. [21]. Since we are only interested in a proof of concept we leave voltage loss reduction for future research.

10.4 Numerical Simulations

In this section we investigate the impact of controlling the local power factors with respect to transmission losses based on a Monte Carlo simulation with 100 runs. To this end, we first provide some details on implementation and then, illustrate our results.

10.4.1 Details on Implementation

Throughout our numerical simulations we use the parameters as listed in Table 10.1. Note that we do not provide any lower bound on Q_0. The values R'_l, ℓ_l as well as the grid topology have been provided by our industrial partner. The P_i were generated

Table 10.1   Parameters used in the implementation

$\underline{\mu}_i$	0.9	$\underline{V}_i$	0.9 [p.u.]	$\underline{Q}_0$	$-\infty$
$\overline{\mu}_i$	1	$\overline{V}_i$	1.1 [p.u.]	$\overline{\overline{Q}}_0$	-100 [kvar]

Table 10.2   `fmincon` setting

Option	Setting
`Algorithm`	`interior-point`
`MaxFunctionEvaluations`	`1e5`
`StepTolerance`	`1e-16`

randomly using Gaussian distribution with expected value $\mathbb{E}(P) = 2.5$ kW and standard deviation $\sigma(P) = 0.5$ kW. We use the MATLAB package `matpower` to solve the PFE (10.2) and MATLAB embedded function `fmincon` to solve (10.4), which provides several state-of-the-art solvers for constraint non-convex optimization. The `fmincon` setting is displayed in Table 10.2. For an introduction to interior-point methods and an advanced version we refer to [19, 25], respectively.

Remark 2 Note that constraint (10.4c) is non-smooth and in our simulations $\left\| |Y^f \cdot V| - |Y^t \cdot V| \right\|_\infty \approx 10^{-12}$ holds. For computational efficiency one could replace the maximum by the mean value yielding a smooth formulation without distorting the result drastically. However, since numerical solvers do not distinguish between smooth and non-smooth functions but only evaluate them at certain points and therefore see a basically smooth function we choose to stick with the more precise max formulation.

We distinguish three scenarios with respect to controllability of the power factors:

1. $\cos(\varphi_i) = 0.9$ for all $i > 0$ (no optimization)
2. $\cos(\varphi_i) = \mu^*$ for all $i > 0$ (1-dimensional optimization)
3. $\cos(\varphi_i) = \mu_i^*$ for all $i > 0$ (n-dimensional optimization).

The first one serves as a reference scenario. In the second one we investigate the impact of setting all power factors μ_i to the same (optimal) value. Then, in the third scenario we study the improvement of the overall performance if the power factors within the grid may differ. The corresponding objective function value is denoted by P_L^{ref}, P_L^{1D}, and P_L^{nD}, respectively. Note that we only consider the stationary problem at one fixed time step.

For the computation of the bus admittance matrices Y, Y^f and Y^t we use the open-source software PandaPower [22].

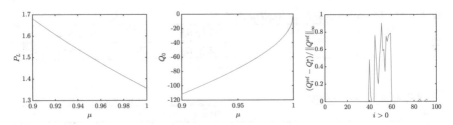

Fig. 10.2 Impact of the choice of μ on P_L (left) and Q_0 (middle) for the 1-dimensional unconstraint optimization problem and comparison of reactive power Q_i^{ref} of the reference scenario and $Q_i^\star$ of the solution of the n-dimensional problem

10.4.2 Results

Ignoring the constraints in (10.4) for a moment, Fig. 10.2 (left) shows that in the 1-dimensional case the objective function is decreasing in μ. Hence, the optimum $\mu^\star$ is attained at the upper bound of the feasible set. Furthermore, the absolute value of the reactive power Q_0 at the slack node is decreasing as well. We tested numerically, that solving (10.4) without considering constraint (10.4g) the optimal solution for both Scenario 2 and 3 is the trivial solution $\mu_i^\star = 1$ yielding $Q_i = 0$ for all $i > 0$, i.e. no constraint is active. If we incorporate (10.4g), i.e. the bounds on Q_0 set by the DSO, the solution becomes non-trivial. In particular, the lower bound $\underline{\mu}_i \leq \mu_i$ becomes active, i.e. $\mu_i^\star = \underline{\mu}_i$, for some $i > 0$, see e.g. Fig. 10.4 (top right). Then, the mean value $\bar{\mu}^\star = \sum_{i>0} \mu_i^\star$ in the n-dimensional case, however, is close to the 1-dimensional solution $\mu_{1D}^\star$. Note that $\mu_{1D}^\star$ can be obtained by truncating the feasible set $[\underline{\mu}, \overline{\mu}]$ at $Q_0^{-1}(\overline{Q}_0)$, where $Q_0 : [\underline{\mu}, \overline{\mu}] \to \mathbb{R}$ is considered as function in $\mu^\star$ as depicted in Fig. 10.2 (middle).

Furthermore, Fig. 10.2 (right) shows the relative difference between the reactive power Q_i^{ref} of the reference scenario and the optimal reactive power $Q_i^\star$. There are two things to be mentioned. First of all, $Q_i^{\mathrm{ref}} \geq Q_i^\star$ holds for all $i > 0$, i.e. reactive power is reduced. Second, only a particular region is affected by the optimization, namely the following nodes of node 28, see also Fig. 10.1. In other words, for the remaining buses the lower bound $\mu_i^\star = 0.9$ is active, see also Fig. 10.3 (top right).

In Fig. 10.3 the impact of slightly changing the active power at a single node is visualized. To this end, we set $P_i = 5$ kW, $i \in \{6, 12, 60, 61\}$, see Fig. 10.1 for the location of these nodes in the grid (green triangles). We make the following observations. (Slightly) Changing the active power at a single node changes the reactive power accordingly while the voltage and the optimal power factor do not change drastically. The corresponding (optimal) objective function values are listed in Table 10.3. The objective function itself does not change drastically. However, it can be improved by approximately 80 and 110 W in average in the 1-dimensional and n-dimensional case, respectively, if one bus is manipulated. Furthermore, the active power at the slack node does slightly change, if the active power at one bus is manipulated, while in all (tested) cases the reactive power takes the value of its

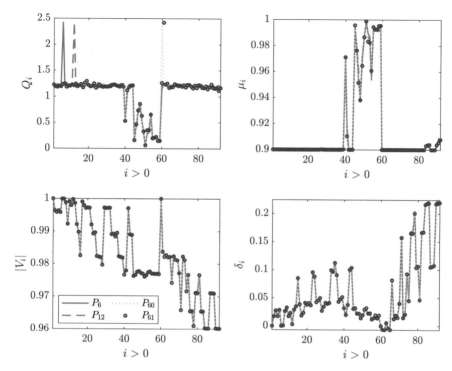

Fig. 10.3 Impact of changing the active power at one node i to $P_i = 5$ kW (minor disturbance)

Table 10.3 Impact of manipulating the active power at a single bus or increasing the variance of the active power within the whole grid on the objective function value. All values are given in kW

Setting	P_L^{ref}	P_L^{1D}	P_L^{nD}
No manipulation	1.72	1.65	1.62
$P_6 = 5$	1.75	1.67	1.64
$P_{12} = 5$	1.76	1.68	1.65
$P_{60} = 5$	1.72	1.64	1.61
$P_{61} = 5$	1.74	1.66	1.63
$P_{63} = 5$	1.73	1.66	1.63
$P_{63} = 10$	1.76	1.68	1.64
$P_{63} = 20$	1.87	1.75	1.71
$P_{63} = 50$	2.16	1.96	1.93
std $= 1$	1.74	1.67	1.64
std $= 2$	1.79	1.71	1.68
std $= 4$	1.82	1.73	1.69
std $= 8$	1.88	1.78	1.73

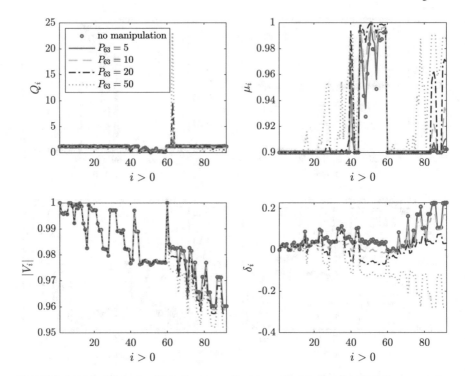

Fig. 10.4 Impact of changing the active power P_i at one node i (major disturbance)

upper bound $\overline{Q}_0$. These numerical tests all show that even the simple one dimensional optimization problems improves the objective value.

Figure 10.4 illustrates what happens if there is a strong disturbance at bus 63, see Fig. 10.1 (red triangle). First of all the corresponding reactive power also changes drastically while the others do not. Furthermore, it affects not only the voltage at node 63 and its following nodes but also the neighbouring subtree rooted at node 62, see Fig. 10.4 (bottom). Moreover, the corresponding active power P_0 at the slack changes significantly. Additionally, Figs. 10.5 and 10.6 illustrate the impact of a strong disturbance ($P_{63} = 50$ kW) on the line losses in the corresponding subtree and the whole grid, respectively. Figure 10.5 shows how the line losses in the subtree change due to the disturbance from (top) to (middle). For the sake of clarity we only denoted the lines where the loss is larger than 5 W. In particular the loss along transmission line (60, 61) increases drastically from 208.74 W (top) to 537.43 W (middle). Solving (10.4) leads clearly reduced losses as depicted in Fig. 10.5c. Furthermore, we subtracted the losses after (see Fig. 10.5c) from the losses before optimization (see Fig. 10.5b) for the whole grid and plotted the results in Fig. 10.6. One can see that the optimization routine mitigate the losses along each line, i.e. the difference is positive at each line. Note that losses are reduced in all three subtrees, see e.g. lines (9, 15) and (57, 59). Furthermore, there is a major reduction along line (2, 4).

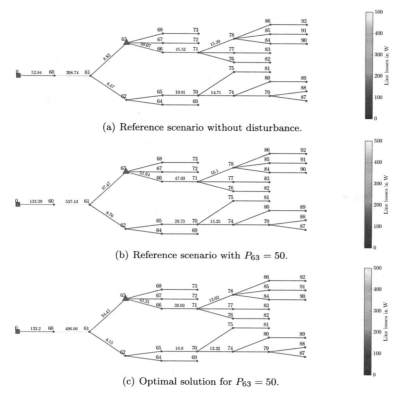

(a) Reference scenario without disturbance.

(b) Reference scenario with $P_{63} = 50$.

(c) Optimal solution for $P_{63} = 50$.

Fig. 10.5 Reference scenario, i.e. $P_i \sim \mathcal{N}(2.5, 0.5)$, (top), reference scenario, i.e. $\mu_i = 0.9, i > 0$, with strong manipulation $P_{63} = 50$ kW (middle), and solution of n-dimensional scenario with $P_{63} = 50$ kW (bottom)

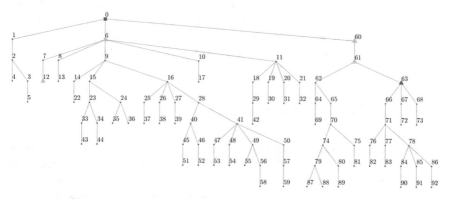

Fig. 10.6 Difference between line losses before and after (n-dimensional) optimization with $P_{63} = 50$ kW

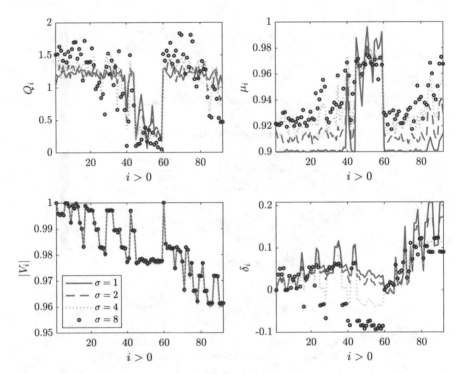

Fig. 10.7 Impact of increasing the variation of the P_i. Here, σ denotes the corresponding standard deviation of $P_i \sim \mathcal{N}(\mathbb{E}(P_i), \sigma)$ in kW

Moreover, Fig. 10.7 shows what happens if the P_i's are more widely spread, i.e. the variance is increased. With increasing variance of the active power distribution, the variance of the reactive power, the control value and the voltage angle increase as well. However, the impact on the voltage magnitude is negligible. The corresponding results in Table 10.3 confirm that the resulting losses can be reduced.

10.5   Conclusions and Outlook

We considered an optimization problem aiming to reduce losses online by governing the power factor in electrical grids. Our numerical results show that controlling the power factor can reduce these losses. We investigated numerically, how the performances are affected by (random) disturbances in the active power demand.

So far, we assumed the active power P_i, $i > 0$, to be given as an exogenous input. The model introduced in [26] can be used to describe the local (active) power demand P_i. Here, the authors assume the load and generation at each bus to be predictable. Furthermore, some of the buses are equipped with energy storage devices which can be controlled in a distributed manner [16] in order to manipulate the local active power demand. Combining this model with the presented approach yields more flexibility and hence, seems promising in order to improve the overall performance.

Another point left for future research is the frequency of the power factor update. In this paper, we addressed the problem of optimizing the power factor for a single time instant. However, in real-life applications one has to deal with time series, e.g. with 96 instances per day. In that case it is worth investigating whether the optimization problem has to be solved every 15 minutes or if it suffices to solve it once a day.

Acknowledgements The authors acknowledge funding by the Federal Ministry of Education and Research within the project KONSENS (BMBF; grants 05M18SIA and 05M18EVA). Karl Worthmann is also indebted to the German Research Foundation (DFG; grant WO 2056/6-1).

References

1. Abido, M.: Optimal power flow using particle swarm optimization. Int. J. Electr. Power Energy Sys. **24**, 563–571 (2002)
2. Aschenbruck, T., Baumann, M., Esterhuizen, W., Filipecki, B., Grundel, S., Helmberg, C., Ritschel, T.K.S., Sauerteig, P., Streif, S., Worthmann, K.: Optimization and Stabilization of Electrical Networked Systems. In: Göttlich, S., Herty, M., Milde, A. (eds.) Mathematical MSO for Power Engineering and Management, Mathematics in Industry. Springer (2021)
3. Bakirtzis, A.G., Biskas, P.N., Zoumas, C.E., Petridis, V.: Optimal power flow by enhanced genetic algorithm. IEEE Trans. Power Syst. **17**(2), 229–236 (2002)
4. Baran, M., Wu, F.F.: Optimal sizing of capacitors placed on a radial distribution system. IEEE Trans. Power Deliv. **4**(1), 735–743 (1989)
5. Carrasco, J.M., García Franquelo, L., Bialasiewicz, J.T., Galván, E., Portillo Guisado, R.C., Martín Prats, M.A., Moreno-Alfonso, N.: Power-electronic systems for the grid integration of renewable energy sources: a survey. IEEE Trans. Ind. Electron. **53**(4), 1002–1016 (2006)
6. Chiradeja, P.: Benefit of distributed generation: a line loss reduction analysis. In: 2005 IEEE/PES Transmission Distribution Conference Exposition: Asia and Pacific, pp. 1–5 (2005)
7. Conejo, A.J., Arroyo, J.M., Alguacil, N., Guijarro, A.L.: Transmission loss allocation: a comparison of different practical algorithms. IEEE Trans. Power Syst. **17**(3), 571–576 (2002)
8. Dommel, H.W., Tinney, W.F.: Optimal power flow solutions. IEEE Trans. Power Appar. Syst. **87**(10), 1866–1876 (1968)
9. Engelmann, A., Jiang, Y., Mühlpfordt, T., Houska, B., Faulwasser, T.: Towards distributed OPF using ALADIN. IEEE Trans. Power Syst. **34**(1), 584–594 (2019)
10. Esmin, A.A.A., Lambert-Torres, G., Zambroni de Souza, A.C.: A hybrid particle swarm optimization applied to loss power minimization. IEEE Trans. Power Syst. **20**(2), 859–866 (2005)
11. Fang, X., Misra, S., Xue, G., Yang, D.: Smart grid - the new and improved power grid: a survey. IEEE Commun. Surv. Tutor. **14**(4), 944–980 (2012)
12. Gao, J., Xiao, Y., Liu, J., Liang, W., Chen, C.P.: A survey of communication/networking in Smart Grids. Futur. Gener. Comput. Syst. **28**(2), 391–404 (2012)
13. García, O., Cobos, J.A., Prieto, R., Alou, P., Uceda, J.: Single phase power factor correction: a survey. IEEE Trans. Power Electron. **18**(3), 749–755 (2003)
14. Goswami, S.K., Basu, S.K.: A new algorithm for the reconfiguration of distribution feeders for loss minimization. IEEE Trans. Power Deliv. **7**(3), 1484–1491 (1992)
15. Grainger, J.J., Stevenson, W.D.: Power System Analysis, 1 edn. McGraw-Hill, Inc. (1994)
16. Jiang, Y., Sauerteig, P., Houska, B., Worthmann, K.: Distributed optimization using ALADIN for model predictive control in smart grids. IEEE Trans. Control Syst. Technol. https://doi.org/10.1109/TCST.2020.3033010 (early access)

17. Leeton, U., Uthitsunthorn, D., Kwannetr, U., Sinsuphun, N., Kulworawanichpong, T.: Power loss minimization using optimal power flow based on particle swarm optimization. In: ECTI-CON2010: The 2010 ECTI International Confernce on Electrical Engineering/Electronics, Computer, Telecommunications and Information Technology, pp. 440–444 (2010)

18. Liserre, M., Sauter, T., Hung, J.Y.: Future energy systems: integrating renewable energy sources into the smart power grid through industrial electronics. IEEE Ind. Electron. Mag. **4**(1), 18–37 (2010)

19. Nocedal, J., Wright, S.J.: Numerical Optimization. Springer (2006)

20. Prasad, A.R., Ziogas, P.D., Manias, S.: An active power factor correction technique for three-phase diode rectifiers. IEEE Trans. Power Electron. **6**(1), 83–92 (1991)

21. Sauerteig, P., Worthmann, K.: Towards multiobjective optimization and control of smart grids. Optim. Control Appl. Methods (OCAM) **41**, 128–145 (2020)

22. Thurner, L., Scheidler, A., Schafer, F., Menke, J.H., Dollichon, J., Meier, F., Meinecke, S., Braun, M.: pandapower - an open source python tool for convenient modeling, analysis and optimization of electric power systems. IEEE Trans. Power Syst. (2018). https://doi.org/10.1109/TPWRS.2018.2829021

23. Tripathy, M., Mishra, S., Lai, L.L., Zhang, Q.P.: Transmission Loss Reduction Based on FACTS and Bacteria Foraging Algorithm. In: Runarsson, T.P., Beyer, H.G., Burke, E., Merelo-Guervós, J.J., Whitley, L.D., Yao, X. (eds.) Parallel Problem Solving from Nature - PPSN IX, pp. 222–231. Springer, Berlin Heidelberg, Berlin, Heidelberg (2006)

24. Vardakas, J.S., Zorba, N., Verikoukis, C.V.: A survey on demand response programs in smart grids: pricing methods and optimization algorithms. IEEE Commun. Surv. Tutor. **17**(1), 152–178 (2015)

25. Wächter, A., Biegler, L.T.: On the implementation of a primal-dual interior point filter line search algorithm for large-scale nonlinear programming. Math. Program. **106**(1), 25–57 (2006)

26. Worthmann, K., Kellett, C., Braun, P., Grüne, L., Weller, S.: Distributed and decentralized control of residential energy systems incorporating battery storage. IEEE Trans. Smart Grid **6**(4), 1914–1923 (2015)

Chapter 11
MathEnergy – Mathematical Key Technologies for Evolving Energy Grids

Tanja Clees, Anton Baldin, Peter Benner⬤, Sara Grundel⬤,
Christian Himpe⬤, Bernhard Klaassen, Ferdinand Küsters,
Nicole Marheineke, Lialia Nikitina, Igor Nikitin, Jonas Pade, Nadine Stahl,
Christian Strohm, Caren Tischendorf, and Andreas Wirsen⬤

Abstract For a sustainable and CO_2 neutral power supply, the entire energy cycles for power, gas and heating grids have to be taken into account simultaneously. Despite rapid progress, the energy industry is insufficiently equipped for the superordinate planning, monitoring and control tasks, based on increasingly large and coupled network simulation models. The German MathEnergy project aims to overcome these shortcomings by developing selected mathematical key technologies for energy networks and respective software. This chapter gives an overview of MathEnergy by discussing selected new developments related to model order reduction for gas networks, state estimation for gas and power networks, as well as cross-sectoral

T. Clees · A. Baldin · B. Klaassen · L. Nikitina · I. Nikitin
SCAI: Fraunhofer SCAI, 53754 Sankt Augustin, Germany

T. Clees (✉)
Bonn-Rhine-Sieg University of Applied Sciences, 53754 Sankt Augustin, Germany
e-mail: tanja.clees@h-brs.de

P. Benner · S. Grundel · C. Himpe
MPI: Max Planck Institute for Dynamics of Complex Technical Systems, Sandtorstraße 1, 39106
Magdeburg, Germany
e-mail: benner@mpi-magdeburg.mpg.de

S. Grundel
e-mail: grundel@mpi-magdeburg.mpg.de

C. Himpe
e-mail: himpe@mpi-magdeburg.mpg.de

F. Küsters · A. Wirsen
ITWM: Fraunhofer Institute for Industrial Mathematics ITWM, Fraunhofer-Platz 1, 67663
Kaiserslautern, Germany
e-mail: andreas.wirsen@itwm.fraunhofer.de

N. Marheineke · N. Stahl
UTrier: Universität Trier, Universitätsring 15, 54296 Trier, Germany
e-mail: marheineke@uni-trier.de

J. Pade · C. Strohm · C. Tischendorf
HUB: Humboldt University of Berlin, Unter den Linden 6, 10099 Berlin, Germany
e-mail: caren.tischendorf@math.hu-berlin.de

© Springer Nature Switzerland AG 2021
S. Göttlich et al. (eds.), *Mathematical Modeling, Simulation and Optimization
for Power Engineering and Management*, Mathematics in Industry 34,
https://doi.org/10.1007/978-3-030-62732-4_11

233

modeling, simulation and ensemble analysis. Several new theoretical results as well as related software prototypes are introduced. Results for selected gas and power networks are presented, including a first version of the partDE-Hy demonstrator for analysis of power-to-gas scenarios.

11.1 Overview

For a sustainable and CO_2 neutral power supply, cross-sectoral approaches have to be developed, taking into account the entire energy cycles for power, gas and heating grids. Not only vertical communication between grid levels, but also horizontal communication between the energy sources is needed. The necessary planning, monitoring and control tasks have to be based on increasingly large and coupled network simulation models. Hence, respective theoretical investigations, numerical methods and related software modules for solving arising systems of (partial) differential-algebraic equations are needed.

The German MathEnergy project (2016–2021; for a reference, see Acknowledgements) aims to overcome related shortcomings by developing selected mathematical key technologies and respective software, see also Fig. 11.1. The prototypical software modules shall support an efficient solution of hierarchical, parametric, nonlinear, switched, dynamic grid models with stochastically varying parameters. Moreover, the project contributes to the development of workflows and standardizations for the integrated simulation and analysis of cross-sectoral scenarios. Several demonstrators are developed for presenting and discussing the newly developed workflows.

This chapter gives an overview of MathEnergy by discussing selected new developments related to model order reduction for gas networks (Sect. 11.2), state estimation for gas networks (Sect. 11.3) and power networks (Sect. 11.6) as well as cross-sectoral modeling, simulation and analysis (Sects. 11.4 and 11.5). Several new theoretical results as well as related software prototypes are introduced. Results for selected gas, power and coupled networks are presented, including a first version of the partDE-Hy demonstrator for analysis of power-to-gas scenarios.

In other chapters of this volume, more MathEnergy-related developments are discussed, namely *Applying Stein's two-stage procedure to uncertainty determination for gas quality tracking systems* and *Improving the accuracy of compressor characteristic diagrams under scarcity of data points and a given computing power*.

11.2 Modeling and Model Order Reduction

The *Model Order Reduction (for Gas Transport Networks)* sub-project of the *Math-Energy* project investigates the modeling, simplification and dimension reduction of the entailing numerical gas network representations. While the model has to incorporate all relevant physical aspects of the gas transport process, it also needs to be

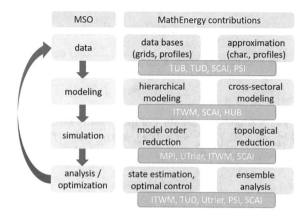

Fig. 11.1 Overview on MathEnergy: on the left, the typical steps of modeling-simulation-optimization (MSO) workflows are listed; on the right, decisive parts of the project and partners contributing to them (to a larger amount) are shown and linked to the MSO steps. In green, the project partners are listed: TUB=Techn. Univ. Berlin, TUD=Techn. Univ. Dortmund, PSI=PSI AG, others: see list of authors

suitable for repeated numerical simulation on short time horizons. To this end, a partial differential-algebraic equation model is developed, spatially discretized, simplified via analytic index reduction or numerical decoupling to a system of ordinary differential equations, and finally reduced in dimensionality via data-driven model reduction.

11.2.1   Modeling

The modeling is exhaustively described in [10] and subsequently presented in abbreviated form. We start with a graph $G = (\mathcal{E}, \mathcal{N})$ representing the pipe network. The set of edges $\mathcal{E}$ corresponds to the pipes and the set of nodes $\mathcal{N}$ either to inlets, outlets or junctions. The pipes are assumed to be circular with diameter d_k, cross-sectional area $a_k = \frac{\pi}{4}d_k^2$, length L_k and a given geodesic incline $\partial_x h_k(x)$ for k indexing $\mathcal{E}$. The time dependent variables are the pressure $p_k(x, t)$, the density $\rho_k(x, t)$ and the mass flow $q_k(x, t)$, where $x \in [0, L_k]$ is the one dimensional spatial coordinate and $t \in [t_0, t_1]$ the time. To simplify the equations, the temperatures T_k are assumed to be constant. The gas flow in each pipe is then modeled by the isothermal Euler equations. Conservation of mass yields the continuity equation (11.1a) and conservation of momentum yields the pressure loss equation (11.1b). Here, g is the gravitational acceleration and λ is a friction factor combining all resistance forces acting on the gas inside the pipe. Furthermore, the relation between the density and the pressure is described by the state equation of a gas (11.1c),

$$\partial_t \rho_k + \frac{1}{a_k} \partial_x q_k = 0, \tag{11.1a}$$

$$\partial_t q_k + a_k \partial_x p_k + a_k g \rho_k \partial_x h_k(x) = -\frac{\lambda(q_k)}{2 d_k a_k} \frac{q_k |q_k|}{\rho_k}, \tag{11.1b}$$

$$p_k = T_k R_S z(p_k, T_k) \rho_k, \tag{11.1c}$$

for the specific gas constant R_S, and the compressibility factor z.

We define by $q_L^k := q_k(0, t)$, $p_L^k := p(0, t)$, and $q_R^k := q_k(L_k, t)$, $p_R^k := p_k(L_k, t)$ the flux and the pressure at the left and right end points of the pipe. With that and the sets

$$\mathcal{I}_R^i = \{k \in \mathcal{E} | \text{ the right end of } k \text{ is } i\},$$
$$\mathcal{I}_L^i = \{k \in \mathcal{E} | \text{ the left end of } k \text{ is } i\},$$

which are defined for every node $i \in \mathcal{N}$, the coupling conditions at the nodes are given by:

$$\sum_{k \in \mathcal{I}_R^i} q_R^k(t) - \sum_{k \in \mathcal{I}_L^i} q_L^k(t) = d_q^i(t), \qquad i \text{ demand or interior node}, \tag{11.2}$$

$$p^i(t) = p_s^i(t), \qquad i \text{ pressure supply}, \tag{11.3}$$

for given demand flux d_q^i and supply pressures s_p^i, where $d_q^i \equiv 0$ defines an interior node. Discretizing equation (11.1) via finite differences with respect to the spatial derivatives leads to

$$\partial_t p_*^k + \frac{\gamma z}{a_k} \frac{q_R^k - q_L^k}{L_k} = 0, \tag{11.4a}$$

$$\partial_t q_*^k + a_k \frac{p_R^k - p_L^k}{L_k} + a_k g \frac{p_*^k}{\gamma z} \partial_x h_k = -\frac{\lambda(q_*^k) \gamma z}{2 d_k a_k} \frac{q_*^k |q_*^k|}{p_*^k}. \tag{11.4b}$$

There are many possible ways to define p_*^k and q_*^k. In [34], we pick them both to be the midpoint, and in [33], we pick them as the endpoints $q_*^k = q_L^k$ and $p_*^k = p_R^k$. In both cases, the discrete equations (11.4) together with the coupling conditions result in a differential-algebraic equation (DAE) first, which can be decoupled yielding an ordinary differential equation (ODE). In order to write the equations more concisely, we introduce the vectors p, q_L, q_+, where p collects the pressures at all interior and demand nodes, while q_L, q_+ collect all fluxes at the left end of a pipe and the average flux of the left and right flux of each pipe, respectively. Furthermore, the vectors d_q and s_p collect the supply pressures and demand fluxes. For the endpoint discretization, the resulting ODEs read:

$$\dot{p} = D_c(Aq_L - q_{\text{set}}), \tag{11.5a}$$

$$\dot{q}_L = -D_\alpha(A^{\mathsf{T}}p + A_S^{\mathsf{T}}p_{\text{set}}) - g_1(p, p_{\text{set}}, q_L), \tag{11.5b}$$

with a sparse matrix D_c, a diagonal matrix D_α, and $g_1(p, p_S, q_L)$ a nonlinear function representing the friction effect from Eq. (11.4b), while for the midpoint discretization, the resulting ODEs are given by:

$$|A|C_\beta|A^{\mathsf{T}}|\dot{p} = 4Aq_+ - 4q_{set} - |A|C_\beta|A_S^{\mathsf{T}}|\dot{p}_{set}, \tag{11.6a}$$

$$\dot{q}_+ = -C_\alpha(A^{\mathsf{T}}p + A_S^{\mathsf{T}}p_{\text{set}}) - g_2(p, p_{\text{set}}, q_+), \tag{11.6b}$$

with a sparse matrix D_c, diagonal matrices D_α, C_α, C_β, and $g_2(p, p_{\text{set}}, q_+)$ nonlinear functions representing the friction effect from Eq. (11.4b). The matrices A and A_S^{T} are the incidence matrix of the graph for the demand and interior nodes and for the supply nodes, respectively.

We see here, that the final model of a pipeline-only network in the two above presented discretizations could be rewritten as an ODE. In both cases, this decoupling was done analytically, which is feasible due to the specific structure of the individual DAEs. However, using other discretizations, for example, a discretization with Riemann invariants from [31], it is not directly obvious if such a decoupling exists. In light of this, [6, 7] present an algorithm which assembles projections for arbitrary differential-algebraic systems with the Euler equation structure, such that we can decouple the system into an ODE and an uni-directionally coupled algebraic equation.

11.2.2 Model Order Reduction

The gas network model is a two-dimensional, hyperbolic, nonlinear, parametric (in temperature and specific gas constant, encoded in θ) system of partial DAEs. After the previous spatial discretization and index-reduction or decoupling, a high-dimensional (input-output) system of stiff, nonlinear, parametric ODEs remains to be simulated to obtain the sought quantities of interest s_q, d_p (mass-flux at inlets and pressure at outlets) for given boundary conditions s_p, d_q (pressure at inlets and mass-flux at outlets):

$$\begin{pmatrix} E_p & 0 \\ 0 & I_q \end{pmatrix}\begin{pmatrix} \dot{p} \\ \dot{q}_* \end{pmatrix} = \begin{pmatrix} 0 & A_p(\theta) \\ A_q(\theta) & 0 \end{pmatrix}\begin{pmatrix} p \\ q_* \end{pmatrix} + \begin{pmatrix} 0 & B_p(\theta) \\ B_q(\theta) & 0 \end{pmatrix}\begin{pmatrix} s_p \\ d_q \end{pmatrix} + \begin{pmatrix} 0 \\ g(p, q_*, s_p, \theta) \end{pmatrix},$$

$$\begin{pmatrix} s_q \\ d_p \end{pmatrix} = \begin{pmatrix} 0 & C_q \\ C_p & 0 \end{pmatrix}\begin{pmatrix} p \\ q_* \end{pmatrix}. \tag{11.7}$$

Yet, since many simulations for varying parameters and boundary values need to be performed to ensure safe operations and guaranteed delivery of denominations, their computation time becomes prohibitive, especially in a time-constrained setting, such as forecasting supply and demand for the next 24 hours.

To accelerate the simulation run-time, model order reduction (MOR) is utilized to reduce the dimensionality of the input-output system (11.7), which determines a surrogate system approximating the input-output (boundary-value/quantity-of-interest) mapping. The underlying MOR technique used is *structured projection-based MOR* [9, 34]. This means that (Petrov-)Galerkin projections U_p, U_q, V_p, V_q are constructed associated to p and q, and applied as follows:

$$\begin{cases} \tilde{p} := V_p p, & p \approx U_p \tilde{p}, & V_p U_p = \text{id}, \\ \tilde{q}_* := V_q q_*, & q_* \approx U_q \tilde{q}_*, & V_q U_q = \text{id}, \end{cases}$$

yielding a low-dimensional reduced state $(\tilde{p} \ \ \tilde{q}_*)^{\mathsf{T}}$ with $\dim(\tilde{p}(t)) \ll \dim(p(t))$ and $\dim(\tilde{q}_*(t)) \ll \dim(q_*(t))$. Given this approximation of the state, the vector field and output functional are adapted by applying the reconstructing projections U_p, U_q, to the pre-image of the operators, and the reducing projections V_p, V_q to their image. For the linear components, a reduced operator is pre-computed (denoted by $\tilde{\cdot}$), i.e., $\tilde{E}_p := V_p E_p U_p$, while for the nonlinear term the composition remains:

$$\begin{pmatrix} \tilde{E}_p & 0 \\ 0 & \tilde{I}_q \end{pmatrix} \begin{pmatrix} \dot{\tilde{p}} \\ \dot{\tilde{q}}_* \end{pmatrix} = \begin{pmatrix} 0 & \tilde{A}_p(\theta) \\ \tilde{A}_q(\theta) & 0 \end{pmatrix} \begin{pmatrix} \tilde{p} \\ \tilde{q}_* \end{pmatrix} + \begin{pmatrix} 0 & \tilde{B}_p(\theta) \\ \tilde{B}_q, (\theta) & 0 \end{pmatrix} \begin{pmatrix} s_p \\ d_q \end{pmatrix} + \begin{pmatrix} 0 \\ V_q f_q(U_p \tilde{p}, U_q \tilde{q}_*, s_p, \theta) \end{pmatrix},$$

$$\begin{pmatrix} s_q \\ d_p \end{pmatrix} = \begin{pmatrix} 0 & \tilde{C}_q \\ \tilde{C}_p & 0 \end{pmatrix} \begin{pmatrix} \tilde{p} \\ \tilde{q}_* \end{pmatrix}.$$

In a subsequent hyper-reduction step, the nonlinear term can also be fully represented by low-dimensional quantities.

In the scope of *MathEnergy*, system-theoretic MOR methods, based on the concepts of *controllability* and *observability*, are evaluated in terms of fitness for MOR of gas network models. Yet classically, system-theoretic methods require a *linear* input-output system. A linearization is not sufficiently accurate, since the nonlinearity models the physically important friction term. Hence, data-driven approaches, based on so-called empirical system Gramian matrices [35], are used to obtain the projections. The advantage of data-driven MOR is that linear methods are generalized to nonlinear systems without explicit analytical handling of the nonlinear operators, but rather their action encoded in discrete trajectory data for p, q_* and output data s_d, d_p, which is obtained by systematic perturbations of boundary values s_p, d_q, as well as steady-states $\bar{p}, \bar{q}_*$.

As a baseline, the proper orthogonal decomposition (POD) method is tested [9]. Among others, the novel method of empirical-cross-Gramian-based dominant subspaces is evaluated [11], which, as demonstrated in [32], produces smaller reduced models for a linear hyperbolic system than POD. To further accelerate the computation of the projections themselves, and to enable compression for big data-sets, the *Hierarchical Approximate Proper Orthogonal Decomposition* (HAPOD) [37] was developed, that also enables the computation of low-rank empirical Gramians

[38]. Practically, the empirical Gramians are computed via the empirical Gramian framework `emgr` [36], that is also advanced as part of the *MathEnergy* project. Additionally, system identification is a possible ansatz for reduced-order data-driven modeling of gas transport networks, for example by *Input-Output Dynamic Mode Decomposition* [12].

Exemplary, we demonstrate the reducibility by seven MOR techniques for the pipeline model from [14], also used as a benchmark in [10, Sect. 5.1]. Pipeline "networks" are a standard test case for transient gas transmission simulations. While pipelines have a trivial network topology, for MOR this is still a challenging set up, due to the lack of reducible redundancy. The considered pipeline has a length of 363 km, a diameter of 1.422 m, and an inside wall roughness of 0.0000015 m, which is incorporated through the *Hofer* friction factor model. Here, the ambient temperature is fixed to 20°C, and the specific gas constant is set to $530 \frac{J}{kg\,K}$, whereas generally, these quantities are parameters and hence, parametric reduced-order models will be implemented. After spatial discretization using the midpoint method and refinement by inserting virtual junctions every 1000 m, the resulting ODE is of order $\dim(p) + \dim(q_+) = 728$.

Since the reduced-order model (ROM) is desired to be flexible for a wide range of boundary values, the demonstrated MOR methods use generic step functions for training. Furthermore, independent from the test scenario time horizon, the training time horizon is fixed at one hour. This means, the simulations providing the data for the considered MOR methods are less complex than real (intraday) simulations which typically have a time horizon of 24 hours or more.

Specifically, three classes of MOR algorithms are tested: POD, *empirical balanced truncation* and *empirical dominant subspaces*, for details see [11]. While all methods are used in a structured variant, meaning separate ROMs for the pressure and mass-flux variables, the latter two methods are evaluated in three versions utilizing different system-theoretic operators, respectively. To test the ROMs of dimensions up to $\dim(\tilde{p}) + \dim(\tilde{q}_+) = 100$, the boundary values from [14] are used with a 24 hour time horizon. The MOR error is empirically evaluated between full and reduced model's output quantities of interest in the relative L_2-norm. Figure 11.2 shows the reduced dimension versus MOR error for the seven considered algorithms. All in all, the numerical results show that for increasing reduced order, empirical dominant subspaces (in each version) produces better ROMs than POD, while empirical balanced truncation (in each version) assembles ROMs of varying accuracy and stability.

In an upcoming work, the utilized simulation and MOR platform, developed as part of the *MathEnergy* project, will be described in detail and tested on realistic gas networks featuring parametric MOR.

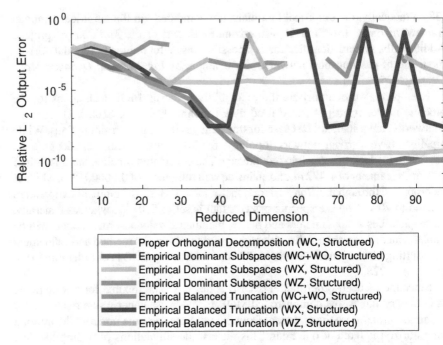

Fig. 11.2 Relative L_2 MOR error of the pipeline model from [10] by seven MOR techniques for ROMs with different state dimensions. Missing line sections mean, that no stable ROM could be computed for this reduced dimension. The results show that a reduced system of around dimension 55 would likely be sufficiently accurate

11.3 State Estimation Using Reduced-Order Models

In view of model predictive control of energy systems, the necessary state estimation usually suffers from the high computational effort required when dealing with large-scale network systems. To establish more efficient filtering variants, we adopt the ideas of model order reduction (MOR) as in the previous section. Aiming for a state reconstruction of a gas pipeline network, we consider here a linear model. Proceeding from the well-known Kalman Filter which is the optimal estimator in a mean-square error sense for linear systems, we investigate low-rank versions of the Kalman Filter in comparison to a Kalman Filter that uses low-rank models derived by MOR. The performance results are informative and show the potential for expansion in the nonlinear case.

11.3.1 Problem Setting

Given some noisy measurements and a mathematical model, the state estimation problem consists of reconstructing the whole state of the dynamical network system

considered. In our setting, we want to reconstruct the pressure and mass flux for a
gas pipeline network.

As model we use a simplified linear version of the Euler equations (11.1), namely
the linear damped wave equations, for pressure p_k and mass flux q_k on every pipe
$k \in \mathcal{E}$:

$$c_1^k \partial_t p_k = -\partial_x q_k, \qquad\qquad c_2^k \partial_t q_k = -\partial_x p_k - c_3^k q_k \qquad (11.8)$$

with constant pipe parameters c_1^k, c_2^k and c_3^k and the coupling conditions as described
in (11.2) and (11.3). A spatial semi-discretization of (11.8) with mixed finite elements
yields a linear time-invariant MIMO (multi-input, multi-output) descriptor system
for the discrete pressure and mass flux values that we summarize in the state vector x,

$$E\,\dot{x} = A\,x + B\,u, \qquad\qquad y = C\,x, \qquad (11.9)$$

with system matrices E, A, B and C, for details see [24]. The input u takes into
account the boundary conditions, whereas y denotes the output of the descriptor
system. In the spirit of Sect. 11.2, a model hierarchy can be established by applying one-sided projection-based MOR to (11.9) with a projection matrix V. For the
reduced-order model, it holds that $x \approx V\tilde{x}$ with $\tilde{x}$ being the reduced state.

The model obviously does not describe the reality exactly — the "truth" is assumed
to be given through a stochastically perturbed model. Therefore, we firstly add an
uncorrelated centered Gaussian process (white noise) to the system (11.9). Additionally, we consider stochastic boundary data, which e.g. reflect market strategies or the
idea of Power-to-Gas, i.e., using electrical energy mostly from renewable energy
plants to generate gas from water and carbon dioxide. So, our input is composed of a
deterministic part and a stochastic part, the last is described by an Ornstein-Uhlenbeck
process with constant parameters κ^v, μ^v, σ^v and a standard Wiener process W_t for
every boundary node v:

$$du_S^v = \kappa^v(\mu^v - u_S^v)dt + \sigma^v dW_t.$$

Both stochastic processes together form the so-called state noise ε which itself is an
uncorrelated Gaussian process. The output corresponding to some measurements is
also assumed to be noisy due to inaccuracies of the measurement devices. Therefore,
we add to the output equation some vector-valued uncorrelated centered Gaussian
process η.

Time discretization with a θ-scheme and equidistant time intervals yields the
time-discrete model:

$$x_{j+1} = \Phi x_j + \Psi u_j + \varepsilon_j, \qquad y_{j+1} = H x_{j+1} + \eta_j \qquad (11.10)$$

for times t_j, $j = 1, \ldots, J$. Here, Φ, Ψ and H are made up from the system matrices
of the semi-discretized model. The normal distributed centered noises are assumed

to behave as $\varepsilon_j \sim \mathcal{N}(0, Q)$ and $\eta_j \sim \mathcal{N}(0, R)$ with constant covariance matrices Q and R, respectively.

11.3.2 State Estimation

Regarding the state estimation of the linear system (11.10), the state at time t_j is denoted by $\bar{x}_{j|l}$ and its error covariance matrix $\text{Cov}[x_j - \bar{x}_{j|l}] := P_{j|l}$. Here, the first index j represents the actual time point the state is considered at, whereas the second index stands for the time up to which measurement data is considered. As all the following Kalman filtering techniques consist of a prediction and a correction step to calculate $\bar{x}_{j|j}$, we have $l = j - 1$ or $l = j$, respectively.

The original *Kalman Filter* algorithm [41] considers the state estimate at time t_j to be given as $\bar{x}_{j|j}$. The forecast or prediction step computes the predicted state estimation $\bar{x}_{j+1|j}$ for the next time point t_{j+1} as:

$$\bar{x}_{j+1|j} = \Phi \bar{x}_{j|j} + \Psi u_j, \qquad P_{j+1|j} = \Phi P_{j|j} \Phi^T + Q.$$

Afterwards, the update step corrects the forecasted distribution using the newly available measurements y_{j+1} which then gives us the corrected distribution $x_{j+1|j+1}$ through its mean and covariance:

$$\bar{x}_{j+1|j+1} = \bar{x}_{j+1|j} + K_{j+1}(y_{j+1} - H\bar{x}_{j+1|j}), \quad P_{j+1|j+1} = (\text{id} - K_{j+1}H)P_{j+1|j},$$

where the Kalman gain K_{j+1} is defined as

$$K_{j+1} = P_{j+1|j} H^T (H P_{j+1|j} H^T + R)^{-1}.$$

Due to the high computational complexity of the above algorithm originating from many matrix-matrix multiplications, we propose low-rank approximations. Making use of MOR we apply the Kalman Filter to a reduced-order model. We refer to this procedure as the *Reduced Kalman Filter*. With the assumption $x \approx V\tilde{x}$ and due to the linearity of the expectation value it holds $P_{j|l} \approx V\tilde{P}_{j|l}V^T$ with $\tilde{P}$ being the covariance matrix of the reduced state variable such that all the information of the estimated low-rank state can be prolongated to the full dimension.

The *Compressed State Kalman Filter Algorithm* presented in [48], in contrast, is based on a factorization of the covariance matrix, i.e., $P_{j|l} \approx V\hat{P}_{j|l}V^T$ where $\hat{P}_{j|l}$ is a low-rank covariance matrix and V a projection matrix with $V^TV = \text{id}$. With this assumption, all calculations for the covariance matrices can be done in lower dimension and the costly matrix-matrix multiplications are avoided. The state itself is calculated on the full dimension. The projection matrix V has to be chosen adequately. In the following, we use the same V as for the Reduced Kalman Filter.

Fig. 11.3 Exemplary result for the mass flux at one node plotted over time. Solution of the deterministic full-order model (blue dashed), sample path of the stochastically perturbed full-order model (red), estimation with the Reduced Kalman Filter (yellow)

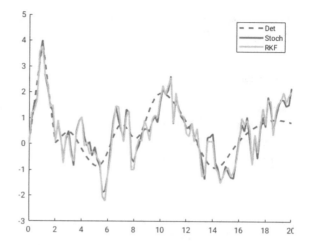

As additional filtering technique, we consider the *Particle Filter* [3] in this work. In contrast to the Kalman Filter, the Particle Filter can handle even nonlinear and non-Gaussian settings. The Particle Filter uses samples which are propagated in each time point according to the mathematical model. These samples are weighted and updated by help of an estimated distribution of the state.

11.3.3   Numerical Results and Discussion

For the performance study of the filtering techniques, we use the setting (network and parameters) as in [55]. Our full-order model, which has $N = 7011$ unknowns, is reduced to $n = 24$ degrees of freedom using the moment-matching method as described in [24]. Figure 11.3 shows exemplarily the solution of the deterministic full-order model and one sample path of the stochastically perturbed full-order model as well as an estimation with the Reduced Kalman Filter. The other filtering methods yield qualitatively similar results.

In Table 11.1, the average of the L^2-errors over time and the computational time needed for one estimation are listed for the different filters. The speed-up is taken in reference to the full Kalman Filter, as this is the theoretically proven optimal filter for a linear system with Gaussian noise. It can be seen that the speed-up for filters using reduced-order models is enormous and also the Compressed State Kalman Filter has a notable speed-up. As the Particle Filter uses sample paths in each time step, one has to generate a large vector of random variables which unfortunately takes a huge amount of time. Here, every run was taken with 100 samples. Concerning the error behavior, the numerical results confirm the theoretical statements for the full Kalman Filter: it is indeed the best one. However, since the Compressed State Kalman Filter does not differ very much from the traditional one, the approximations are also very

Table 11.1 Relative L^2-error of estimations and computational times for the estimation. For the speed-up, the time needed for the full Kalman Filter is used as reference time

Filter	$\text{Mean}_j \dfrac{\|\mathbb{E}[x_j - \bar{x}_{j|j}]\|_{L^2}}{\|\mathbb{E}[x_j]\|_{L^2}}$	$t[s]$	Speed-up
Kalman full	0.0197	7915.00	–
Kalman compressed	0.0233	13.25	60.00
Kalman reduced	0.4830	0.11	70 044.00
Particle full	0.3619	51 077.00	0.15
Particle reduced	0.6695	3.06	2 586.00

good. The Reduced Kalman Filter yields here larger errors, but increasing the size of the reduced-order model improves the errors notably, while the computation is still much faster than the Compressed State Kalman Filter.

Summarizing, already this academic problem illustrates that an enormous computational speed-up in the state estimation can be achieved by using reduced-order models instead of full-order ones, providing results of acceptable accuracy. This is the case for the Kalman Filter as well as the Particle Filter. In the underlying linear setting, the Kalman Filter is obviously superior to the Particle Filter due to its optimality properties. Comparing the low-rank approximations, the Reduced Kalman Filter shows a remarkable performance. The Compressed State Kalman Filter is more accurate, but much slower and has no error bounds in literature. For the Reduced Kalman Filter, error bounds for parabolic problems exist (see [21]). It is future work to adapt existing error bound approaches, like e.g., [26], to the damped wave equations.

11.4 Efficient MSO for Gas Networks with Hydrogen Injection

Hydrogen plays an important role in approaches for transforming our overall energy system. However, many of these approaches need intense planning and optimization based on simulations for several reasons. First, there are just a few larger power-to-gas projects. Second, technical, economical and ecological impacts of injecting larger amounts of hydrogen to systems which were originally designed for transporting and compressing natural gas are not fully understood. Third, larger technical and economical changes for the consumers of gas products have to be planned.

11.4.1 MYNTS

For simulation and analysis of gas networks, we developed a software called MYNTS (e.g. [16]). MYNTS solves Euler equations with advanced models for compressor thermodynamics and mixing of currently 21 gas components. To be more specific, Eq. (11.1) are iterated with the following equation for conserving energy

$$
\partial_t \left(\rho_k \left(\frac{1}{2} v_k^2 + e_k \right) \right) + \partial_x \left(\rho_k \left(\frac{1}{2} v_k^2 + e_k \right) v_k + p_k v_k \right) = -\frac{c_h}{d_k}(T_k - T_S),
$$

$$(11.11)$$

where v_k is the velocity, e_k the specific inner energy (can be reformulated in terms of the specific enthalpy), c_h a heat transfer coefficient and T_S the soil temperature.

Here, T_k is not assumed to be constant anymore. A multi-step workflow is used for solving the resulting set of equations. It starts from a linearization as an initial step, uses a simplified compressor model, adds advanced compressor modeling, and in the final phase also gas mixing and temperature dependency. In this final phase, a number of iterations n_f is performed till convergence (see Sect. 11.4.3 below).

MYNTS is not open-source. It has been compared to a widely used simulator intensively. More information on MYNTS can be found in the following articles: in particular, we developed a theoretical background (see [17, 19] for basics, and [18] for advanced compressor modeling) and constructive practical solution approach which guarantees existence and uniqueness of solutions in relevant situations. For speeding up the simulation process, several levels of topological reduction have been developed and built into MYNTS [5, 20]. Exemplary results for the scenario considered here can be found in Table 11.2.

Particularly with the goal to model cross-sectoral applications (see [22]), we developed a so-called universal translator (UT) [4]. UT takes a description of the network and its scenario data (settings of compressors etc., profiles) and a translation matrix (TM) in order to form the system to be solved by a nonlinear and/or DAE solver. TM contains formulas in a readable fashion. Output can be in NL (for e.g. IPOPT, used here), Mathematica, or MATLAB format currently.

Table 11.2 Results of topological reduction: basic layer. Even for this already rather coarsely modeled network, a speedup can be obtained

Net	No. of nodes	No. of edges	Simulation time
partDE	508	644	5.64 sec
Coarsened	328	375	4.66 sec
Factor	1.55	1.72	1.21

Fig. 11.4 partDE-Hy
demonstrator, as displayed
by the MYNTS OSM View
module

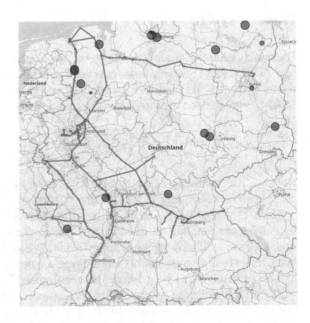

11.4.2 partDE-Hy Demonstrator

Based on [57] and an own study on 21 power-to-gas (PtG) projects with (planned) injection of larger amounts of hydrogen, we assembled a first demonstrator called partDE-Hy (see Fig. 11.4) for analyzing gas networks with hydrogen injection. The partDE data set is a coarse version of a decisive part of the long-distance gas network of Western Germany (cf. dashboard of OGE, www.open-grid-europe.com). Out of the 21 PtG plants, 7 are directly connected to the net (see Fig. 11.5). The remainder should be connected to a different network in Eastern Germany.

For analyzing impacts of different methane-hydrogen mixtures with up to 100% hydrogen and different amounts of injection at the 7 PtG plants, an ensemble analysis was set up (see Table 11.6). The injection starts at minimum values representing a bit less than the respective project plans to inject in the near future and is stepped up simultaneously ("group 2"), the mixture is varied independent from that.

11.4.3 Gas Laws - Comparisons and Results of a First Ensemble Analysis

In order to describe the behaviour of hydrogen in a gas network, the compressibility factor z has to be modeled accurately, in particular. Among the typical models for natural gas compositions $\mathbf{x}$ are Papay, AGA8-DC92 ("DC92") and GERG2008 ("GERG"). Papay (cf. [10]) can be written as

Fig. 11.5 partDE-Hy
demonstrator: colors
represent absolute difference
in power values for Papay
versus GERG for MYNTS
runs with 100% hydrogen
and ten times the max. power
setpoint, see Fig. 11.6

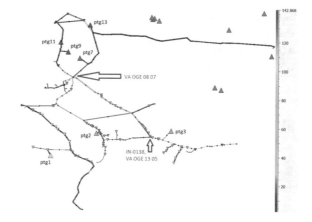

Fig. 11.6 Ensemble setup.
prop. = property, gr. =
group, pr.h. = fraction of
hydrogen, pow.=power
setpoint in MW. Results in
Fig. 11.8

where	prop.	min	max	gr.
gas	pr.h.	0	100	1
ptg1	pow.	-400	-4	2
ptg2	pow.	-400	-4	2
ptg3	pow.	-100	-1	2
ptg7	pow.	-100	-1	2
ptg9	pow.	-700	-7	2
ptg11	pow.	-300	-3	2
ptg13	pow.	-700	-7	2

$$z(p, T, \mathbf{x}) = 1 - 3.52\, p_r e^{-2.26 T_r} + 0.247\, p_r^2 e^{-1.878 T_r} \qquad (11.12)$$

with $p_r = p/p_c$ being the reduced, p_c the critical pressure, $T_r = T/T_c$ the reduced, and T_c the critical temperature. They depend on $\mathbf{x}$. Using the DC92 equation of state (Starling and Savidge (1992), [40]), z is the result of a virial-type equation:

$$z(\rho, T, \mathbf{x}) = 1 + B\rho_m - \rho_r \sum_{n=13}^{18} C_n^* + \sum_{n=13}^{58} C_n^* \left(b_n - c_n k_n \rho_r^{k_n} \right) \rho_r^{b_n} e^{-c_n \rho_r^{k_n}}, \quad (11.13)$$

where B is the second virial coefficient, ρ_m the molar density, ρ_r the reduced density, b_n, c_n, k_n constants, and C_n^* are functions of T and $\mathbf{x}$. The experimentally up-to-date GERG [43] has a much more complex description. Here, z is modeled based on the Helmholtz free energy, split into parts for ideal gas and residual fluid.

Table 11.3 Range of applicability for DC92 and GERG

Gas law	Reference	Pressure	Temperature	x_{CH4}	x_{H2}
AGA8-DC92	[40]	Up to 12 MPa	263–338 K	70%–100%	up to 10%
GERG2008	[43]	Up to 35 MPa	90–450 K	0%–100%	0–100%

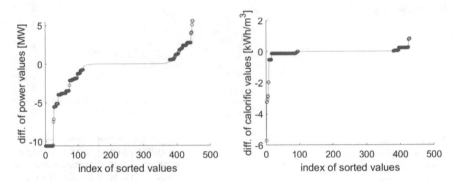

Fig. 11.7 Further analysis of gas laws for the scenario of Fig. 11.5: difference between DC92 and GERG for (left) absolute power values, (right) calorific values

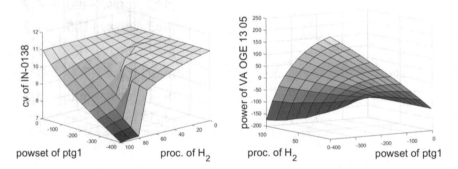

Fig. 11.8 Ensemble analysis: exemplary results (calorific values, power) at decisive nodes in partDE-Hy,showing substantial changes w.r.t. the amount of hydrogen and direction of gas flow

Applicability is compared in Table 11.3. Differences in power and calorific values for MYNTS runs are shown in Figs. 11.5, 11.7, and 11.9. In [15], deviations in pressure drop for a long pipe and influences on the dynamics for a transient partDE-Hy scenario are analyzed. Results indicate that for longer and/or more involved networks, differences can sum up substantially in parts of the networks. For instance, in the South of Germany (e.g. at "VA OGE 13 05"), not only the amount, but also the direction of flow depends on the gas law for the scenarios considered here and in [15].

An experimental convergence analysis indicates that for DC92, the final number of iterations n_f is at least as high as for GERG, but quite often higher, making runs

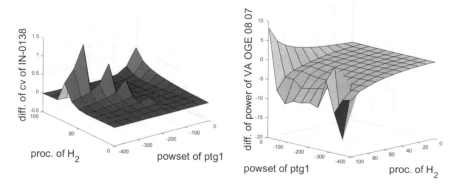

Fig. 11.9 Analysis of gas laws: differences of the results of the ensemble analysis with DC92 versus GERG: calorific values (left) and (right) power at decisive nodes in partDE-Hy

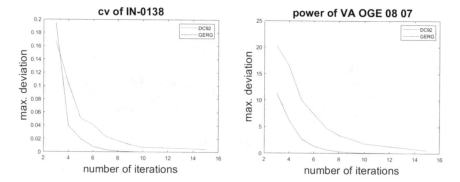

Fig. 11.10 Convergence analysis of DC92 versus GERG, deviations from converged states: calorific values (left) and (right) power at decisive elements in partDE-Hy

with GERG (slightly) faster. Exemplary results are shown in Figs. 11.10, 11.11, and 11.12. An explanation might be given by the quality of the solution processes for the nonlinear equations inside the gas laws. Both use Newton-type iterations. The older DC92 (implemented according to its ISO norm [40]) seems to use a less accurate approximation. This shall be investigated further.

To summarize, Papay cannot be used for the hydrogen scenarios; DC92 can show substantial differences to the up-to-date GERG. Runs with GERG can even be slightly faster for the same level of accuracy. Overall, GERG should be used.

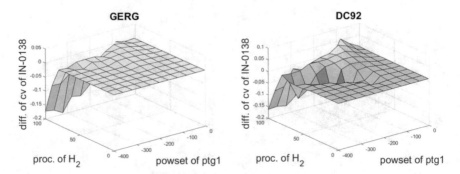

Fig. 11.11 Convergence analysis, deviations of run with $n_f = 3$ final iterations from converged states: results with GERG (left) and DC92 (right) for node IN-0138

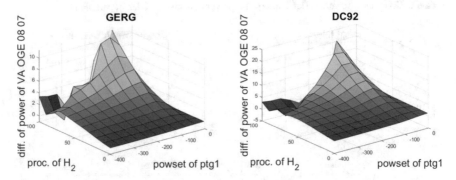

Fig. 11.12 Convergence analysis, deviations of run with $n_f = 3$ final iterations from converged states: results with GERG (left) and DC92 (right) for edge "VA OGE 08 07"

11.5 Coupled Transient Modeling and Simulation of Power and Gas Networks

11.5.1 Transient Modeling of Gas Networks

Here, we consider gas networks with pipes, resistors, valves, controlled valves and compressors. They can be described as a coupled system of element equations of the form (11.1a)–(11.1c) for pipes [34] together with [39, 42, 58]

$$p_r - p_l = c_5 \frac{z(p_r)}{p_r} q|q|, \quad g(p_r, p_l, q, y, t) = 0 \quad \text{for resistors and compressors,}$$

$$s(p_r - p_l) + (1 - s)q = 0, \quad p_r - p_l = p_{ctrl} \quad \text{for switching and controlled valves.}$$

Here, s is a switching function with values in $[0, 1]$ and g is a nonlinear function with $\partial_y g$ non-singular. For pipes, the pressure p and the mass flow q depend on space x and time t. For the other elements, the pressures p_r and p_l at the right and left end as well as the flow q depend only on time. For compressors, we have additionally the time-dependent variable y containing the adiabatic enthalpy, the volumetric flow rate, the speed and the efficiency. For shortness and simplicity, we use an isothermal model for the pipes, and the resistor model is based on the *Darcy-Weisbach* model [25, 49]. The system is completed with the boundary conditions $p(x_l) = p_l$ and $p(x_r) = p_r$ for the left and right end of the pipes as well as the flow balance equations (11.2) at each node and the pressure supplies (11.3). The system of equations represents a hyperbolic partial differential algebraic system [39]. Applying a spatial discretization (cf. (11.4)), we obtain a differential-algebraic equation (DAE) of the form

$$A_G z'_G + b_G(z_G, t) = 0 \tag{11.14}$$

with z_G containing the node pressures, the element flows and the compressor variables y. Notice that z'_G denotes the time derivative of z_G. The index of this system depends on the choice of the spatial discretization. Variants of IMEX schemes usually lead to index-2 DAEs, cf. [33], whereas network topologically adapted schemes imply index-1 DAEs, cf. [9, 39]. DAEs of index 1 are well-posed whereas DAEs of higher index are ill-posed [46]. Hence, spatial discretizations yielding index-1 DAEs are preferable in order to avoid stability problems.

11.5.2   Transient Modeling of Power Networks

We consider power networks containing generators, transmission lines and loads. For shortness and simplicity, we neglect transformers. Power networks with transformers can be modeled as a combination of sub-networks (e.g., high and low voltage grid) coupled by transformers. They can be treated by co-simulation as described in Sect. 11.5.4. Using equivalent circuits for transmission lines and loads, the power network can be described as a coupled system of element equations of the form [27]

$$M\omega' = D\omega - iv + P_{mech}(t), \quad \theta' = \omega, \quad v = v_{max} \sin(\theta) \quad \text{for generators,}$$
$$Cv' = i, \quad Li' = v, \quad i = g(v) \quad \text{for capacitances, inductances and resistances,}$$

together with the Kirchhoff laws $Ai_* = 0$, $v_* = A^\top e_*$, where all element currents, element voltages and node potentials are summarized in the vectors i_*, v_* and e_*. Collecting the unknowns, i. e. the element angles θ, the element angular velocities ω, the element voltages v, the element currents i and the nodal potentials e in z_P, we can express the power network as a DAE of the form

$$A_P(z_P)z'_P + b_P(z_P, t) = 0. \tag{11.15}$$

Following the decoupling approach in [50], we can show that the DAE index is at most 2.

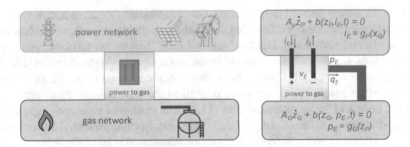

Fig. 11.13 The power network and the gas network are coupled by an electrolyser. The electrolyser uses power from the power network to generate hydrogen from water. The hydrogen is methanated before it enters the gas network

11.5.3 Model Coupling of Gas and Power Networks

We consider the coupling of gas and power networks by an electrolyser (see Fig. 11.13). As additional unknowns, we have a current i_E flowing from the cathode to the anode and the voltage v_E between the poles as well as the pressure p_E and the mass flow q_E of the gas H_2/CH_4 at the outflow point of the electrolyser. In order to model the coupling of the systems, we use first the power balance equation

$$q_E(\alpha p_E + \beta) = \mu v_E i_E$$

with the efficiency μ. The second equation is given by Faraday's law $q_E = \gamma i_E$, where γ includes a chemical reaction factor caused by methanation. Furthermore, we exploit that we can express v_E in terms of the node potentials, i.e. u_E can be expressed in terms of z_E. Analogously, we can express q_E as mass flow entering the pipe connected to the electrolyzer, i.e. q_E belongs to z_G. Using Faraday's law as an expression for i_E and the power balance equation to express p_E, we obtain a coupled model of the form

$$A_G \dot{z}_G + f_G(z_G, z_P, t) = 0$$
$$A_P \dot{z}_P + f_P(z_P, z_G, t) = 0.$$

11.5.4 Transient Co-Simulation for Coupled System

In order to simulate the coupled system, we want to use different solvers for the subsystems. It allows us to use well-established as well as newly developed tools for each of the subsystems that have a complete different physical behavior. Therefore, we study a co-simulation of both sub-systems in the following way. We start from an initial guess z_P^0 for the gas network and then iterate as follows:

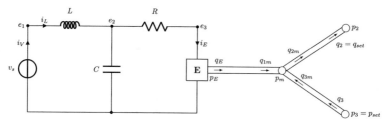

Fig. 11.14 Benchmark coupling of a simple power network with a simple gas network by an electrolyzer E

$$A_G(z_G^k)\dot{z}_G^k + f_G(z_G^k, z_P^{k-1}, t^k) = 0$$
$$A_P(z_P^k)\dot{z}_P^k + f_P(z_P^k, z_G^k, t^k) = 0.$$

Here, z_G^k and z_P^k denote the k-th solution iterates approximating the exact solutions z_G and z_P. Notice that all solution iterates are functions on the whole time interval. This approach is well-known as Gauß–Seidel waveform relaxation [47]. Since the sub-systems are DAEs and not ODEs, such a co-simulation may fail, see e.g. [2, 8, 54]. To the best of the authors' knowledge, there are no general convergence results for a coupling of index-1 and index-2-DAEs. However, following the convergence proof of [50] and the network topological decoupling approaches in [39, 50], we are able to prove the following convergence result.

Theorem 1 *Let the gas network be spatially discretized in such a way that the resulting DAE is of index 1 and the pressure p_E at the electrolyzer is given by an ODE resulting from the discretization of the pipe leaving the electrolyzer. Let, furthermore, the electrolyzer be located in the power network in such a way that anode and cathode are connected by a capacitance that does not belong to a loop of generators and capacitances. Then, the co-simulation via Gauß-Seidel waveform relaxation converges to the exact solution of the coupled system.*

Figure 11.14 shows an example that couples a simple power network model with a simple H_2 gas network by an electrolyzer E in the middle. Power is supplied by the voltage source on the left. The gas network has the supply node p_3, where the pressure is set and a demand node p_2, where the flow q_2 is set.

Figure 11.15 shows the simulation results for the node potentials and the currents of the power network and the pressures and flows of the gas network. Here, the demand q_2 changes from 4 kg/s to 2 kg/s after $600\,\mathrm{s} = 10\,\mathrm{min}$ and the power supply changes from 2 V to 1.5 V after $1200\,\mathrm{s} = 20\,\mathrm{min}$.

The co-simulation via Gauß–Seidel iteration converges after a few iterations as the error plots in Fig. 11.16 show.

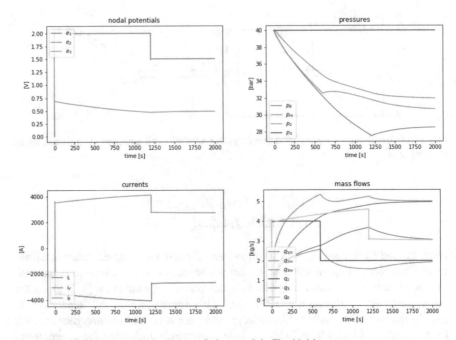

Fig. 11.15 Simulation results for the coupled network in Fig. 11.14

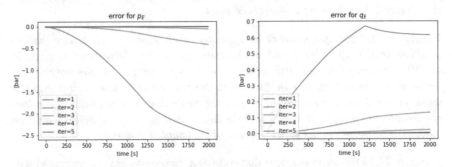

Fig. 11.16 Error for co-simulation via Gauß-Seidel iteration for the pressure p_E and the flow e_E at the electrolyzer for the coupled network in Fig. 11.14. The counter "iter" reflects the iteration number

11.6 State Estimation of the Power Grid

Within our work in MathEnergy, we also address the topic of state estimation of power grids. This topic already plays an important role in observing and controlling the power grid over a long time period. However, in available simulation tools, state estimation is referred to determining the states of the load flow model every 15 minutes applying a weighted least squares approach. Within MathEnergy, we focus on certain approaches like hybrid state estimation as well as on dynamic observers, e.g. *Constraint Extended Kalman Filters* for nonlinear DAEs, which are not applied in control centers of the Transmission and Distribution System Operators (TSO and DSO, respectively) yet. In addition to pure state observation of the power grid model, we also address the topic of estimating switches in the grid topology, i.e. determining the active mode after switching off, e.g. a power line. W.r.t. this task, most algorithms in the literature try to separate the problem of mode detection and state observation. We consider problems where the aforementioned is not possible and instead separate the problems of switching time detection and switch observation [44]. This also makes sense from a practical point of view as they represent fault detection and identification, respectively. The corresponding switch observer is useful, for example in situations where one wants to identify faults that have already been noticed. For this purpose, we model the power network as a linear DAE. Switches are modeled by a change in the algebraic constraint, which might lead to jumps or impulses in the system state and output [44, 45].

11.6.1 Observability

Using linear DAEs to model the individual modes of a power grid, the switched system can be described by a switched DAE:

$$E_\sigma \dot{x} = A_\sigma x + B_\sigma u, \quad x(0^-) = x_0$$

$$y = C_\sigma x + D_\sigma u, \tag{11.16}$$

with a right-continuous function $\sigma : \mathbb{R} \to \{1, \dots, N\}$, $N \in \mathbb{N}_{\geq 2}$, describing the switching signal, an initial value x_0, an input function $u \in C^\infty(\mathbb{R}, \mathbb{R}^{n_u})$, and system matrices $E_p, A_p \in \mathbb{R}^{n \times n}$, $B_p \in \mathbb{R}^{n \times n_u}$, $C_p \in \mathbb{R}^{n_y \times n}$ for some $n_u, n, n_y \in \mathbb{N}$ describing the different system modes p. The matrix pairs (E_p, A_p) are assumed to be regular for all $p \in \{1, \dots, N\}$. A distributional solution theory is required to solve (11.6.1), see [56].

Observability is a classical concept of control theory that has already been studied for various system classes. It deals with recovering the system states x from input u and output data y. For inhomogeneous systems ($(B_\sigma, C_\sigma) \neq (0, 0)$), one usually distinguishes between weak, generic and strong observability, i.e. observability for some, almost all or all input functions u, respectively.

For the switched DAE (11.6.1), we are interested in recovering both the state x and the switching signal σ. Switch observability ($[\sigma_1]$-observability) describes the observability of state and switching signal under the assumption that the switching signal σ is not constant. It is strictly weaker than mode and state observability as one can make use of the behavior during the switch, e.g. jumps or sudden changes in the dynamics of the output. An even weaker concept is that of switching time observability ($[t_S]$-observable), i.e. determining the time of the switch, which corresponds to the concept of fault detection. The concepts of (weak/generic/strong)

$[t_S]$-observablity and $[\sigma_1]$-observablity have been developed at Fraunhofer ITWM [44] within MathEnergy.

11.6.2   Switch Observer for Mode Detection

For mode detection, we use a linear DAE model of the power grid. To reduce the computational costs, we furthermore assume that the dwell time is sufficiently large for the local observer of the correct mode to "converge", i.e. to sufficiently approach the correct state. A switch in a $[t_S]$-observable system causes an impulse or a change in the dynamics. To observe this, the results of an impulse-observer are combined with those of N parallel-running Luenberger observers - one for each possible mode. Denote by $P^{curr}(t)$ the set of currently reasonable modes, i.e. of those modes for which the output error estimation of the Luenberger observer is sufficiently small. An impulse-free switch of a $[t_S]$-observable system causes that – at least locally – the set $P^{curr}(t)$ becomes empty for some time. This criterion allows to determine the switching times.

The switch observer developed at Fraunhofer ITWM allows for $[\sigma_1]$-observable systems to identify the system state and switching signal [44]. Its basic outline is given in the following algorithm:

Algorithm Switch Observer

1. On the pre-switch interval, classical observers are used for each mode. As the individual modes do not have to be observable, this might not give a full state estimate, but only a state estimate on a subspace, i.e. a partial state estimate. Most of the observers will not converge, as the measured output of the system does not fit to the dynamics of the corresponding modes. Those modes that give well-behaving observers and their partial state estimations are stored.
2. The same is done for the post-switch interval.
3. For the impulsive part, it is checked which modes can cause the measured impulse. For those modes that might cause the measured impulse, corresponding state information is computed.
4. After computing these local results, all pairs of remaining mode candidates for the pre-switch and impulsive/post-switch mode are checked. For each candidate pair, an estimation of the overall state is computed from the partial state estimations.
5. To select the correct mode pair, one can simulate the results for each possible mode pair together with the corresponding state estimation and compare these outputs to the measured one. (Strong) $[\sigma_1]$-observability guarantees that this procedure works properly given correct results for the partial states.

The switch observer relies on classical observers for the local estimates on pre- and post-switch interval. This idea can also be extended to the case of multiple switches [44].

11.6.3   Example Switch Observer

We now apply the switch observer to a switched version of the 9 bus benchmark system from Matpower [59]. The network is modeled by the linear swing equation which is a simplified power network model. The swing equation [13] is a common model for power systems [28, 52]. It does not take voltage variation or reactive power into account, yet the model and its linearization are widely used, see [1, 28, 51, 53].

The swing equation can be derived from first principles [13] or as a simplifica-tion of a nonlinear system consisting of the power flow equations and a generator model [29]. In the derivation [29], certain assumptions were made. Most impor-tantly: (1) voltage and current are sinusoidal with constant frequency, (2) the voltage amplitude is constant, (3) the transmission lines are lossless. The last assumption is a realistic approximation only for high voltages, i.e. for the transmission system. Uses and limitations of the swing equation are discussed in [1, 13]. The authors argue that the swing equation is often used beyond its scope, as the approximations are valid only in a short time-interval. As we are interested only in the system dynamics close to a switch, this limitation is not an issue here. The following simple extensions of the model are possible:

- One can add dynamic loads as an additional type of nodes. Dynamic loads can be used, among others, to model droop-controlled inverters, see e.g. [23].
- The generator has been modeled by a single rotating mass. It can easily be extended to multiple rotating masses, see [28].
- One can include transient reactance at the generators as in [30].

Within our example, the system consists of three generator nodes, six load nodes and nine transmission lines. Details on the parametrization of the swing equations can be found in [44]. Power infeed and extraction are modeled as additional constant states. The following modes for switching are considered:

1. Mode: All lines are active.
2. Mode: The line connecting nodes 4 and 5 is switched off.
3. Mode: The line connecting nodes 8 and 9 is switched off.
4. Mode: The line connecting nodes 9 and 4 is switched off.

The lines that might be switched off are highlighted in Fig. 11.17. As shown in [44], this system is not strongly $[t_S]$-observable, not even for full state measure-ment, and the switched DAE is not $[\sigma_1]$-observable. For certain inputs (which do not describe realistic load curves), the switching signal cannot be determined even if a full state measurement is available. However, it is generically $[\sigma]$-observable, i.e. $[\sigma]$-observable for almost all inputs. To identify the current mode on the pre-switch interval [0, 1], a Luenberger observer is used for each mode. The output of the first two local observers converge to the measured output, meaning that both modes rea-sonably describe the pre-switch dynamics. On the post-switch interval, the output estimation of all four local observers converge to the measured output. Hence, the observer yields $P^{pre} = \{1, 2\}$ and $P^{post} = \{1, 2, 3, 4\}$. In particular, $|P^{pre}| > 1$ and

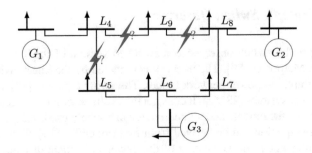

Fig. 11.17 9 bus benchmark system from Matpower [59]

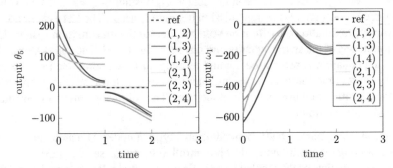

Fig. 11.18 Estimated output corresponding to different mode pairs and homogeneous reference solution

$|P^{post}| > 1$ imply that one cannot determine the switching signal (i.e. which fault occurred) if only the information before or after the switch is considered. As there is no impulse in the output (see Fig. 11.19), all modes are plausible for the impulsive part, i.e. $P^{imp} = \{1, 2, 3, 4\}$. While it is not possible to determine the correct switching signal using only the information on the pre-switch or the post-switch interval, the switch observer allows to uniquely determine the switching signal σ: Fig. 11.18 shows the simulation for the different possible mode pairs. The estimation for some pairs is way off. Figure 11.19 shows that only the output estimations for the mode pairs $(1, 2)$ and $(2, 1)$ are of the same magnitude as the measured output. Only the correct mode pair $(1, 2)$ gives an output estimation reasonably close to the measured one.

Beside this simulation-based plausibility test for the candidate mode pairs, one can also use the state projection error ([44, Condition (6.14)]) or the output estimation error ([44, Condition (6.15)]) to determine the correct switching signal. For the given example, the error term of the correct mode pair is for all three conditions by at least 1-2 orders of magnitude smaller than the error terms of all other mode pairs. Hence, the correct switching signal can easily be determined.

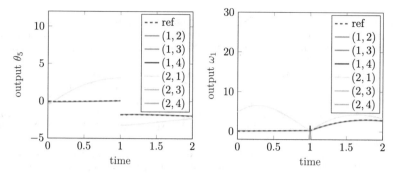

Fig. 11.19 Estimated output corresponding to different mode pairs and homogeneous reference solution for the system (Fig. 11.18 zoomed in)

11.7   Outlook

A main focus of the remaining project work is the development of efficient approaches for solving large coupled gas and power grids. In particular, further extensions of the existing demonstrators, among them partDE-Hy, shall be developed, the software prototypes be combined further to efficient workflows, and, finally, their efficiency be shown by means of demonstrators.

Acknowledgements This work is supported by the German Federal Ministry for Economic Affairs and Energy, in the joint project: "**MathEnergy** – Mathematical Key Technologies for Evolving Energy Grids" with its subprojects 0324019A, 0324019B, 0324019E, and 0324019F. The authors thank Philipp Spelten (Fraunhofer SCAI) for providing locations and dimensions of the PtG plants.

References

1. Anderson, P.M., Fouad, A.A.A.: Power System Control and Stability, 1st edn. Iowa State Univ. Press, Ames, Iowa (1977)
2. Arnold, M., Günther, M.: Preconditioned dynamic iteration for coupled differential-algebraic systems. BIT **41**, 1–25 (2001). https://doi.org/10.1023/A:1021909032551
3. Arulampalam, M.S., Maskell, S., Gordon, N., Clapp, T.: A tutorial on particle filters for online nonlinear/non-Gaussian Bayesian tracking. IEEE Trans. Signal Process. **50**(2), 174–188 (2002). https://doi.org/10.1109/78.978374
4. Baldin, A., Cassirer, K., Clees, T., Klaassen, B., Nikitin, I., Nikitina, L., Torgovitskaia, I.: Universal translation algorithm for formulation of transport network problems. In: 8th International Conference on Simulation and Modeling Methodologies, Technologies and Applications, SIMULTECH 2018. Proceedings, pp. 315–322 (2018). https://doi.org/10.5220/0006831903150322
5. Baldin, A., Clees, T., Fuchs, B., Klaassen, B., Nikitin, I., Nikitina, L., Torgovitskaia, I.: Topological reduction of gas transport networks. In: INFOCOMP 2019, the Ninth International Conference on Advanced Communications and Computation, pp. 15–20 (2019). http://www.thinkmind.org/download.php?articleid=infocomp_2019_2_10_68001

6. Banagaaya, N., Benner, P., Grundel, S.: Index-preserving model order reduction for differential-algebraic systems arising in gas transport networks. In: Progress in Industrial Mathematics at ECMI 2018, Mathematics in Industry, vol. 30, pp. 291–297 (2019). https://doi.org/10.1007/978-3-030-27550-1_36

7. Banagaaya, N., Grundel, S., Benner, P.: Index-aware MOR for gas transport networks with many supply inputs. IUTAM Symposium on Model Order Reduction of Coupled Systems, IUTAM Bookseries **36**, 191–207 (2019). https://doi.org/10.1007/978-3-030-21013-7_14

8. Bartel, A., Brunk, M., Günther, M., Schöps, S.: Dynamic iteration for coupled problems of electric circuits and distributed devices. SIAM J. Sci. Comput. **35**(2), B315–B335 (2013). https://doi.org/10.1137/120867111

9. Benner, P., Braukmüller, M., Grundel, S.: A direct index 1 DAE model of gas networks. In: Keiper, W., Milde, A., Volkwein, S. (eds.) Reduced-Order Modeling (ROM) for Simulation and Optimization, pp. 99–119. Springer, Cham (2018). https://doi.org/10.1007/978-3-319-75319-5_5

10. Benner, P., Grundel, S., Himpe, C., Huck, C., Streubel, T., Tischendorf, C.: Gas network benchmark models. In: Applications of Differential Algebraic Equations: Examples and Benchmarks, Differential-Algebraic Equation Forum, pp. 171–197. Springer, Cham (2018). https://doi.org/10.1007/11221_2018_5

11. Benner, P., Himpe, C.: Cross-Gramian-based dominant subspaces. Adv. Comput. Math. **45**(5), 2533–2553 (2019). https://doi.org/10.1007/s10444-019-09724-7

12. Benner, P., Himpe, C., Mitchell, T.: On reduced input-output dynamic mode decomposition. Adv. Comput. Math. **44**(6), 1821–1844 (2018). https://doi.org/10.1007/s10444-018-9592-x

13. Caliskan, S.Y., Tabuada, P.: Uses and abuses of the swing equation model. In: 2015 54th IEEE Conference on Decision and Control (CDC), pp. 6662–6667. IEEE (15.12.2015 - 18.12.2015). https://doi.org/10.1109/CDC.2015.7403268

14. Chaczykowski, M.: Sensitivity of pipeline gas flow model to the selection of the equation of state. Chem. Eng. Res. Des. **87**, 1596–1603 (2009). https://doi.org/10.1016/j.cherd.2009.06.008

15. Clees, T., Baldin, A., Klaassen, B., Nikitina, L., Nikitin, I., Spelten, P.: Efficient modeling and simulation of long-distance gas transport networks with large amounts of hydrogen injection. In: SWEDES 2020, Procs. 15th Conf. Sustainable Development of Energy, Water and Environment Systems, Cologne, Sep 1-5, 2020. Accepted as archival paper

16. Clees, T., Cassirer, K., Hornung, N., Klaassen, B., Nikitin, I., Nikitina, L., Suter, R., Torgovitskaia, I.: MYNTS: Multi-phYsics NeTwork Simulator. In: SIMULTECH 2016, 6th International Conference on Simulation and Modeling Methodologies, Technologies and Applications. Proceedings, pp. 179–186 (2016). https://doi.org/10.5220/0005961001790186

17. Clees, T., Hornung, N., Nikitin, I., Nikitina, L.: A globally convergent method for generalized resistive systems and its application to stationary problems in gas transport networks. In: SIMULTECH 2016, 6th International Conference on Simulation and Modeling Methodologies, Technologies and Applications. Proceedings, pp. 64–70 (2016)

18. Clees, T., Nikitin, I., Nikitina, L.: Advanced modeling of gas compressors for globally convergent stationary network solvers. In: Seventh International Conference on Advanced Communications and Computation, INFOCOMP 2017, pp. 52–57 (2017)

19. Clees, T., Nikitin, I., Nikitina, L.: Making network solvers globally convergent. In: Simulation and Modeling Methodologies, Technologies and Applications. International Conference, SIMULTECH 2016, pp. 140–153 (2018). https://doi.org/10.1007/978-3-319-69832-8_9

20. Clees, T., Nikitin, I., Nikitina, L., Segiet, L.: Modeling of gas compressors and hierarchical reduction for globally convergent stationary network solvers. Int. J. Adv. Syst. Meas. **11**(1-2), 61–71 (2018). https://www.thinkmind.org/download.php?articleid=sysmea_v11_n12_2018_6

21. Dihlmann, M., Haasdonk, B.: A reduced basis Kalman filter for parametrized partial differential equations. ESAIM: COCV **22**(3), 625–669 (2016). https://doi.org/10.1051/cocv/2015019

22. Doetsch, C., Clees, T.: Systemansätze und -komponenten für cross-sektorale Netze, pp. 311–340 (2017). https://doi.org/10.1007/978-3-658-15737-1_17

23. Dörfler, F.A.: Dynamics and control in power grids and complex oscillator networks. University of California, Santa Barbara, [Santa Barbara, Calif.] (2013)
24. Egger, H., Kugler, T., Liljegren-Sailer, B., Marheineke, N., Mehrmann, V.: On structure-preserving model reduction for damped wave propagation in transport networks. SIAM J. Sci. Comput. **40**, A331–A365 (2017). https://doi.org/10.1137/17M1125303
25. Finnemore, E., Franzini, J.: Fluid Mechanics with Engineering Applications, 10th edn. Asia Higher Education Engineering/Computer Science Civil Engineering, McGraw-Hill Higher Education (2001)
26. Gerner, A.L., Veroy, K.: Certified reduced basis methods for parametrized saddle point problems. SIAM J. Sci. Comput. **34**(5), A2812–A2836 (2012). https://doi.org/10.1137/110854084
27. Grainger, J., Stevenson, W.: Power System Analysis. McGraw-Hill (1994)
28. Groß, T., Trenn, S., Wirsen, A.: Solvability and stability of a power system DAE model. Syst. Control Lett. **97**, 12–17 (2016). https://doi.org/10.1016/j.sysconle.2016.08.003
29. Groß, T.B.: DAE-Modellierung und mathematische Stabilitätsanalyse von Energieversorgungsnetzen. Fraunhofer IRB Verlag, Stuttgart (2016)
30. Gross, T.B., Trenn, S., Wirsen, A.: Topological solvability and index characterizations for a common DAE power system model. In: 2014 IEEE Conference on Control Applications (CCA), pp. 9–14. IEEE (08.10.2014 - 10.10.2014). https://doi.org/10.1109/CCA.2014.6981321
31. Grundel, S., Herty, M.: Hyperbolic discretization of the isothermal Euler equation via Riemann invariants. Cornell University (2019). 2005.12158. Math.NA
32. Grundel, S., Himpe, C., Saak, J.: On empirical system Gramians. Proc. Appl. Math. Mech. **19**(1), e201900006 (2019). https://doi.org/10.1002/PAMM.201900006
33. Grundel, S., Jansen, L.: Efficient simulation of transient gas networks using IMEX integration schemes and MOR methods. In: 54th IEEE Conference on Decision and Control (CDC), Osaka, Japan, pp. 4579–4584 (2015). https://doi.org/10.1109/CDC.2015.7402934
34. Grundel, S., Jansen, L., Hornung, N., Clees, T., Tischendorf, C., Benner, P.: Model order reduction of differential algebraic equations arising from the simulation of gas transport networks. In: Progress in Differential-Algebraic Equations, Differential-Algebraic Equations Forum, pp. 183–205. Springer Berlin Heidelberg (2014). https://doi.org/10.1007/978-3-662-44926-4_9
35. Himpe, C.: emgr - the Empirical Gramian Framework. Algorithms **11**(7), 91 (2018). https://doi.org/10.3390/a11070091
36. Himpe, C.: emgr – EMpirical GRamian framework (version 5.7) (2019). https://gramian.de. https://doi.org/10.5281/zenodo.2577980
37. Himpe, C., Leibner, T., Rave, S.: Hierarchical approximate proper orthogonal decomposition. SIAM J. Sci. Comput. **40**(5), A3267–A3292 (2018). https://doi.org/10.1137/16M1085413
38. Himpe, C., Leibner, T., Rave, S., Saak, J.: Fast low-rank empirical cross Gramians. Proc. Appl. Math. Mech. **17**(1), 841–842 (2017). https://doi.org/10.1002/pamm.201710388
39. Huck, C.: Perturbation analysis and numerical discretisation of hyperbolic partial differential algebraic equations describing flow networks. Ph.D. thesis, Humboldt-Universität zu Berlin, Mathematisch-Naturwissenschaftliche Fakultät (2018). https://doi.org/10.18452/19596
40. International Standard: Iso 12213-2: Natural gas - calculation of compression factor - part 2: Calculation using molar-composition analysis (2nd ed. 2006-11-15)
41. Kalman, R.E.: A new approach to linear filtering and prediction problems. J. Basic Eng. **82**(1), 35–45 (1960). https://doi.org/10.1115/1.3662552
42. Koch, T., Hiller, B., Pfetsch, M., Schewe, L.: Evaluating Gas Network Capacities. Society for Industrial and Applied Mathematics, Philadelphia, PA (2015). https://doi.org/10.1137/1.9781611973693
43. Kunz, O., Wagner, W.: The GERG-2008 wide-range equation of state for natural gases and other mixtures: an expansion of GERG-2004. J. Chem. Eng. Data **57**, 3032–3091 (2012). https://doi.org/10.1021/je300655b
44. Küsters, F.: Switch Observability for Differential-Algebraic Systems: Analysis. Observer Design and Application to Power Networks. Fraunhofer Verlag, Stuttgart (2018)
45. Küsters, F., Trenn, S., Wirsen, A.: Switch-observer for switched linear systems. In: Proceedings of the IEEE Conference on Decision and Control. IEEE (2017). https://doi.org/10.1109/CDC.2017.8263903

46. Lamour, R., März, R., Tischendorf, C.: Differential-Algebraic Equations. A Projector Based Analysis. Springer, Berlin (2013)
47. Lelarasmee, E.: The waveform relaxation method for time domain analysis of large scale integrated circuits: theory and applications. College of Engineering, University of Berkeley, Electronics Research Laboratory (1982)
48. Li, J.Y., Kokkinaki, A., Ghorbanidehno, H., Darve, E.F., Kitanidis, P.K.: The compressed state Kalman filter for nonlinear state estimation: application to large-scale reservoir monitoring. Water Resour. Res. 51(12), 9942–9963 (2015). https://doi.org/10.1002/2015WR017203
49. Lurie, M.: Modeling of Oil Product and Gas Pipeline Transportation. Wiley-VCH Verlag GmbH Co. KGaA (2008). https://doi.org/10.1002/9783527626199
50. Pade, J., Tischendorf, C.: Waveform relaxation: a convergence criterion for differential-algebraic equations. Numer. Algorithms 81, 1327–1342 (2019). https://doi.org/10.1007/s11075-018-0645-5
51. Pasqualetti, F., Dorfler, F., Bullo, F.: Attack detection and identification in cyber-physical systems. IEEE Trans. Autom. Control 58(11), 2715–2729 (2013). https://doi.org/10.1109/TAC.2013.2266831
52. Sastry, S., Varaiya, P.: Hierarchical stability and alert state steering control of interconnected power systems. IEEE Trans. Circuits Syst. 27(11), 1102–1112 (1980). https://doi.org/10.1109/TCS.1980.1084747
53. Scholz, E.: Observer-based monitors and distributed wave controllers for electromechanical disturbances in power systems. Ph.d. thesis, Massachusetts Institute of Technology (2004). https://dspace.mit.edu/bitstream/handle/1721.1/26723/59669742-MIT.pdf?sequence=2
54. Schöps, S.: Multiscale modeling and multirate time-integration of field/circuit coupled problems. Ph.D. thesis, Universität Wuppertal, Fakultät für Mathematik und Naturwissenschaften (2018)
55. Stahl, N., Marheineke, N.: Filtering and model reduction of PDAEs with stochastic boundary data. Proc. Appl. Math. Mech. 19(1), e201900130 (2019). https://doi.org/10.1002/pamm.201900130
56. Trenn, S.: Distributional differential algebraic equations. <Ilmenau>|Universitätsverlag Ilmenau (2009)
57. TUB-ER: partDE data set, Technical University of Berlin (2019). https://www.er.tu-berlin.de/menue/home/parameter/en/
58. Walther, T., Hiller, B., Saitenmacher, R.: Polyhedral 3d models for compressors in gas networks. In: Kliewer, N., Ehmke, J.F., Borndörfer, R. (eds.) Operations Research Proceedings 2017, pp. 517–523. Springer International Publishing, Cham (2018). https://doi.org/10.1007/978-3-319-89920-6_69
59. Zimmerman, R.D., Murillo-Sánchez, C.E.: Matpower: Users manual (2016)

Chapter 12
Modeling and Simulation of Sector-Coupled Energy Networks: A Gas-Power Benchmark

Eike Fokken, Tillmann Mühlpfordt, Timm Faulwasser, Simone Göttlich, and Oliver Kolb

Abstract In this contribution, we present a gas-power benchmark problem tailored to simulation and optimization of coupled electrical grids and gas networks in a time-varying setting. Based on realistic data sets from the IEEE database and the GasLib suite, we describe the full set up of the underlying equations and motivate the choice of parameters. The illustrative simulation results demonstrate the applicability of the proposed approach and also allow for a clear visualization of gas-power conversion. Moreover, they show that the proposed benchmark problem warrants further investigations.

12.1 Introduction

The current transformation of the energy system is driven by at least three trends: decarbonization and defossilization of energy supply, growing concerns about climate change, and transformative political decisions, e.g. the phase-out of nuclear energy [1] and scheduled phase-out of coal and lignite in Germany. Phase out of all

E. Fokken · S. Göttlich (✉) · O. Kolb
Department of Mathematics, University of Mannheim, Mannheim, Germany
e-mail: goettlich@uni-mannheim.de

E. Fokken
e-mail: fokken@uni-mannheim.de

O. Kolb
e-mail: kolb@uni-mannheim.de

T. Mühlpfordt
Institute for Automation and Applied Informatics, Karlsruhe Institute of Technology, Karlsruhe, Germany
e-mail: tillmann.muehlpfordt@kit.edu

T. Faulwasser
Institute for Energy Systems, Energy Efficiency and Energy Economics, TU Dortmund University, Dortmund, Germany
e-mail: timm.faulwasser@ieee.org

© Springer Nature Switzerland AG 2021
S. Göttlich et al. (eds.), *Mathematical Modeling, Simulation and Optimization for Power Engineering and Management*, Mathematics in Industry 34,
https://doi.org/10.1007/978-3-030-62732-4_12

fossil energies is, in principle, doable as averaged over long time spans renewable sources (wind, solar, etc.) provide sufficient amounts of energy to achieve decarbonization. Yet, renewable generation and demand are not synchronized in time and space and thus energy storage as well as energy distribution/transport are of crucial importance.[1] Currently, neither a readily and widely usable storage technology to buffer the quantities of electrical energy required for decarbonization exists, nor is there scientific consensus on which large-scale storage technologies will be available in near-to-mid future. Hence, it comes at no surprise that the coupling of energy domains and sectors is gaining increasing research attention. For example, the economic viability of future power-to-X pathways (where X can be Hydrogen, Methane, or synthetic bio-fuels, for instance) has been investigated in several studies, see e.g. [4–6]. These investigations are driven by the fact that natural gas can be stored in sufficient quantities in dedicated installations and to a certain extend directly in the gas grid itself. In other words, coupling of electrical grids and gas networks is currently considered a very promising road towards a high share of renewables.

Historically, however, the critical energy system infrastructure for gas and power grids has been separated in terms of operation and control. Hence, there do not exist established standards for joint operation and control of multi-energy grids. In turn multi-energy systems arising from sector-coupling pose control and optimization challenges, many of which are still open or are subject to ongoing research efforts, see [7–9], or more recently [10–13].

In the context of electrical grids, the Optimal Power Flow (OPF) problem is of vital importance for safe, reliable, and economic operation, see [14–16] for tutorial introductions and overviews. In its plain form, OPF means to solve a non-convex Nonlinear Program (NLP) of finite dimension. This problem can scale up easily to several thousand nodes, hence several thousand decision variables (4· number of nodes). Among the challenges arising are distributed optimization [17–19], convex relaxations [20, 21], and the consideration of stochastic disturbances [22, 23].

In principle, similar to electrical grids, gas grids are typically described by a coupled system of conservation laws, see e.g. [24–28] for an overview. However, due to the compressibility of gas, the arising models differ as one has to consider nonlinear and time-dependent hyperbolic partial differential equations (PDEs) for the gas flow through pipelines supplemented with appropriate coupling conditions at intersections.[2] Since these dynamics allow for discontinuous solutions, a careful theoretical and numerical treatment is needed to master the challenges of simulation, optimization, and control for complex networks. Hence, in the context of gas grids, already the deterministic case is numerically challenging. In the context of the so-called energy transition, there is a strong need to couple the different energy supply systems, in particular power-to-gas, to compensate the differences between supply and demand in the electricity system. Hence several recent papers investigate the

---

[1] For example, the International Energy Agency (IEA) predicts a global need for substantial growth of energy storage capacity [2, 3].

[2] Note that already nowadays so-called linepack flexibility is used for different purposes by operators of gas grids, see for example [29].

coupling of gas and power grids, see [7, 8, 30], or more recently [10–12]. The resulting mathematical problems require the design of appropriate numerical solution methods since the transport of electricity happens on a shorter time scale compared to gas.

It is worth to be remarked that for electrical grids and the corresponding optimization problems, there exists a large number of established benchmarks. This includes IEEE test cases, CIGRE benchmarks [31] and, in case of the German system, the scigrid model [32]. Likewise for gas networks, the GasLib suite[3] offers test cases for simulation and optimization purposes, respectively.

However, when it comes to coupled gas-power systems, there do not exist widely accepted benchmarks. One of the few exceptions appears to be the case study presented in [9], which comprises the IEEE RTS96 One Area 24 node electrical grid and a 24 pipeline gas network. Hence, the present chapter takes further steps towards a more realistic optimization benchmark for multi-energy systems. Specifically, we couple a model of the Greek gas network — the gaslib-134 model [33] which includes 86 pipelines — with the IEEE 300-bus system under AC conditions. We formulate a combined simulation framework, wherein the key actions are taken for the gas network and the electrical side translates into the solution of the AC power flow equations. In particular, we observe a significant pressure drop for the gas-plant nodes while the gas consumed varies over time.

12.2 Model and Algorithm

We model the combined power-and-gas network over a time horizon $t \in [0, T]$ by a graph whose nodes and edges represent certain components of the respective networks. A small example of such a graph is given in Every component contributes variables, parameters and/or model equations, see Sect. 12.2.1 for the gas model, and Sect. 12.2.4 for the power model. We model the coupling between the power grid and the gas network via gas-power-conversion (i.e. a gas power) plants, which are represented by certain edges of the graph. Starting from a continuous-time formulation we then discretize the time interval $[0, T]$ to time steps $t_k = k\frac{T}{N}$, hence formulating the model equations as an algebraic system of nonlinear equations, which is solved with Newton's method. The following two sections describe the gas and power model, and Sect. 12.2.5 describes the coupling of both (Fig. 12.1).

12.2.1 Model of the Gas Network

For the gas network we use the definitions of gaslib [33]. The following theory is also presented in [12]. In our application, a gas edge is one of the following: a

---

[3]http://gaslib.zib.de/.

Fig. 12.1 A graph of a
combined gas and power
network. In the lower part is
a small gas network, the bold
arrow represents a
gas-power-conversion plant,
the upper network consists of
different power components

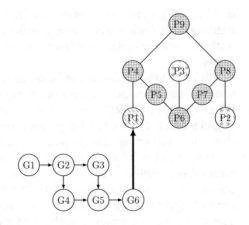

pipeline, a so-called short pipe, a (controlled) valve, a compressor station, or a gas-power-conversion plant, where both valves and compressors act as short pipes in our setting. The nodes represent sources (where gas is injected into the network), sinks (where it is withdrawn) and inner nodes (where no gas is injected or withdrawn).

12.2.1.1 Pipelines

The gas flow in each pipe is modeled by the isentropic Euler equations

$$\begin{pmatrix} \rho_l \\ q_m \end{pmatrix}_t + \begin{pmatrix} q_m \\ p(\rho_l) + \frac{q_m^2}{\rho_l} \end{pmatrix}_x = \begin{pmatrix} 0 \\ S(\rho_l, q_m) \end{pmatrix}, \tag{12.1}$$

where ρ_l is the line density (kg/m) of the gas, q_m is its mass flow (kg/s) and p is the pressure function. As is often done we employ the (possibly space-dependent) density and the volumetric flow are instead

$$\rho = \frac{\rho_l}{A}, \quad q = \frac{q_m}{\rho_0}, \tag{12.2}$$

where A is the cross section of the pipe and ρ_0 is the density at standard conditions. This is relevant for the coupling of pipes with possibly different cross sections, see Sect. 12.2.3. The subscript indices indicate the partial derivatives where $t \in [0, T]$ is time, and $x \in [0, L]$ is the position along the pipe of length L; S is a friction term given by

$$S(\rho, q) = -\frac{\lambda(q)}{2d_{\text{pipe}}} \frac{q|q|}{\rho}. \tag{12.3}$$

The friction factor $\lambda(q)$ is defined by the Prandtl-Colebrook formula

Table 12.1 Gas net constants for (12.8)

ρ_0 [kg/m^3]	p_0 [bar]	z_0	T_0 [k]	T [K]	α [1/bar]
0.785	1.01325	1.005	273.15	283.15	-0.00224

$$\frac{1}{\sqrt{\lambda}} = -2 \log_{10} \left(\frac{2.51}{\text{Re}(q)\sqrt{\lambda}} + \frac{k}{3.71 d_{\text{pipe}}} \right) \tag{12.4}$$

with Reynolds number

$$\text{Re}(q) = \frac{d_{\text{pipe}}}{\eta} q , \tag{12.5}$$

roughness of the pipes $k = 8 \times 10^{-6}$ m and the dynamic viscosity $\eta = 10^{-5}$ kg/(ms) and the pipe diameter d_{pipe}.

We use the isothermal pressure function with compressibility factor

$$p(\rho) = \frac{c_{\text{vac}}^2 \rho}{1 - \alpha c_{\text{vac}}^2 \rho}. \tag{12.6}$$

The pressure function can be inverted, yielding

$$\frac{p}{c_{\text{vac}}^2 z(p)} = \rho, \tag{12.7}$$

where c_{vac} is the limit of the speed of sound in the vacuum limit (that is, for $\rho \to 0$) and $z(p)$ is the compressibility factor. These parameters are given by

$$c_{\text{vac}} = \sqrt{\frac{p_0}{z_0} \frac{T}{T_0} \frac{1}{\rho_0}}, \tag{12.8}$$

$$z(p) = 1 + \alpha p.$$

The numerical values for parameters ρ_0, p_0, z_0, T_0, T, α are listed in Table 12.1. According to Table 12.1 we have that $\alpha < 0$. We define $\beta = -\alpha$ and obtain for the pressure function (12.6).

$$p(\rho) = \frac{c_{\text{vac}}^2 \rho}{1 + \beta c_{\text{vac}}^2 \rho} = \frac{1}{\beta} + \frac{1}{\beta^2 c_{\text{vac}}^2} \left(\frac{1}{\rho + \frac{1}{\beta c_{\text{vac}}^2}} \right). \tag{12.9}$$

At first glance it is not clear whether the isentropic Euler equations are well-posed with this pressure function. Note that, according to [12, Proposition C], well-posedness[4] of the isentropic Euler equations is unaltered under addition or multi-

---

[4]By well-posedness we mean unique solvability.

plication of a positive constant with the pressure function. Therefore (12.9) leads to well-posedness if the innermost bracket $\left(\rho + \frac{1}{\beta c_{vac}^2}\right)^{-1}$ is fulfills the prerequisites of the cited proposition. This can easily be checked to be true, again with [12, Proposition C].

As PDEs, the model equations (12.1) for the pipelines are infinite-dimensional. To use them in our algorithm we need to choose a discretization in both time and space. To this end we employ the implicit box scheme [34], which is of the form

$$\frac{U_{j-1}^* + U_j^*}{2} = \frac{U_{j-1} + U_j}{2} - \frac{\Delta t}{\Delta x}\left(F(U_j^*) - F(U_{j-1}^*)\right) + \Delta t \frac{G(U_j^*) + G(U_{j-1}^*)}{2}.$$
(12.10)

where $\Delta t = \frac{T}{N}$ is the time step-size, $\Delta x = \frac{L}{M}$ the space step-size, and

$$U_j = \begin{pmatrix} \rho(k\Delta t, j\Delta x) \\ q(k\Delta t, j\Delta x) \end{pmatrix} \text{ and } U_j^* = \begin{pmatrix} \rho((k+1)\Delta t, j\Delta x) \\ q(k\Delta t, j\Delta x) \end{pmatrix}, \qquad (12.11)$$

are the states at the last time step and the current time step, respectively. Variables with superscript $(\cdot)^*$ are the unknowns to be computed. In the box scheme (12.10), the flux term F, and the source term G are given by

$$F(U) = \begin{pmatrix} q \\ p(\rho) + \frac{q^2}{\rho} \end{pmatrix} \text{ and } G(U) = \begin{pmatrix} 0 \\ S(\rho, q) \end{pmatrix}. \qquad (12.12)$$

Boundary conditions are prescribed by the nodes in Sect. 12.2.3.

12.2.2 Other Edges

All other edges—apart from the gas-power-conversion plants—act like a short pipe in our setting. A short pipe has no physical properties and exactly two states, corresponding to the beginning and the end of it respectively. Their model equations set these states to be equal. That is, for a short pipe with incoming state (ρ_{in}, q_{in}) and outgoing state (ρ_{out}, q_{out}) there holds $\rho_{in} = \rho_{out}$ and $q_{in} = q_{out}$. Short pipes can be used to separate boundary conditions from coupling conditions on the computational level by inserting the artificial short pipe in between a node with multiple attached pipes and a source/sink node.

12.2.3 Nodes

The nodes of the gas network comprise the algebraic equations coupling the states of the corresponding edges. In addition, the source nodes and the sink nodes entail the boundary conditions describing inflow and outflow, respectively. We use two kinds of boundary conditions, which share the same structure: On the one hand, we enforce equality of the pressure at a node. On the other hand, we demand Bernoulli invariant coupling conditions to be satisfied as introduced in [35]. These coupling conditions yield entropy reduction at the nodes. Specifically, at a node with $l \in \mathbb{N}$ attached edges, the coupling conditions are of the form

$$\sum_{k=1}^{l} q_k = 0$$

$$H(\rho_k, q_k) = H(\rho_{k-1}, q_{k-1}) \text{ for } k = 2, \ldots, l.$$

(12.13)

The pressure coupling condition reads

$$H_p(\rho, q) = p(\rho),$$

(12.14)

while the Bernoulli coupling condition is

$$H_b(\rho, q) = \frac{1}{2}\left(\frac{\rho_0 q}{\rho A}\right)^2 + \int_{\rho_0}^{\rho} \frac{p'(\hat{\rho})}{\hat{\rho}} \, d\hat{\rho}.$$

(12.15)

Note that we use the space-dependent density, not the line density.[5] Because $\rho \mapsto p(\rho)$ is one-to-one (12.15) can be written as

$$H_b(p, q) = \frac{1}{2}\left(\frac{\rho_0 q}{\rho(p) A}\right)^2 + \int_{p_0}^{p} \frac{1}{\rho(\hat{p})} \, d\hat{p}.$$

(12.16)

Note that $v = \frac{\rho_0 q}{\rho A}$ is simply the flow velocity of the gas. If we were to omit the first part of H_b, this would be equivalent to the usual condition of pressure equality. The integral in (12.16) can be solved when $\rho(p)$ is inserted from (12.7). This yields

$$H_b(p, q) = \frac{1}{2}\left(\frac{\rho_0 q}{\rho(p) A}\right)^2 + c_{\text{vac}}^2 \left[\ln\left(\frac{p}{p_0}\right) + \alpha(p - p_0)\right].$$

(12.17)

As just remarked H_b behaves differently from H_p only because of the first part, which involves the velocity. We will see in our simulations in Sect. 12.4.1 that the contribution of this first part is negligible in all pipelines we considered at all times. This seems plausible in case the velocity is much smaller than the speed of sound, which is true for realistic pipeline settings.

Remark 1 (*Implementing Bernoulli coupling conditions*) Although H_b represents the better physical model, it brings about implementation issues: At a node where

---

[5]The line density is usually discontinuous over a node, whereas the three-dimensional space-dependent density is not in case of H_p.

Fig. 12.2 Components at
bus $i \in \mathcal{N}$, and its
connection to the remaining
grid [39]

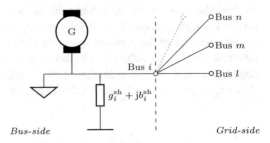

in addition to pipelines a short pipe or any other connection type (like a compressor
or a valve) is attached, the term H_b cannot be easily evaluated, as this requires to
know the pipe cross section; a quantity often not available for these components. A
workaround is to set the pressure for short pipes arbitrarily, and only set the Bernoulli
invariants for all edges that have a known cross section area.

Let us now turn to the model of the electrical grid.

12.2.4 Model of the Power Grid

We study a connected electrical grid under steady state conditions in terms of its
single-phase equivalent; as is commonly done, we model electrical lines by algebraic relations derived from the so-called Π-line equivalent [14, 16, 36–38]. These
standard assumptions simplify the mathematical model of the electrical grid tremen-
dously: instead of partial differential equations for three-phase systems it suffices
to study a system of nonlinear algebraic equations, the so-called power flow equations [14, 36] (Fig. 12.2).

We represent the electrical grid by the triple $(\mathcal{N}, \mathcal{L}, Y)$: $\mathcal{N} = \{1, \ldots, n_{\text{bus}}\}$ is
the non-empty set of buses, $\mathcal{L} = \{1, \ldots, n_{\text{line}}\}$ is the non-empty set of lines, and
$Y = G + iB \in \mathbb{C}^{n_{\text{bus}} \times n_{\text{bus}}}$ is the so-called bus-admittance matrix that contains both
topological and physical information—such as shunts and line impedances—about
the grid [36].[6] Each bus $i \in \mathcal{N}$ is characterized by its voltage phasor $V_i \exp(i\varphi_i)$, and
its net apparent power $P_i + iQ_i$, with i being the imaginary unit. We call V_i voltage
magnitude, φ_i voltage angle, P_i active power, and Q_i reactive power at bus i,
respectively. The power flow equations—here given in polar form—relate the bus
voltages to the powers according to

---

[6]In graph-theoretic terms, $\mathcal{N}$ contains the nodes, $\mathcal{L}$ contains the edges, and Y corresponds to the
weighted Graph Laplacian in the absence of phase-shifting transformers. For the construction of
the bus admittance matrix Y there exist explicit formulas, see e.g. [14, 36]. Alternatively, power
systems software packages such as Matpower [40] provide the functionality to generate the bus
admittance matrix Y from a given case file (via `makeYbus(casefile)`).

Bus name	Fixed quantities
PQ	Active power P_i and Reactive power Q_i
PV	Active power P_i and Voltage magnitude V_i
Slack	Voltage angle φ_i and Voltage magnitude V_i

Table 12.2 Common bus specifications for bus $i \in \mathcal{N}$ [14, 36, 37]

$$P_i = \sum_{j \in \mathcal{N}} V_i V_j \left(G_{ij} \cos(\varphi_i - \varphi_j) + B_{ij} \sin(\varphi_i - \varphi_j) \right), \qquad (12.18a)$$

$$Q_i = \sum_{j \in \mathcal{N}} V_i V_j \left(G_{ij} \sin(\varphi_i - \varphi_j) - B_{ij} \cos(\varphi_i - \varphi_j) \right), \qquad (12.18b)$$

for all buses $i \in \mathcal{N}$. The PF equations (12.18) constitute $2n_{\text{bus}}$ nonlinear and non-convex equations in $4n_{\text{bus}}$ variables. The remaining $2n_{\text{bus}}$ degrees of freedoms are fixed by introducing so-called bus specifications, which are listed in Table 12.2. A PQ represents a load, a PV represents a generator, and the slack bus is a specific generation node that provides an angle reference. The angle reference is needed to overcome the rotational degeneracy of the PF equations (12.18).

A power flow study is to solve the nonlinear system of $4n_{\text{bus}}$ equations built from the $2n_{\text{bus}}$ PF equations (12.18) together with the $2n_{\text{bus}}$ bus specifications from Table 12.2. We concisely write

$$g(x) = 0, \qquad (12.19)$$

where x contains the voltage magnitude, the voltage angle, the active power, and the reactive power of every bus $i \in \mathcal{N}$. These are the model equations provided by the power network. Arguably, there exists a plethora of methods to solve the system (12.19), the Newton method, which is also used by us, being perhaps the most prevalent one [14, 36, 37].

12.2.5 Gas-Power-Conversion

Having covered the modeling of both the gas and the electrical side, how are they connected? Each gas-power-conversion plant is modeled as an edge between the gas network nodes and the power grid nodes listed in Table 12.4. These operate in two modes, namely Gas-to-Power (GtP), that is as a gas power plant and Power-to-Gas (PtG), where surplus electric power is converted to natural gas, e.g. by electrolysis and methanisation. The simplified model equations are taken from [12], where conditions for their applicability is discussed. These are all fulfilled here. In particular the flow in attached pipes must be subsonic, which is the case in realistic pipeline settings. The equations have the same form in both cases, namely

$$q = E_{\text{mode}}(\text{sign}(P))P, \tag{12.20}$$

where P is the power demand (positive) or supply (negative) of the connected power node, q is the outflow of the sink, and E_{mode} a conversion factor of PtG and GtP conversion, respectively. This piece-wise linear model serves as an approximation of the heat rate of a power plant, respectively the efficiency of a PtG-plant. To overcome the non-differentiability of (12.20) at $P = 0$, we employ an interpolating function S

$$S(x, a, b, \epsilon) = x \left(\frac{1}{2}(a + b) - \frac{3}{4}(b - a)\frac{x}{\epsilon} + \frac{(b - a)}{4} \left(\frac{x}{\epsilon}\right)^3 \right), \tag{12.21}$$

with properties

- S is a polynomial of degree 4 in x,
- $S(0, a, b, \epsilon) = 0$,
- $S(\epsilon, a, b, \epsilon) = a \cdot \epsilon$,
- $S(-\epsilon, a, b, \epsilon) = b \cdot (-\epsilon)$,
- $\frac{\partial S}{\partial x}(\epsilon, a, b, \epsilon) = a$,
- $\frac{\partial S}{\partial x}(-\epsilon, a, b, \epsilon) = b$.

Hence, we replace the conversion (12.20) by

$$q = \begin{cases} E_{\text{PtG}} \cdot P & \text{for} & P < -\epsilon \\ S(P, E_{\text{GtP}}, E_{\text{PtG}}, \epsilon) & \text{for} & -\epsilon < P < \epsilon \\ E_{\text{GtP}} \cdot P & \text{for} & \epsilon < P \end{cases}, \tag{12.22}$$

which makes $P \mapsto q(P) \in C^1(\mathbb{R})$ and $q(0) = 0$, so that no gas is taken from or injected into the gas network if electrical power is neither drawn nor supplied. In this way, we don't have to account for switching times beforehand. Instead the switching happens automatically by virtue of the conversion function (12.22). We used $\epsilon = 0.01$ in our examples, which worked well.

12.3 Network Data

12.3.1 Gas Network

We use the gaslib-134 model [33] with inactive compressor and inactive valve. This is a network with 90 sink nodes, 3 source nodes and 86 inner nodes. As connections, there are 86 pipes, 45 short pipes, one compressor and one valve. The pipes have a total length of approximately 1500 km. As valve and compressor are inactive, they just let gas flow through them. We let them act like short pipes with the following exception. Although gaslib-134 doesn't provide one we attach a cross section to these components so they can partake in the Bernoulli coupling (12.15). The compressor

Table 12.3 Volumetric inflow and outflow at source and sink nodes of the gas network

(a) Volumetric inflow for gas network.   (b) Volumetric outflow at sinks other than conversion plants.

Node ID	Inflow [m³/s]
node_1	58.993631
node_20	190.815287
node_80	61.866242

Gas network ID	Outflow [m³/s]
node_ld1	0.000000
node_ld3	0.000000
node_ld4	0.121019
node_ld5	0.000000
node_ld7	1.490446
node_ld8	2.089172
node_ld9	0.000000
node_ld11	5.490446
node_ld14	0.452229
node_ld15	0.280255
node_ld16	0.076433
node_ld17	4.617834
node_ld18	4.617834
node_ld19	0.802548
node_ld20	0.445860
node_ld21	0.286624
node_ld22	7.592357
node_ld23	0.082803
node_ld25	0.802548
node_ld26	0.000000
node_ld27	0.012739
node_ld28	0.000000
node_ld30	1.426752
node_ld32	0.000000
node_ld33	1.101911
node_ld34	0.000000
node_ld35	0.000000
node_ld37	7.732484
node_ld38	0.000000
node_ld39	0.000000
node_ld40	7.732484
node_ld41	1.528662
node_ld43	0.000000
node_ld44	0.000000
node_ld45	0.000000

begins at the end of a single pipe and the valve ends at a single pipe. Therefore we
endow them with the cross section of their respective attached pipes. We do so in
order to have the coupling reach through the entire network. Otherwise, there would
be three distinct parts, one before the compressor, one after the valve and one in
between, that are not coupled through the Bernoulli coupling. The inflow of gas into
the three source nodes and outflow at (non-gas plant) sink nodes of the network is
chosen constant. A list of both can be found in Table 12.3.

Table 12.4 Connection of
gas nodes and power nodes
for conversion

Power grid ID	Gas network ID
213	node_ld31
221	node_ld24
230	node_ld13
7001	node_ld36
7017	node_ld2
7024	node_ld12
7039	node_ld42
7057	node_ld6
7061	node_ld29
7071	node_ld10

12.3.2 Power Network

For the power model we adapt the IEEE 300-bus test case that is part of the Matpower
software [40]. Originally, this system has a total of $n_{bus} = 300$ buses (1 slack bus,
68 PV buses, 231 PQ buses), and $n_{line} = 411$ lines. We modify the grid such that the
original slack bus is now a PV bus, and the nodes listed in Table 12.4 are all slack
buses. These are linked to sinks of the gas network. At these buses, gas and electricity
can be converted into each other. As gaslib-134 is inspired by the Greek gas network
we aim to have a plausible number of gas power plants for Greece. Unfortunately
the authors were unable to find reliable data on the total number of gas power plants
in Greece, although a choice of ten gas power plants seems to be reasonable.[7] The
IDs of the connected sinks in the gas network are given in the table.

Therefore we have a total of 10 slack buses, 59 PV buses and 231 PQ buses. The
nominal total active power generation of the grid is about 24, 000 MW. For the slack
buses and the PV buses, we use the bus specifications from the original case file. For
PQ buses we use time-dependent bus specifications given by

$$P(t) = P_{300} \left(0.9 + 0.4 \sin \left(\frac{2\pi t}{24h} \right) \right),$$
$$Q(t) = Q_{300} \left(0.9 + 0.4 \sin \left(\frac{2\pi t}{24h} \right) \right),$$

(12.23)

where P_{300} and Q_{300} are the active and reactive power demand from the original case
file.

---

[7]See e.g. https://de.wikipedia.org/wiki/Liste_von_Kraftwerken_in_Griechenland and https://en.
wikipedia.org/wiki/List_of_power_stations_in_Greece.

12.3.2.1 Parameters of Gas-Power Conversion

We need to specify values for the conversion factors E_{GtP} and E_{PtG} in (12.20). For the operation as a gas power plant we choose an efficiency of $\eta_{GtP} = 0.4$ with respect to the lower heating value of the gas. This is a realistic value, given that there are gas power plants with efficiencies of up to 60 % [41]. The lower heating value L of natural gas is usually in the range of $36\text{MJ/kg} \le L \le 50\text{MJ/kg}$, depending on the gas composition [42]. We choose $L = 40\text{MJ/kg}$. The parameter E_{GtP} is then obtained from

$$E_{GtP} = \frac{1}{\rho_0 L \eta_{GtP}} \approx 0.0796 \frac{m^3}{MJ}. \tag{12.24}$$

For PtG conversion we choose an efficiency of $\eta_{PtG} = 0.8$, this time with respect to the upper heating value according to [43]. The upper heating value U of natural gas is given by $U = 1.11L$. Therefore we obtain

$$E_{PtG} = \frac{\eta_{PtG}}{\rho_0 U} \approx 0.0229 \frac{m^3}{MJ} \tag{12.25}$$

12.4 Numerical Results

Using the network data from Sect. 12.3 we simulate the combined network over a time horizon of 24 h. Each simulation run took about 20 s on an *Intel(R) Core(TM) i7-7700 CPU @ 3.60GHz* with 32 Gigabytes of RAM.

12.4.1 Comparison of Coupling Conditions

In order to quantify the difference of the two coupling conditions H_p and H_b, we simulate once (simulation "p") with the pressure coupling constant H_p, and once (simulation "b") with the Bernoulli coupling constant H_b. We compare the values of the pressures and the volumetric flows on the whole gas network at every timestep. Let $p_p(t, x)$ be the pressure obtained from simulation "p" at time t and at some position in the network (x ranges over all pipes and all pipe lengths). Further, let $p_b(t, x)$ be the analogue for simulation "b" and let $q_p(t, x)$ and $q_b(t, x)$ the corresponding values for the volumetric flow. Table 12.5 shows our findings with regard to the different coupling constants.

For the relative differences in the flow we used different cut-off values, because although the relative error grows when approaching $q = 0$, the absolute values are very small and hence probably of little significance. In contrast to [35] we find little difference for the two coupling conditions. The key difference is the absence of a friction term in [35], which allows errors to accumulate. In our case artificial energy

Table 12.5 Absolute and relative differences for the two coupling constants
(a) Absolute and relative difference of pressure.

$\max\lvert p_p - p_b\rvert$	$\max \frac{\lvert p_p - p_b\rvert}{p_p}$
0.1828 bar	0.0041

(b) Absolute and relative difference of flow for different cut-off values.

range [m^3/s]	$\max\lvert q_p - q_b\rvert$	$\max \frac{\lvert q_p - q_b\rvert}{\lvert q_p\rvert}$
$10^{-3} < \lvert q_p\rvert < 10^{-2}$	$0.0004\,m^3/s$	0.3270
$10^{-2} < \lvert q_p\rvert < 10^{-1}$	$0.0180\,m^3/s$	0.32670
$10^{-1} < \lvert q_p\rvert < 10^{0}$	$0.0225\,m^3/s$	0.1105
$10^{0} < \lvert q_p\rvert < 10^{1}$	$0.0227\,m^3/s$	0.0207
$10^{1} < \lvert q_p\rvert$	$0.0570\,m^3/s$	0.0029

produced at the nodes is consumed by friction and cannot cause much error. In light of the small size of the error introduced by using the physically unsound pressure coupling constant practitioners should trade-off carefully the need for more accuracy against the practical hurdles mentioned in Remark 1.

12.4.2 Gas-Power-Conversion

We now present the results of Gas-to-Power conversion and Power-to-Gas conversion. Over a day all of the plants go through a cycle of high power demand during which gas is consumed to power a generator. During the second half of the day much less power is needed and so the Gas-to-Power mode is used to convert power back to gas. All the data of pressure and flow in the conversion plants is found in Tables 12.6 and 12.8 on pages xxx and xxx. The total volume of gas consumed by the power plants is obtained by integrating the outflow over time: using the trapezoidal rule, in our case it is $2.3098 \times 10^7\,m^3$, the total volume of gas generated is $2.0522 \times 10^7\,m^3$. Figure 12.3 shows the pressure evolution at the conversion nodes. It shows that the gas network cannot provide the peak power demand indefinitely as the pressure drops considerably during power generation (in the first 12 h). But it is suitable to counter balance high and low power demand over the course of a day as it recuperates during low power demand when gas is injected into the pipeline network by power-to-gas operation.

Table 12.6 Pressure values at the gas-power conversion plants (time in hours, pressure in bar)

t	N7071	N7024	N230	N221	N7061	N7017	N213	N7001	N7039	N7057
0.5000	71.0455	68.7749	69.1475	53.7685	51.8813	74.8464	51.1480	51.0639	50.1773	73.8533
1.0000	70.7183	67.5806	68.3356	53.2087	51.7221	74.6727	51.1268	50.8318	49.9303	73.5722
1.5000	70.2439	66.0861	67.3185	52.1056	51.2539	74.3291	50.8509	50.2887	49.4061	73.1693
2.0000	69.6307	64.2225	66.0704	50.4439	50.4842	73.8227	50.2854	49.4100	48.5786	72.6254
2.5000	68.9075	62.2002	64.7148	48.4085	49.4746	73.1821	49.4661	48.2694	47.5002	71.9598
3.0000	68.0585	59.8037	63.1425	45.8429	48.2067	72.4049	48.3963	46.8280	46.1479	71.1605
3.5000	67.1256	57.4434	61.5661	43.0930	46.7655	71.5286	47.1358	45.2112	44.6087	70.2682
4.0000	66.0892	54.7383	59.8057	39.8636	45.1188	70.5470	45.6883	43.3498	42.8498	69.2680
4.5000	65.0069	52.3600	58.1767	36.7211	43.3824	69.5084	44.1217	41.4267	40.9891	68.2175
5.0000	63.8572	49.7330	56.4249	33.2189	41.5144	68.4044	42.4360	39.3420	38.9853	67.0974
5.5000	62.7105	47.8000	54.9567	30.1795	39.6577	67.2927	40.7078	37.3300	36.9822	65.9799
6.0000	61.5441	45.7813	53.4489	27.0182	37.7607	66.1634	38.9373	35.2621	34.9319	64.8411
6.5000	60.4340	44.7646	52.3513	24.7939	35.9811	65.0778	37.2053	33.4050	32.9930	63.7592
7.0000	59.3556	43.7928	51.2905	22.8096	34.2578	64.0244	35.5113	31.6069	31.1123	62.7058
7.5000	58.3800	43.8553	50.6880	22.1493	32.7428	63.0595	33.9318	30.1305	29.4455	61.7526
8.0000	57.4819	43.9650	50.1644	21.9610	31.3714	62.1701	32.4653	28.8127	27.9376	60.8702
8.5000	56.7205	44.8620	50.0599	22.9462	30.2741	61.4023	31.1786	27.8732	26.7254	60.1165
9.0000	56.0720	45.7336	50.0430	24.1956	29.3906	60.7433	30.0709	27.1528	25.7559	59.4658
9.5000	55.5805	47.0555	50.3491	26.0187	28.8166	60.2267	29.1967	26.7986	25.1315	58.9597
10.0000	55.2262	48.2887	50.7319	27.7734	28.4977	59.8413	28.5574	26.6718	24.8009	58.5786
10.5000	55.0366	49.6904	51.3252	29.5911	28.4850	59.6077	28.1880	26.8397	24.8182	58.3477
11.0000	54.9968	50.9792	51.9798	31.1656	28.7349	59.5170	28.0627	27.2097	25.1356	58.2543
11.5000	55.1181	52.2577	52.7481	32.5462	29.2599	59.5771	28.1738	27.7955	25.7606	58.3085
12.0000	55.3914	53.4322	53.5678	33.6339	30.0288	59.7817	28.5369	28.5618	26.6578	58.5035
12.5000	55.8117	54.5168	54.4360	34.3566	30.9620	60.1269	29.2231	29.5118	27.8101	58.8370
13.0000	56.3744	55.5234	55.3512	35.1781	32.0166	60.6090	30.1445	30.5516	29.1532	59.3059
13.5000	57.0377	56.4177	56.2633	36.1206	33.1627	61.2019	31.2350	31.7098	30.6045	59.8902
14.0000	57.7644	57.2086	57.1477	37.1864	34.3806	61.8273	32.4753	33.0018	32.1458	60.5551
14.5000	58.5205	57.9972	57.9750	38.3290	35.6358	62.4940	33.8232	34.3851	33.7216	61.2679
15.0000	59.3027	58.8135	58.7796	39.5603	36.9585	63.1991	35.2322	35.8317	35.2923	62.0055
15.5000	60.0868	59.6250	59.5729	40.8282	38.3172	63.9286	36.6674	37.2919	36.8079	62.7516
16.0000	60.8857	60.4567	60.3846	42.1504	39.7089	64.6803	38.1296	38.7701	38.3199	63.5143
16.5000	61.6906	61.2860	61.2007	43.4672	41.1107	65.4437	39.6022	40.2453	39.8220	64.2869
17.0000	62.5088	62.1313	62.0320	44.8103	42.5271	66.2192	41.0866	41.7276	41.3251	65.0711
17.5000	63.3278	62.9626	62.8593	46.1200	43.9358	66.9971	42.5654	43.1923	42.8105	65.8588
18.0000	64.1539	63.8019	63.6946	47.4359	45.3425	67.7787	44.0400	44.6500	44.2854	66.6514
18.5000	64.9731	64.6178	64.5172	48.7006	46.7247	68.5546	45.4923	46.0749	45.7290	67.4402
19.0000	65.7912	65.4336	65.3394	49.9551	48.0882	69.3264	46.9234	47.4773	47.1472	68.2263
19.5000	66.5942	66.2208	66.1419	51.1469	49.4114	70.0848	48.3157	48.8328	48.5196	69.0008
20.0000	67.3871	67.0006	66.9357	52.3142	50.6996	70.8313	49.6701	50.1506	49.8515	69.7646

(continued)

Table 12.7 (continued)

t	N7071	N7024	N230	N221	N7061	N7017	N213	N7001	N7039	N7057
20.5000	68.1570	67.7502	67.7045	53.4132	51.9333	71.5570	50.9705	51.4093	51.1240	70.5093
21.0000	68.9080	68.4857	68.4565	54.4761	53.1169	72.2629	52.2177	52.6164	52.3416	71.2351
21.5000	69.6079	69.1644	69.1454	55.4625	54.2182	72.9363	53.3731	53.7289	53.4334	71.9165
22.0000	70.2464	69.7832	69.7330	56.3825	55.2219	73.5701	54.4081	54.7299	54.3815	72.5376
22.5000	70.7852	70.2768	70.1534	57.1860	56.0680	74.1509	55.3028	55.5863	55.1608	73.0851
23.0000	71.1909	70.5080	70.3734	57.8705	56.7468	74.5951	56.0517	56.2940	55.7739	73.5244
23.5000	71.4407	70.4238	70.3619	58.4119	57.2346	74.8895	56.6179	56.8014	56.1775	73.8243
24.0000	71.5473	70.0346	70.1380	58.7883	57.4952	75.0361	56.9596	57.0237	56.3332	73.9787

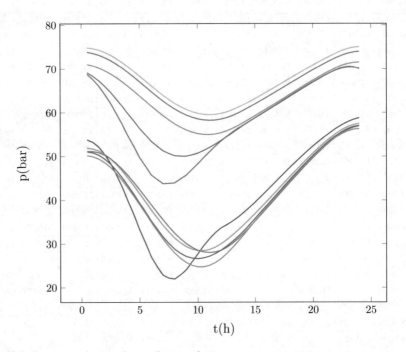

Fig. 12.3 Pressure at the gas-plant nodes over time

Table 12.8 Flow values at the gas-power conversion plants (time in hours, in m^3/s). Positive flow means gas is converted to power, negative flow means gas is generated with power

t	N7071	N7024	N230	N221	N7061	N7017	N213	N7001	N7039	N7057
0.5000	8.3239	28.8649	24.8018	12.4990	28.1027	25.0951	19.4250	29.8060	44.6346	13.4619
1.0000	10.9487	35.7187	28.9484	25.9047	34.5863	32.1913	22.8247	42.6988	52.1582	15.8722
1.5000	13.4409	42.2100	32.8427	38.5824	40.7160	39.0234	26.0325	54.8326	59.2368	18.1753
2.0000	15.9830	48.8202	36.7720	51.4521	46.9399	46.0882	29.2849	67.0847	66.3854	20.5409
2.5000	18.2112	54.6092	40.1799	62.6711	52.3712	52.3627	32.1192	77.7048	72.5857	22.6303
3.0000	20.4881	60.5241	43.6260	74.0623	57.8987	58.8562	34.9989	88.4243	78.8518	24.7838
3.5000	22.2743	65.1663	46.3020	82.9349	62.2208	64.0105	37.2452	96.7253	83.7131	26.4890
4.0000	24.1016	69.9199	49.0121	91.9391	66.6334	69.3407	39.5298	105.1035	88.6320	28.2512
4.5000	25.2758	72.9781	50.7371	97.6769	69.4672	72.7977	40.9895	110.4176	91.7613	29.3956
5.0000	26.4741	76.1035	52.4825	103.4846	72.3621	76.3526	42.4712	115.7769	94.9279	30.5756
5.5000	26.8893	77.1876	53.0833	105.4835	73.3669	77.5910	42.9824	117.6172	96.0183	30.9879
6.0000	27.3082	78.2823	53.6873	107.4927	74.3824	78.8442	43.4970	119.4648	97.1151	31.4061
6.5000	26.8893	77.1876	53.0833	105.4835	73.3669	77.5910	42.9824	117.6172	96.0183	30.9879
7.0000	26.4741	76.1035	52.4825	103.4846	72.3621	76.3526	42.4712	115.7769	94.9279	30.5756
7.5000	25.2758	72.9781	50.7371	97.6769	69.4672	72.7977	40.9895	110.4176	91.7613	29.3956
8.0000	24.1016	69.9199	49.0121	91.9391	66.6334	69.3407	39.5298	105.1035	88.6320	28.2512
8.5000	22.2743	65.1663	46.3020	82.9349	62.2208	64.0105	37.2452	96.7253	83.7131	26.4890
9.0000	20.4881	60.5241	43.6260	74.0623	57.8987	58.8562	34.9989	88.4243	78.8518	24.7838
9.5000	18.2112	54.6092	40.1799	62.6711	52.3712	52.3627	32.1192	77.7048	72.5857	22.6303
10.0000	15.9830	48.8202	36.7720	51.4521	46.9399	46.0882	29.2849	67.0847	66.3854	20.5409
10.5000	13.4409	42.2100	32.8427	38.5824	40.7160	39.0234	26.0325	54.8326	59.2368	18.1753
11.0000	10.9487	35.7187	28.9484	25.9047	34.5863	32.1913	22.8247	42.6988	52.1582	15.8722
11.5000	8.3239	28.8649	24.8018	12.4990	28.1027	25.0951	19.4250	29.8060	44.6346	13.4619
12.0000	5.7461	22.1108	20.6834	−0.2049	21.7107	18.2245	16.0640	17.0461	37.1834	11.1087
12.5000	3.2120	15.4428	16.5889	−3.9583	15.4078	11.5666	12.7370	4.4213	29.8041	8.8081
13.0000	0.7195	8.8495	12.5149	−7.6573	9.1946	5.1122	9.4404	−2.3249	22.4972	6.5572
13.5000	−0.4518	2.7640	8.7345	−11.0546	3.4888	−0.2088	6.3931	−5.6415	15.7555	4.5029
14.0000	−1.1012	−0.9434	4.9679	−14.4022	−0.6141	−1.8424	3.3676	−8.9201	9.0817	2.4899
14.5000	−1.6511	−2.4281	1.7433	−17.2346	−1.9831	−3.2062	0.7856	−11.7036	3.4081	0.7944

(continued)

Table 12.9 (continued)

t	N7071	N7024	N230	N221	N7061	N7017	N213	N7001	N7039	N7057
15.0000	−2.1938	−3.9052	−0.4246	−20.0245	−3.3294	−4.5335	−0.5138	−14.4560	−0.6368	−0.2506
15.5000	−2.6052	−5.0343	−1.1344	−22.1333	−4.3450	−5.5267	−1.0792	−16.5457	−1.8675	−0.6122
16.0000	−3.0121	−6.1605	−1.8428	−24.2108	−5.3433	−6.4974	−1.6424	−18.6152	−3.0871	−0.9677
16.5000	−3.2655	−6.8672	−2.2874	−25.4986	−5.9609	−7.0957	−1.9952	−19.9052	−3.8476	−1.1879
17.0000	−3.5170	−7.5734	−2.7314	−26.7704	−6.5697	−7.6846	−2.3472	−21.1865	−4.6031	−1.4055
17.5000	−3.6024	−7.8142	−2.8826	−27.2001	−6.7753	−7.8832	−2.4670	−21.6215	−4.8595	−1.4790
18.0000	−3.6875	−8.0549	−3.0338	−27.6276	−6.9796	−8.0807	−2.5867	−22.0555	−5.1153	−1.5522
18.5000	−3.6024	−7.8142	−2.8826	−27.2001	−6.7753	−7.8832	−2.4670	−21.6215	−4.8595	−1.4790
19.0000	−3.5170	−7.5734	−2.7314	−26.7704	−6.5697	−7.6846	−2.3472	−21.1865	−4.6031	−1.4055
19.5000	−3.2655	−6.8672	−2.2874	−25.4986	−5.9609	−7.0957	−1.9952	−19.9052	−3.8476	−1.1879
20.0000	−3.0121	−6.1605	−1.8428	−24.2108	−5.3433	−6.4974	−1.6424	−18.6152	−3.0871	−0.9677
20.5000	−2.6052	−5.0343	−1.1344	−22.1333	−4.3450	−5.5267	−1.0792	−16.5457	−1.8675	−0.6122
21.0000	−2.1938	−3.9052	−0.4246	−20.0245	−3.3294	−4.5335	−0.5138	−14.4560	−0.6368	−0.2506
21.5000	−1.6511	−2.4281	1.7433	−17.2346	−1.9831	−3.2062	0.7856	−11.7036	3.4081	0.7944
22.0000	−1.1012	−0.9434	4.9679	−14.4022	−0.6141	−1.8424	3.3676	−8.9201	9.0817	2.4899
22.5000	−0.4518	2.7640	8.7345	−11.0546	3.4888	−0.2088	6.3931	−5.6415	15.7555	4.5029
23.0000	0.7195	8.8495	12.5149	−7.6573	9.1946	5.1122	9.4404	−2.3249	22.4972	6.5572
23.5000	3.2120	15.4428	16.5889	−3.9583	15.4078	11.5666	12.7370	4.4213	29.8041	8.8081
24.0000	5.7461	22.1108	20.6834	−0.2049	21.7107	18.2245	16.0640	17.0461	37.1834	11.1087

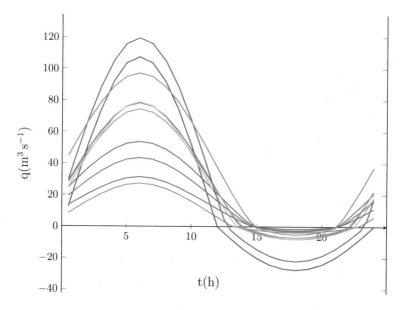

Fig. 12.4 Gas flow for power conversion over time

Figure 12.4 shows the amount of gas consumed ($q > 0$) by power generation and generated ($q < 0$) by power-to-gas operation. Note the kink at the switching times, where $q = 0$ which is due to the difference in efficiencies of the two processes.

12.5   Conclusion

We have presented a model for a combined gas and power network and simulated the operation of it over a time horizon of 24 h. We evaluated the impact of two different coupling conditions, namely pressure coupling and Bernoulli coupling, and found them to be negligible for practical purposes of pipeline simulation. In addition our simulation results showed the given gas network to be able to provide enough power to counter balance power demand peaks and our tools provide visualization and quantification of gas consumed and produced. Our data can be used to benchmark similar tools. The data can be found under https://bitbucket.org/efokken/gas-power-benchmark

Future work includes the solution of corresponding optimal control problems and the inclusion of stochastic effects.

References

1. Bundesgesetzblatt. Tech. rep. G 5702 Teil I, Nr. 43. Bundesanzeiger Verlag GmbH (2011)
2. Tracking Energy Integration. Tech. rep. Paris, France: International Energy Agency (IEA) (2019). https://www.iea.org/reports/tracking-energy-integration
3. Zablocki, A.: Fact Sheet: Energy Storage. Tech. rep. Environmental and Energy Study Institute (EESI) (2019). https://www.eesi.org/papers/view/energy-storage-2019 (visited on 05/27/2020)
4. Schiebahn, S., Grube, T., Robinius, M., Tietze, V., Kumar, B., Stolten, D.: Power to gas: technological overview, systems analysis and economic assessment for a case study in Germany. Int. J. Hydrog. Energy **40**(12), 4285–4294 (2015)
5. Brown, T., Schlachtberger, D., Kies, A., Schramm, S., Greiner, M.: Synergies of sector coupling and transmission reinforcement in a cost-optimised, highly renewable European energy system. Energy **160**, 720–739 (2018)
6. Heinen, S., Burke, D., O'Malley, M.: Electricity, gas, heat integration via residential hybrid heating technologies - an investment model assessment. Energy **109**, pp. 906–919 (2016)
7. Chertkov, M., Backhaus, S., Lebedev, V.: Cascading of fluctuations in interdependent energy infrastructures: gas-grid coupling. Appl. Energy **160**, 541–551 (2015). https://doi.org/10.1016/j.apenergy.2015.09.085
8. Zeng, Q., Fang, J., Li, J., Chen, Z.: Steady-state analysis of the integrated natural gas and electric power system with bi-directional energy conversion. Appl. Energy **184**(C), 1483–1492 (2016). https://EconPapers.repec.org/RePEc:eee:appene:v:184:y:2016:i:c:p:1483-1492
9. Zlotnik, A., Roald, L., Backhaus, S., Chertkov, M., Andersson, G.: Coordinated scheduling for interdependent electric power and natural gas infrastructures. IEEE Trans. Power Syst. **32**(1), 600–610 (2016)
10. Herty, M., Müller, S., Sikstel, A.: Coupling of compressible Euler equations. In: Vietnam Journal of Mathematics (2019). issn: 2305-2228. https://doi.org/10.1007/s10013-019-00353-7
11. Fokken, E., Göttlich, S., Kolb, O.: Optimal control of compressor stations in a coupled gas-to-power network. In: Bertsch, V., Ardone, A., Suriyah, M., Fichtner, W., Leibfried, T., Heuveline, V. (eds.) Advances in Energy System Optimization, pp. 67–80. Springer International Publishing, Cham (2020). isbn: 978-3-030-32157-4
12. Fokken, E., Göttlich, S., Kolb, O.: Modeling and simulation of gas networks coupled to power grids. In: Journal of Engineering Mathematics (2019). issn: 1573-2703. https://doi.org/10.1007/s10665-019-10026-6
13. O'Malley, C., Delikaraoglou,S., Roald, L., Hug, G.: Natural gas system dispatch accounting for electricity side flexibility. Electr. Power Syst. Res. **178**, 106038 (2020)
14. Frank, S., Rebennack, S.: An introduction to optimal power flow: theory, formulation, and examples. IIE Trans. **48**(12), 1172–1197 (2016). https://doi.org/10.1080/0740817X.2016.1189626
15. Capitanescu, F.: Critical review of recent advances and further developments needed in AC optimal power flow. Electr. Power Syst. Res. **136**, 57–68 (2016)
16. Faulwasser, T., Engelmann, A., Mühlpfordt, T., Hagenmeyer, V.: Optimal power flow: an introduction to predictive, distributed and stochastic control challenges. Automatisierungstechnik **66**(7), 573–589 (2018). https://doi.org/10.1515/auto-2018-0040
17. Engelmann, A., Jiang, Y., Mühlpfordt, T., Houska, B., Faulwasser, T.: Toward distributed OPF using ALADIN. IEEE Trans. Power Syst. **34**(1), 584–594 (2019). issn: 0885-8950. https://doi.org/10.1109/TPWRS.2018.2867682
18. Molzahn, D., Dörfler, F., Sandberg, H., Low, S.H., Chakrabarti, S., Baldick, R., Lavaei, J.: A survey of distributed optimization and control algorithms for electric power systems. IEEE Trans. Smart Grid **8**(6), 2941–2962 (2017)
19. Kim, B.H., Baldick, R.: A comparison of distributed optimal power flow algorithms. IEEE Trans. Power Syst. **15**(2), 599–604 (2000)
20. Low, S.H.: Convex relaxation of optimal power flow-Part I: formulations and equivalence. IEEE Trans. Control Netw. Syst. **1**(1), 15–27 (2014)

21. Low, S.H.: Convex relaxation of optimal power flow-Part II: exactness. IEEE Trans. Control Netw. Syst. **1**(2), 177–189 (2014)
22. Mühlpfordt, T., Roald, L., Hagenmeyer, V., Faulwasser, T., Misra, S.: Chance-constrained AC optimal power flow - a polynomial chaos approach. IEEE Trans. Power Syst. (2019). arXiv:1903.11337. https://doi.org/10.1109/TPWRS.2019.2918363
23. Bienstock, D., Chertkov, M., Harnett, S.: Chance-constrained optimal power flow: risk-aware network control under uncertainty. Siam Rev. **56**(3), 461–495 (2014)
24. Banda, M., Herty, M., Klar, A.: Coupling conditions for gas networks governed by the isothermal Euler equations. Netw. Heterog. Media **1**, 295–314 (2006). https://doi.org/10.3934/nhm.2006.1.295
25. Banda, M., Herty, M., Klar, A.: Gas flow in pipeline networks. Netw. Heterog. Media **1**(1), 41–56 (2006). https://doi.org/10.3934/nhm.2006.1.41
26. Bressan, A., Canic, S., Garavello, M., Herty, M., Piccoli, B.: Flow on networks: recent results and perspectives. Eur. Math. Soc.-Surv. Math. Sci. **1**(1), 47–111 (2014). https://doi.org/10.4171/EMSS/2
27. Brouwer, J., Gasser, I., Herty, M.: Gas pipeline models revisited: model hierarchies, nonisothermal models, and simulations of networks. Multiscale Model. Simul. **9**(2), 601–623 (2011). https://doi.org/10.1137/100813580
28. Colombo, R., Garavello, M.: On the Cauchy problem for the p-system at a junction. SIAM J. Math. Anal. **39**(5), 1456–1471 (2008). https://doi.org/10.1137/060665841
29. NTS linepack. Tech. rep. NationalGrid UK (2019). https://www.nationalgrid.com/uk/gas-transmission/balancing/nts-linepack (visited on 05/27/2020)
30. Zlotnik, A., Roald, L., Backhaus, S., Chertkov, M., Andersson, G.: Control policies for operational coordination of electric power and natural gas transmission systems. In: 2016 American Control Conference (ACC). July 2016, pp. 7478–7483 (2016). https://doi.org/10.1109/ACC.2016.7526854
31. Strunz, K.: Developing benchmark models for studying the integration of distributed energy resources. In: 2006 IEEE Power Engineering Society General Meeting. IEEE, 2-pp (2006)
32. Matke, C., Medjroubi, W., Kleinhans, D., Sager, S.: Structure analysis of the German transmission network using the open source model SciGRID. In: Advances in Energy System Optimization. Springer, pp. 177–188 (2017)
33. Humpola, J., Joormann, I., Kanelakis, N., Oucherif, D., Pfetsch, M., Schewe, L., Schmidt, M., Schwarz, R., Sirvent, M.: GasLib - A library of gas network instances. Tech. rep. Sept. (2017). http://www.optimization-online.org/DB_HTML/2015/11/5216.html
34. Kolb, O., Lang, J., Bales, P.: An implicit box scheme for subsonic compressible flow with dissipative source term. Numer. Algorithms **53**(2), 293–307 (2010). https://doi.org/10.1007/s11075-009-9287-y
35. Reigsta, G.: Numerical network models and entropy principles for isothermal junction flow. Netw. Heterog. Media **9**(1556–1801), 65 (2014). issn: 1556-1801. https://doi.org/10.3934/nhm.2014.9.65. http://aimsciences.org//article/id/fd4fb8f6-dc83-405c-951d-b45d4114f45f
36. Andersson, G.: Lecture Notes on Power System Analysis. Power Flow Analysis, Fault Analysis, Power System Dynamics and Stability (2015)
37. Grainger, J.J., Stevenson, W.D.: Power System Analysis. McGraw-Hill Education (1994)
38. Wood, A.J., Wollenberg, B.F.: Power Generation, Operation, and Control. Wiley (2012)
39. Mühlpfordt, T.: Uncertainty Quantification via Polynomial Chaos Expansion - Methods and Applications for Optimization of Power Systems. Dissertation. Karlsruhe Institute of Technology
40. Zimmerman, R.D., Murillo-Sanchez, C.E., Thomas, R.J.: MATPOWER: steady-state operations, planning, and analysis tools for power systems research and education. In: IEEE Transactions on Power Systems **26**(1), 12–19 (2011). issn: 0885-8950. https://doi.org/10.1109/TPWRS.2010.2051168
41. Danish Energy Agency. Technology Data - Generation of Electricity and District Heating. Tech. rep. Energistyrelsen (Danish Energy Agency) (2019). https://ens.dk/sites/ens.dk/files/Analyser/technology_data_catalogue_for_el_and_dh.pdf (visited on 01/12/2020)

42. Cerbe, G., Lendt, B.: Grundlagen der Gastechnik: Gasbeschaffung - Gasverteilung - Gasver-
    wendung. Carl Hanser Verlag GmbH & Company KG (2016). isbn: 9783446449664
43. Trimis, D., Anger, S.: Potenzial der thermisch integrierten Hochtemperaturelektrolyse und
    Methanisierung für die Energiespeicherung durch Powerto-Gas (PtG). In: gfw Gas 155.1-2,
    pp. 50–59 (2014)

Chapter 13
Coupling of Two Hyperbolic Systems by Solving Half-Riemann Problems

Michael Herty, Siegfried Müller, and Aleksey Sikstel

Abstract The modelling of gas networks requires the development of coupling techniques at junctions. Recent work on the coupling of hyperbolic systems based on solving two half Riemann problems can be useful also for the coupling issue in gas networks. This strategy is exemplified here for the coupling of a fluid with a solid modelled by the Euler equations supplemented with a stiffened gas equation and a linear elastic model, respectively. This framework may serve as a basis for investigations of coupling conditions on nodes of a gas network.

Keywords Linear elastic equations · Compressible Euler equations · Coupling conditions · Coupled Riemann problem · Lax curves

13.1 Introduction

The modelling of natural gas-fired power generator requires the coupling of two networks: the natural gas transportation and the electric transmission network [1, 19, 23]. It is reasonable to assume that strong time-dependent changes appear according to the demand of the electricity network lead to pressure fluctuations along the natural

This work has been supported in part by the German Research Council (DFG) within the DFG Collaborative Research Center SFB-TR-40, TP A1, and by KI-Net NSF RNMS grant No. 1107444, grants DFG Cluster of Excellence Production technologies for high-wage countries', HE5386/13,14,15- 1, DAAD-MIUR project, Deutsche Forschungsgesellschaft HE5386/13,15 and Bundesministerium für Bildung und Forschung 05M18PAA.

M. Herty (✉) · S. Müller · A. Sikstel
Institut für Geometrie und Praktische Mathematik, RWTH Aachen, Templergraben 55, 52056 Aachen, Germany
e-mail: herty@igpm.rwth-aachen.de

S. Müller
e-mail: mueller@igpm.rwth-aachen.de

A. Sikstel
e-mail: sikstel@igpm.rwth-aachen.de

© Springer Nature Switzerland AG 2021
S. Göttlich et al. (eds.), *Mathematical Modeling, Simulation and Optimization for Power Engineering and Management*, Mathematics in Industry 34,
https://doi.org/10.1007/978-3-030-62732-4_13

gas pipe network. Thus, we are interested in the fine-scale resolution of the resulting flow patterns in the gas network. In particular, we consider the compressible Euler equations that are suitable for fine-scale modelling of the gas dynamics. The relevant regime and the underlying scaling leading to such a model can be found e.g. in [2].

We present a general framework for coupling of hyperbolic conservation laws based on the solutions of Riemann problems. In particular we apply this strategy to coupled linear elastic structures and fluids governed by the compressible Euler equations at a steady interface. We refer to this kind of problems as fluid-structure coupling problem (FSC). For a perfect gas equation of state (EoS) the FSC has been solved previously in [12] by means of coupled RPs. This leads in fact to identifying the unique root of a scalar nonlinear function. Properties of this function, and hence, conditions for the existence of a unique solution of the FSC have been presented in Theorem 3.1, [12].

In this work we present a new result, namely the FSC with a stiffened gas EoS. Those EoS are more suitable for fluids such as water. The stiffened gas EoS models the fluid as an ideal gas under high pressure. For that purpose, pressure and internal energy of an ideal gas EoS are scaled. Although the shift from ideal to stiffened gas is linear, the solution of the FSC with stiffened gas EoS may differ drastically from the ideal fluid case due to the *nonlinear* nature of the fluid. Lax curves corresponding to the compressible Euler equations with stiffened gas EoS have to be determined to obtain states at the coupling interface. A scalar nonlinear function is then constructed. Its root provides a unique solution to the FSC. We state similar results and accompanying proofs as in [12].

Although elastic structures do not appear in common models for gas networks, the framework employed for the solution of the FSC may be applied to fine-scale models for gas networks. In particular, the modelling of turbine effects on high-pressure gas networks proposed in [13] assumed a perfect gas EoS. This model may be extended to general natural gas mixtures in a similar manner. We refer to [17] for an overview of natural gas EoS.

13.2   Riemann Problems for Coupled Conservation Laws

Coupling the dynamics requires to postulate conditions to be fulfilled at the interface for almost all times $t \geq 0$. Let the domain Ω be partitioned into two subdomains $\Omega_- \cup \Omega_+ \cup \Gamma = \Omega$ by a fixed interface plane Γ, see Fig. 13.1. Assume that the dynamics in Ω_- are governed by a system of conservation laws

$$\overline{\mathbf{w}}_t + \mathrm{div}(\overline{\mathbf{f}}(\overline{\mathbf{w}})) = 0, \ \mathbf{x} \in \Omega_-, \ t \geq 0, \tag{13.1}$$

$$\overline{\mathbf{w}}(0, \mathbf{x}) = \overline{\mathbf{w}}_0(\mathbf{x}), \ \mathbf{x} \in \Omega_- \tag{13.2}$$

Fig. 13.1 Domain of the coupled conservation laws

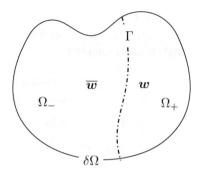

for the conserved quantity $\overline{\mathbf{w}} \colon \mathbb{R}_+ \times \Omega_- \to \overline{\mathcal{D}}$ and differentiable flux function $\overline{\mathbf{f}} \colon \overline{\mathcal{D}} \to \mathbb{R}^{\overline{m}}$ where $\overline{\mathcal{D}} \subset \mathbb{R}^{\overline{m}}$ is the set of admissible values for $\overline{\mathbf{w}}$. The dynamics in the other half of the domain, Ω_+, is governed by the system

$$\mathbf{w}_t + \mathrm{div}(\mathbf{f}(\mathbf{w})) = 0, \ \mathbf{x} \in \Omega_+, \ t \geq 0 \tag{13.3}$$

$$\mathbf{w}(0, \mathbf{x}) = \mathbf{w}_0(\mathbf{x}) \text{ for } \mathbf{x} \in \Omega_+ \tag{13.4}$$

for $\mathbf{w} \colon \mathbb{R}_+ \times \Omega_+ \to \mathcal{D}$ and a possibly different differentiable flux function $\mathbf{f} \colon \mathcal{D} \to \mathbb{R}^m$ where $\mathcal{D} \subset \mathbb{R}^m$ is the set of admissible values for $\mathbf{w}$. At the non-interface boundary of each domain $\delta\Omega_- \backslash \Gamma$ and $\delta\Omega_+ \backslash \Gamma$ we prescribe boundary conditions, cf. [11]. At the interface Γ we prescribe coupling conditions

$$\Psi\left(\overline{\mathbf{w}}(t, \mathbf{x}), \mathbf{w}(t, \mathbf{x})\right) = 0, \ \mathbf{x} \in \Gamma \tag{13.5}$$

where $\Psi \colon \mathbb{R}^{\overline{m}+m} \to \mathbb{R}^\ell$. We refer to [5] for a precise definition in the case of a 2×2 system of conservation laws.

Definition 1 Let the domain Ω be separated by a smooth interface Γ, i.e. $\Omega = \Omega_- \cup \Omega_+ \cup \Gamma$, see Fig. 13.1. As a **weak solution of the coupled problem** we understand a pair of weak entropy solutions $(\overline{\mathbf{w}}(t, \mathbf{x}), \mathbf{w}(t, \mathbf{x}))$ to Eqs. (13.1) and (13.3) for $\mathbf{x} \in \Omega_-$ and $\mathbf{x} \in \Omega_+$, respectively. Further, assuming the traces of $\overline{\mathbf{w}}$ and $\mathbf{w}$ at Γ exist, the solution at $\mathbf{x} \in \Gamma$ fulfils a.e. in t the coupling condition $\Psi = 0$, that is

$$\Psi\left(\overline{\mathbf{w}}(t, \mathbf{x}^-), \mathbf{w}(t, \mathbf{x}^+)\right) = 0, \ t > 0, \ \mathbf{x} \in \Gamma, \tag{13.6}$$

where

$$\begin{aligned} \overline{\mathbf{w}}(\mathbf{x}^-) &:= \lim_{\varepsilon \to 0^+} \overline{\mathbf{w}}(\mathbf{x} - \varepsilon\mathbf{n}), \\ \mathbf{w}(\mathbf{x}^+) &:= \lim_{\varepsilon \to 0^+} \mathbf{w}(\mathbf{x} + \varepsilon\mathbf{n}) \end{aligned} \tag{13.7}$$

denote the limits in the normal direction $\mathbf{n} = \mathbf{n}(\mathbf{x})$ of Γ.

Remark 1 It is reasonable to assume that whenever $\ell = \overline{m} = m, \mathbf{f} = \overline{\mathbf{f}}$ and $\Psi(\overline{\mathbf{v}}, \mathbf{v}) = \overline{\mathbf{f}}(\overline{\mathbf{v}}) - \mathbf{f}(\mathbf{v})$ the solution of the coupling problem coincides with the solution of the single conservation law

$$\mathbf{w}_t + \mathrm{div}(\mathbf{f}(\mathbf{w})) = 0, \ \mathbf{x} \in \Omega, \ t \geq 0, \tag{13.8}$$

$$\mathbf{w}(0, \mathbf{x}) = \overline{\mathbf{w}}_0(\mathbf{x}), \ \mathbf{x} \in \Omega_-, \tag{13.9}$$

$$\mathbf{w}(0, \mathbf{x}) = \mathbf{w}_0(\mathbf{x}), \ \mathbf{x} \in \Omega_+. \tag{13.10}$$

In order to solve the coupled problem (13.1)–(13.5) an approach is to iterate the coupling condition at each time step solving alternately the conservation laws (13.1) and (13.3) as in [8, 14]. Besides being computationally expensive such methods have significant drawbacks. Firstly, it is in general an open question if and how fast the iterative method for the coupling condition converges. Secondly, given a convergent method there is no guarantee that the limit constitutes an entropy solution in Ω_+.

Alternatively, one may use a strategy based on the solution of coupled Riemann Problems (RP) that has been applied and validated e.g. in [4, 12]. To this end, we project the conservation laws (13.1) and (13.3) to the normal direction at Γ and define two (half-)RPs that are coupled by conditions $\Psi = 0$ at Γ. A (half-)RP associated with the conservation law (13.1) on Ω_- consists of the initial data

$$\overline{\mathbf{w}}(0, \mathbf{x}) = \begin{cases} \overline{\mathbf{w}}_L & \text{if } \mathbf{x} \in \Omega_-, \\ \overline{\mathbf{w}}_\Gamma & \text{if } \mathbf{x} \in \Gamma \end{cases} \tag{13.11}$$

where $\overline{\mathbf{w}}_L$ and $\overline{\mathbf{w}}_\Gamma$ are constant. The solution $\overline{\mathbf{w}} = \overline{\mathbf{w}}(t, \mathbf{x})$ of the (half-)RP is the solution *restricted to* Ω_- of the classical RP for the projected conservation law (13.1) and initial data (13.11). Similarly, we consider a (half-)RP associated with the projected conservation law (13.3) on Ω_+ and initial data

$$\mathbf{w}(0, \mathbf{x}) = \begin{cases} \mathbf{w}_\Gamma & \text{if } \mathbf{x} \in \Gamma, \\ \mathbf{w}_R & \text{if } \mathbf{x} \in \Omega_+ \end{cases} \tag{13.12}$$

where again $\mathbf{w}_\Gamma$ and $\mathbf{w}_R$ are constant. In view of the coupling problems, particularly interesting are solutions where the trace of the solution in Ω_- fulfils $\overline{\mathbf{w}}(t, \mathbf{x}^-) = \overline{\mathbf{v}}$, and in Ω_+ the trace of the solution fulfils $\mathbf{w}(t, \mathbf{x}^+) = \mathbf{v}$ such that the coupling conditions $\Psi(\overline{\mathbf{v}}, \mathbf{v}) = 0$ hold true.

Given a constant state $\overline{\mathbf{w}}_L$ we introduce the notion of admissible boundary states as follows. The set

$$\overline{V}(\overline{\mathbf{w}}_L) := \left\{ \overline{\mathbf{v}} \in \overline{\mathcal{D}} : \exists \, (\varepsilon_i)_{i=1}^{\overline{M}} \subset \mathbb{R}^{\overline{M}}, \right.$$

$$\left. \overline{\mathbf{v}} = \overline{L}_{\overline{M}}^+ \left(\varepsilon_{\overline{M}}; \overline{L}_{\overline{M}-1}^+ \left(\varepsilon_{\overline{M}-1}; \ldots; \overline{L}_1^+ (\varepsilon_1, \overline{\mathbf{w}}_L) \ldots \right) \right) \right\} \tag{13.13}$$

consists of all states $\overline{\mathbf{v}} = \overline{\mathbf{w}}(t, \mathbf{x}^-)$ that solve a (half-)RP in Ω_-, i.e. are attainable by a composition of forward Lax curves $\overline{L}_{\overline{m}}^+(\cdot\,;\, \overline{\mathbf{w}}_0^m)$, $\overline{m} \in \{1, \ldots \overline{M}\}$, see [22] for a precise definition, each emerging at some $\overline{\mathbf{w}}_0^{\overline{m}} \in \mathcal{D}$. Furthermore, $\overline{M}$ is chosen such that for the eigenvalues of the Jacobian of the flux $\overline{f}$ in (13.1) $\overline{\lambda}_{\overline{M}} < 0$ and $\overline{\lambda}_{\overline{M}+1} > 0$ holds. Thus, the $\overline{M}$-th Lax curve $\overline{L}_{\overline{M}}^+(\cdot\,;\, \mathbf{w}_0)$ connects $\mathbf{w}_0$ to a state that is located at the interface. Similarly, for a given state $\mathbf{w}_R$ the set of admissible boundary states in Ω_+ is defined by

$$V(\mathbf{w}_R) := \left\{ \mathbf{v} \in \mathcal{D} : \exists\, (\theta_i)_{i=1}^M \subset \mathbb{R}^M, \right.$$

$$\left. \mathbf{v} = L_M^-\left(\theta_M;\, L_{M-1}^-\left(\theta_{M-1};\, \ldots;\, L_1^-(\theta_1, \mathbf{w}_R)\ldots\right)\right) \right\}$$

(13.14)

where M is chosen such that the eigenvalues of the Jacobian of the flux f in (13.3) $\lambda_M > 0$ and $\lambda_{M-1} < 0$.

Definition 2 Having the sets $\overline{V}(\overline{\mathbf{w}}_L)$ and $V(\mathbf{w}_R)$ at hand the solution to the coupled problem is constructed as follows: we need to prove that there exist unique states $\overline{\mathbf{v}} \in \overline{V}(\overline{\mathbf{w}}_L)$ and $\mathbf{v} \in V(\mathbf{w}_R)$ such that $\Psi(\overline{\mathbf{v}}, \mathbf{v}) = 0$. Provided the traces of the solutions in $\Omega_\pm$ are well-defined and unique $\overline{\mathbf{v}}$ and $\mathbf{v}$ are identified the **solution to the coupled RP** is given by the solution of the two half-RPs with the corresponding initial data $(\overline{\mathbf{w}}_L, \overline{\mathbf{v}})$ and $(\mathbf{v}, \mathbf{w}_R)$, respectively.

By definition of the sets $V(\mathbf{w}_R)$, $\overline{V}(\overline{\mathbf{w}}_L)$, the trace of the solution fulfils the coupling condition. The solution of the coupled RP is summarised in the following algorithm.

Algorithm 1 Riemann Solver for Coupled Problems

1: Let $\mathbf{x} \in \Gamma$ be a point at the interface, $\overline{\mathbf{w}}_L = \overline{\mathbf{w}}(t, \mathbf{x}^-)$ and $\mathbf{w}_R = \mathbf{w}(t, \mathbf{x}^+)$ be the attached interface states for the systems in Ω_- and Ω_+ respectively.
2: Project the systems of conservation laws on the normal interface direction $\mathbf{n} = \mathbf{n}(\mathbf{x})$.
3: Solve the coupled RP consisting of two half-RPs. Thus, identify the sets of admissible boundary states $\overline{V}$ and V by compositions of Lax curves. Find parameters $(\varepsilon_i^*)_{i=1}^{\overline{M}}$ and $(\theta_i^*)_{i=1}^M$ corresponding to $\overline{\mathbf{v}}^* \in \overline{V}(\overline{\mathbf{w}}_L)$ and $\mathbf{v}^* \in V(\mathbf{w}_R)$, respectively, satisfying $\Psi(\overline{\mathbf{v}}^*, \mathbf{v}^*) = 0$.
4: Evaluate the Lax curves with respect to ε^* and θ^* to determine the boundary values $\overline{\mathbf{w}} = \overline{\mathbf{w}}(\varepsilon^*)$ and $\mathbf{w} = \mathbf{w}(\theta^*)$ for both systems of conservation laws.
5: Project the systems of conservation laws back on $\Omega_\pm$.

13.3 Fluid-Structure Coupling: Linear Elastic and Compressible Euler Equations

In this section Algorithm 1 is applied to the FSC problem with stiffened gas EoS. After introducing the model for the FSC the sets $\overline{V}$ and V are determined. Finally, the existence of a unique solution of a RP for the FSC is analysed.

13.3.1 Modelling of the Fluid-Structure Coupling Problem

Consider a situation as sketched in Fig. 13.2 where the interface Γ separates a material and a compressible fluid. It is assumed that the interface remains unaffected by the interaction of the fluid flow with a material structure [8], i.e. we do not account for deformation of the structure.

The solid regime and the fluid regime is assumed to be governed by the linear elastic equations

$$\frac{\partial \overline{v}}{\partial t} - \frac{1}{\overline{\rho}} \nabla \cdot \overline{\sigma} = 0, \tag{13.15a}$$

$$\frac{\partial \overline{\sigma}}{\partial t} - \overline{\lambda}(\nabla \cdot \overline{v}) I - \overline{\mu}\left(\nabla\overline{v} + \nabla\overline{v}^{T}\right) = 0. \tag{13.15b}$$

Here, the density of the material is denoted by $\overline{\rho}$ and assumed to be constant. The deformation velocities are $\overline{v} = (\overline{v}_1, \dots, \overline{v}_d)^{T}$, the stress tensor is denoted by $\overline{\sigma} = (\overline{\sigma}_{ij})_{i,j,=1,\dots,d} = \overline{\sigma}^{T}$, and the Lamé constants are $\overline{\lambda}, \overline{\mu} > 0$. Finally, the dilatation wave velocity and the shear wave velocity are $\overline{c}_1^2 := (2\overline{\mu} + \overline{\lambda})/\overline{\rho}$ and $\overline{c}_2^2 := \overline{\mu}/\overline{\rho}$, respectively. Due to the symmetry of the stress tensor $\overline{\sigma}$, the system of equations (13.15) contains redundant equations. Those may be removed and the system

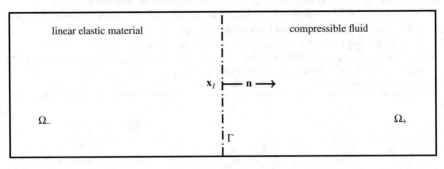

Fig. 13.2 Sketch of the 2D FSC problem with elastic material on the left and compressible fluid on the right. Shown is the interface Γ (dashed) with its normal direction $\boldsymbol{n}$ starting at $\boldsymbol{x}_I \in \Gamma$

can be written in the canonical form of a system of conservation laws, see Eq. (29) in the Appendix A of [12].

The fluid regime is governed by the compressible Euler equations

$$\frac{\partial \rho}{\partial t} + \nabla \cdot (\rho \mathbf{v}) = 0, \tag{13.16a}$$

$$\frac{\partial \rho \mathbf{v}}{\partial t} + \nabla \cdot (\rho \mathbf{v}^T \mathbf{v} + p\mathbf{I}) = \mathbf{0}, \tag{13.16b}$$

$$\frac{\partial \rho E}{\partial t} + \nabla \cdot (\rho \mathbf{v}(E + p/\rho)) = 0, \tag{13.16c}$$

where we use the notation ρ for the gas density, $\mathbf{v} = (v_1, \ldots, v_d)^T$ for its velocity, E for the total energy $E = e + \frac{1}{2}\mathbf{v}^2$, pressure p, internal energy e and total enthalpy $H := E + \frac{p}{\rho}$. The fluid equations have to be supplemented with an EoS, see [7, 15]. Here, we consider a stiffened gas EoS

$$p(\rho, e) = (\gamma - 1)\rho(e - Q) - \gamma\pi, \tag{13.17}$$

where γ, Q and π are constants, see for instance [9]. In particular, $\gamma > 1$ denotes the ratio of specific heats at constant pressure, Q is the formation enthalpy and π is referenced to as the pressure stiffness. Setting Q and π to zero results in the ideal gas EoS.

Across the interface we couple the fluid dynamics model (13.16) to the solid model (13.15). To this end, we again project the system of equations onto the normal direction $\mathbf{n}$ of the interface. For notational convenience we assume the normal direction $\mathbf{n}$ pointing from the solid towards the fluid regime, see Fig. 13.2. The projection of the linear elastic model in $d \in \{1, 2, 3\}$ spatial dimensions is realised by means of the matrices $\mathbf{R}_d$, $\mathbf{S}_d$ and $\mathbf{G}_d$, introduced in the Appendix A of [12], to obtain a quasi–1D model in the normal direction. Similarly, we project the fluid equations (13.16) assuming there is no flow in tangential directions. Since both systems (13.16) and (13.15) are invariant under rotation and reflection, it is sufficient to consider the projection onto direction $\mathbf{n} = \mathbf{e}_1 = (1, 0, \ldots, 0)^T \in \mathbb{R}^d$, as depicted in Fig. 13.2.

The basic problem is now to couple the two projected systems at the interface Γ. Note that the projected linear elastic model is defined in Ω_- and the projected Euler equations in Ω_+. To distinguish the fluid states from the solid states, quantities of the solid shall be denoted by a bar, e.g. $\bar{\mathbf{u}}, \bar{p}, \bar{\lambda}$ and so on.

According to the transition conditions of continuum mechanics at a material interface we model the coupling by requiring the following conditions $\Psi = 0$ to be fulfilled at the interface:

$$\mathbf{n}^T \bar{\sigma} \mathbf{n} \equiv \bar{\sigma}_{nn} \overset{!}{=} -p, \tag{13.18a}$$

$$\bar{\mathbf{v}}^T \mathbf{n} \equiv \bar{v}_n \overset{!}{=} v_n \equiv \mathbf{v}^T \mathbf{n}, \tag{13.18b}$$

where we neglect viscosity and heat conduction in the fluid flow. The conditions prescribe a continuous normal stress and pressure at the interface. Also, we assume that across the interface the normal velocities are equal. The conditions (13.18) are referred to as transition and kinematic conditions or coupling conditions, respectively. These conditions are used to provide boundary condition at some point $\mathbf{x}_I \in \Gamma$ of the interface for both the solid and the fluid. The procedure is described in detail in Algorithm 1, and is realised in the following two sections. For both systems the Lax curves, specified in the next section, are used to prove the existence of a unique solution of the FSC problem provided the initial data in the fluid is subsonic, i.e.

$$-c(\rho_R, e_R) \leq v_R \leq c(\rho_R, e_R) \tag{13.19}$$

where c denotes the speed of sound that is given in terms of density and internal energy as

$$c^2(\rho, e) = \frac{\gamma(p(\rho, e) + \pi)}{\rho} = \gamma(\gamma - 1)\left(e - Q + \frac{\pi}{\rho}\right). \tag{13.20}$$

13.3.2 Riemann Problem for the Fluid-Structure Coupling

Solutions of the classical RP for the projected linear elastic and the projected compressible Euler equations are well-known, see [7, 22]. Figures 13.3 and 13.4 depict the wave structures for both systems of conservation laws in the x-t plane. According to Algorithm 1, admissible states for both systems given by respective Lax curves are required.

Definition 3 Let the domain $\Omega = \mathbb{R}^d$, $d \in \{1, 2, 3\}$, be separated by the interface plane $\Gamma = \{\mathbf{x} \in \mathbb{R}^d : x_1 = 0\}$ into a solid and a fluid subdomain $\Omega_- = \{\mathbf{x} \in \mathbb{R}^d : x_1 < 0\}$ and $\Omega_+ = \{\mathbf{x} \in \mathbb{R}^d : x_1 > 0\}$. Let the dynamics of the solid and of the fluid be governed by the linear elastic equations (13.15) and by the compressible Euler equations (13.16) equipped with stiffened gas EoS (13.17), respectively. Furthermore, let both systems be projected onto $\mathbf{e}_1$, the normal of Γ. Then, the **FSC Riemann problem** (FSC-RP) is defined by the piecewise constant initial data

$$u(0, \mathbf{x}) = \begin{cases} \overline{\mathbf{u}}_L & \text{if } \mathbf{x} \in \Omega_-, \\ u_R & \text{if } \mathbf{x} \in \Omega_+. \end{cases} \tag{13.21}$$

First, we consider the solid, i.e. the linear elastic equations. The wave speeds are constant in $\overline{\mathbf{u}}$ and depend on the material parameters, ρ, $\overline{\lambda}$, $\overline{\mu}$, $\overline{c}_1$ and $\overline{c}_2$, of the solid only. Recall that coupled quantities, i.e. normal velocity $\overline{v}_1$ and stress $\overline{\sigma}_{11}$ in the direction $\mathbf{n} = \mathbf{e}_1$, are constant across the waves corresponding to the zero eigenvalue. The admissible states of the solid at Γ are thus obtained by Lax curves $\overline{L}_i(\,\cdot\,; \overline{\mathbf{u}}_L)$, given in [12], emanating from the initial state $\overline{\mathbf{u}}_L$:

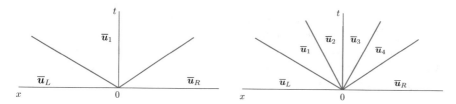

Fig. 13.3 Wave structure of the linear elastic equations for a classical RP with initial data $\overline{\mathbf{u}}_L, \overline{\mathbf{u}}_R$ where $d = 1$ (left) and $d = 3$ (right)

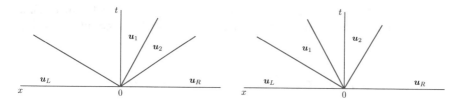

Fig. 13.4 Wave structure of the compressible Euler equations for a classical RP for subsonic initial data $\mathbf{u}_L, \mathbf{u}_R$ where $\lambda_0 \geq 0$ (left) and $\lambda_0 < 0$ (right)

$$d = 1 : \quad \overline{\mathbf{u}}_1 = \overrightarrow{L}_1^+(\varepsilon_1; \overline{\mathbf{u}}_L) = \overline{\mathbf{u}}_L + \varepsilon_1 \, \overline{r}_{1,-}, \tag{13.22a}$$

$$d = 2 : \quad \overline{\mathbf{u}}_2 = \overrightarrow{L}_2^+(\varepsilon_2; \overrightarrow{L}_1^+(\varepsilon_1; \overline{\mathbf{u}}_L)) = \overline{\mathbf{u}}_L + \varepsilon_1 \, \overline{r}_{1,-} + \varepsilon_2 \, \overline{r}_{2,-}, \tag{13.22b}$$

$$d = 3 : \quad \overline{\mathbf{u}}_2 = \overrightarrow{L}_3^+(\varepsilon_3; \overrightarrow{L}_2^+(\varepsilon_2; \overrightarrow{L}_1^+(\varepsilon_1; u_0))) = \overline{\mathbf{u}}_L + \varepsilon_1 \, \overline{r}_{1,-} + \varepsilon_2 \, \overline{r}_{2,-}^1 + \varepsilon_3 \, \overline{r}_{2,-}^2, \tag{13.22c}$$

where $\overline{r}_{i,-}$ denote the eigenvectors of the flux Jacobian as introduced in the Appendix A of [12] for $d = 1, 2, 3$.

We investigate the fluid part, i.e. the compressible Euler equations equipped with stiffened gas EoS and subsonic initial data. The wave structure is the same as in the case of a perfect gas EoS. The coupled quantities, i.e. normal velocity and pressure, are constant across the contact wave, however the density may jump. The contact wave corresponds to the 2-Lax curve, L_2, and the (multiple) flux Jacobian eigenvalue λ_0. In case $\lambda_0 < 0$, admissible fluid states $\mathbf{u}_1$ at Γ are given by the backward 3-Lax curve L_3^- emanating from the initial data $\mathbf{u}_L$. In case $\lambda_0 \geq 0$, the contact wave enters the fluid domain and the composition of L_3^- with the backward 2-Lax curve L_2^- yields the admissible fluid states at Γ, see Fig. 13.4. Thus, admissible interface fluid states $\mathbf{u}$ are given by

$$\begin{cases} \mathbf{u}_2 = \mathbf{L}_3^-(\theta^*; \mathbf{u}_R) & \text{if } \lambda_0 < 0, \\ \mathbf{u}_1 = \mathbf{L}_2^-(\theta_2; \mathbf{L}_3^-(\theta^*; \mathbf{u}_R)) & \text{if } \lambda_0 \geq 0. \end{cases} \tag{13.23}$$

In contrast to the solution of the classical RP for the compressible Euler equations, it is not necessary to reparametrise the Lax curves by the pressure. The straightforward parametrisation, as in the definition of Lax curves for the compressible Euler equations [12], suffices for the solution of the FSC-RP.

The following thermodynamic identities are important for the explicit formulation of the 1,3- Lax curves corresponding to genuinely nonlinear waves. The specific entropy s in terms of density and internal energy

$$s(\rho, e) = c_v \left(\ln \left(\frac{e - Q - \pi/\rho}{c_v} \right) - (\gamma - 1) \ln \rho \right) + Q', \qquad (13.24)$$

where c_v denotes the specific heat at constant volume and Q' is a thermodynamic constant. Finally, internal energy depending on density and specific entropy is given by

$$e(\rho, s) = \exp \left(\frac{s - Q'}{c_v} + (\gamma - 1) \ln \rho \right) + Q + \frac{\pi}{\rho}. \qquad (13.25)$$

For details on the derivation of the above equations, we refer to [6, 9, 18, 21].

Plugging Eqs. (13.20), (13.24), (13.25) and (13.17) into the general definition of the shock and rarefaction curves for $i = 1, 3$ and applying elementary calculus yields the 1- and 3-curves for the compressible Euler equations and stiffened gas EoS. The backward (S^-, R^-) and the forward curves (S^+, R^+) read:

$$S_i^\pm(\theta, \mathbf{u}_R) = \begin{cases} \rho = \pm(i - 2)\theta + \rho_R \\ v = v_R \mp \sqrt{\dfrac{1}{\rho_R} \dfrac{\theta^2 \kappa}{\rho(\theta)} \left(\dfrac{2\gamma \rho_R e_R - Q\kappa(2\rho_R \pm (i - 2)\theta) - 2\gamma\pi}{2\rho_R \pm (2 - i)\kappa\theta} - Q \right)} \\ e = \dfrac{\rho_R}{\rho(\theta)} \dfrac{1}{2 \pm (2 - i)\kappa\theta/\rho_R} \Big((2 \pm (i - 2)(\gamma + 1)\theta/\rho_R)e_R \pm \\ \qquad\qquad (2 - i)\theta \dfrac{1}{\rho_R^2} (Q\kappa(2\rho_R \pm (i - 2)\theta) + 2\gamma\pi) \Big) \end{cases}, \quad (13.26)$$

$$R_i^\pm(\theta, \mathbf{u}_R) = \begin{cases} \rho = \pm(i - 2)\theta + \rho_R \\ v = v_R + 2(2 - i) \sqrt{\dfrac{\gamma}{\gamma - 1} \left(1 - \left(\dfrac{\rho(\theta)}{\rho_R} \right)^{(\gamma-1)/2} \right)} \sqrt{e_R - \left(Q + \dfrac{\pi}{\rho_R} \right)} \\ s = s_R \end{cases},$$

$$(13.27)$$

and we have

$$L_i^\pm = \begin{cases} S_i^\pm & \text{if } \theta < 0 \\ R_i^\pm & \text{if } \theta \geq 0 \end{cases}. \qquad (13.28)$$

In the following we assume the initial data is subsonic, i.e. the condition (13.19) holds. The wave structure of the FSC-RP is illustrated in Fig. 13.5. The initial data and related quantities are marked with a subscript L and R in the solid and fluid part respectively. The admissible fluid states (13.26), (13.27) and solid states (13.22) are used in the coupling conditions (13.18) with $\mathbf{n} = \mathbf{e}_1$. Then, the coupling condi-

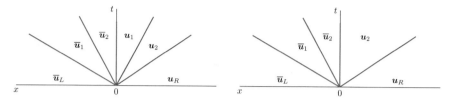

Fig. 13.5 Wave structure of the FSC-RP for $d \geq 2$ where $\lambda_0 \geq 0$ (left) and $\lambda_0 < 0$ (right). Blue and red lines signify the fluid and solid characteristics respectively

tions (13.18) depend on the two parameters $\varepsilon := \varepsilon_1 - \bar{\lambda}_1$ and θ

$$\overline{\sigma}_{11}(\varepsilon) \stackrel{!}{=} -p(\theta), \quad \overline{v}_1(\varepsilon) \stackrel{!}{=} v_1(\theta). \tag{13.29}$$

These are equivalent to

$$\overline{\sigma}_{11,L} + \varepsilon \overline{\rho c_1} \overline{\beta} = -p_3^-(\theta, \mathbf{u}_R), \tag{13.30a}$$

$$\overline{v}_{1,L} + \varepsilon \overline{\beta} = (v_1)_3^-(\theta, \mathbf{u}_R), \tag{13.30b}$$

where $\overline{\beta}$ is a constant defined as following

$$\overline{\beta} := \begin{cases} 1 & \text{if } d = 1 \\ \overline{c}_1^2 - \overline{c}_2^2 & \text{if } d = 2 \ . \\ -1/(\overline{\rho c_1}) & \text{if } d = 3 \end{cases}$$

Hence, the solution of the FSC-RP is equivalent to finding a tuple $(\varepsilon^*, \theta^*)$ that fulfils (13.30). This is equivalent to

$$g(\theta) := \overline{\sigma}_{11,L} + (v_1(\theta; \mathbf{u}_R) - \overline{v}_{1,L})\overline{\rho c_1} + p_3^-(\theta; \mathbf{u}_R) = 0, \tag{13.31a}$$

$$\varepsilon = v_1(\theta; \mathbf{u}_R) - \overline{v}_{1,L}. \tag{13.31b}$$

Since g does not depend on the parameter ε related to the solid part, it is sufficient to determine the roots of g. This is done in the same manner as for the perfect gas case in [12]. The function g is expressed only in terms of the initial data as follows. For the shock branch, i.e. $\theta \leq 0$, g reads

$$g(\theta) = \overline{\sigma}_{11,L} + \overline{\rho c} \left(v_R + \sqrt{\frac{2}{(2\rho_R + (\gamma - 1)\theta)(\rho_R - \theta)}} |\theta| c_R - \overline{v}_{1,L} \right)$$

$$+ \frac{1}{\gamma} \frac{2\rho_R - (\gamma + 1)\theta}{2\rho_R + (\gamma - 1)\theta} \rho_R c_R^2 - \pi, \tag{13.32}$$

while for the rarefaction branch, i.e. $\theta \geq 0$,

$$g(\theta) = \bar{\sigma}_{11,L} + \overline{\rho c}_1 \left(v_R - \frac{2}{\gamma - 1} \left(1 - \left(1 - \frac{\theta}{\rho_R} \right)^{\frac{\gamma-1}{2}} \right) c_R - \bar{v}_{1,L} \right)$$

$$+ (\gamma - 1) \exp \left(\frac{s_R - Q'}{c_v} \right) (\rho_R - \theta)^\gamma - \pi. \tag{13.33}$$

The structure of the solution of $g(\theta) = 0$ will be detailed below.

13.3.3 Entropy Solutions of the Riemann Problem for the Fluid-Structure Coupling

We investigate properties of the scalar, nonlinear function g to conclude on the existence of its roots and, equivalently, on the solution of the FSC-RP. Statements and proofs are very similar to those given in [12] for the perfect gas case. However, for the sake of completeness and in order to show the behaviour of g for the implementation of the FSC-RP, the following proofs are given at full length.

Lemma 1 *Let $\gamma > 1$, $\rho_R, \bar{\rho}, \bar{c}_1 > 0$ and $p > -\pi$. Then there exist $\theta_m < 0$ and $\theta_M > 0$ such that the function $g : (\theta_m, \theta_M] \to \mathbb{R}$ composed of a shock branch (13.32) and a rarefaction branch (13.33) for $\theta \leq 0$ and $\theta \geq 0$, respectively, is differentiable and strictly monotonically decreasing.*

Proof The shock branch $(\theta \leq 0)$ is well-defined provided the discriminant in (13.33) is positive, i.e.

$$\theta_m := -\frac{2\rho_R}{\gamma - 1} < 0.$$

The rarefaction branch $(\theta \geq 0)$ is well-defined provided that the term $(\rho_R - \theta)^{\frac{\gamma-1}{2}}$ is well-defined. Here, two cases have to be distinguished:

$$\theta_M := \begin{cases} \infty & \text{if } \frac{\gamma-1}{2} \in \mathbb{N}, \\ \rho_R & \text{if } \frac{\gamma-1}{2} \in (0, \infty] \backslash \mathbb{N} \end{cases}.$$

The derivatives of g for the shock branch $\theta_m < \theta \leq 0$ and the rarefaction branch $0 \leq \theta \leq \theta_M$ read:

$$g'(\theta) = \bar{\rho}\bar{c}_1 \sqrt{2c} \left(\frac{(\gamma-3)\rho_R\theta - 2(\gamma-1)\theta^2}{2[(2\rho_R + (\gamma-1)\theta)(\rho_R - \theta)]^{\frac{3}{2}}} - \frac{1}{\sqrt{(2\rho_R + (\gamma-1)\theta)(\rho_R - \theta)}} \right)$$

$$- \frac{4\rho^2 c^2}{(2\rho + (\gamma-1)\theta)^2}, \quad \theta \leq 0, \tag{13.34a}$$

$$g'(\theta) = -\frac{\bar{\rho}\bar{c}_1 c}{\rho_R} \left(1 - \frac{\theta}{\rho_R} \right)^{\frac{\gamma-3}{2}} - c^2 \left(1 - \frac{\theta}{\rho_R} \right)^{\gamma-1}, \quad \theta \geq 0. \tag{13.34b}$$

Since the limits at $\theta = 0$ fulfil

$$g(0^-) = g(0^+) = \bar{\sigma}_{11,L} + \bar{\rho}\bar{c}_1 \left(v - \bar{v}_{1,L}\right) + \rho c^2/\gamma - \pi,$$
$$g'(0^-) = g'(0^+) = -\left(\bar{\rho}\bar{c}_1 + \rho c\right) c/\rho < 0$$

and the Lax curves are smooth functions of the parameters, g is continuously differentiable at $\theta = 0$. Next, we show that along the shock branch $(\theta_m < \theta \leq 0)$ the function g is strictly monotonically decreasing. According to (13.34a) this holds true if the inequality

$$(\gamma - 3)\rho\theta - 2(\gamma - 1)\theta^2 - 2(2\rho + (\gamma - 1)\theta)(\rho - \theta) \leq 2\sqrt{2}\frac{\rho^2 c}{\bar{\rho}\bar{c}_1}\sqrt{\frac{(\rho - \theta)^3}{2\rho + (\gamma - 1)\theta}}$$

holds. The left-hand side of the inequality reduces to $-\rho(4\rho + \theta(\gamma - 3))$ that is negative for $\theta_m < \theta \leq 0$, since

$$\begin{cases} 4\rho + \theta(\gamma - 3) \geq 4\rho > 0 & \text{if } 1 < \gamma < 3, \\ 2\rho(\gamma + 1)/(\gamma - 1) > 0 & \text{if } \gamma \geq 3 \end{cases}.$$

Along the rarefaction branch $(0 \leq \theta \leq \theta_M)$ we obtain from (13.34b)

$$g'(\theta) = -c^2 \rho^{-2r} \left(A(\rho - \theta)^{r-1} + (\rho - \theta)^{2r}\right),$$

where $A := c^{-1}\rho^r \bar{\rho}\bar{c}_1 > 0$ and $r := \frac{\gamma - 1}{2}$. It is obvious that g' has a single root in ρ. If $r \notin \mathbb{N}$ or $r \in \mathbb{N}$ is odd, then g' is negative for $\theta \in [0, \theta_M)$. If $r \in \mathbb{N}$ is even, then g' is positive if

$$\rho < \theta < \rho + A^{\frac{1}{r+1}}.$$

Therefore, g is strictly monotonically decreasing, if θ does not exceed θ_M. This concludes the proof. $\qquad\square$

Since $g \in C^1((\theta_m, \theta_M])$ and g is strictly monotone there exists a unique root of g if g has opposite signs at θ_m and θ_M. This is shown in the next theorem that is equivalent to Theorem 3.1 in [12].

Theorem 1 *Let* $\gamma > 1$ *and* $\rho_R, \bar{\rho}, \bar{c}_1 > 0$, $p_R > -\pi$. *The initial data* $(\bar{v}_{1,L}, \bar{\sigma}_{11,L})$ *and* (ρ_R, v_R, p_R) *are assumed to satisfy*

$$v_R \leq \frac{2}{\gamma - 1}c_R + \bar{v}_{1,L} - \frac{\bar{\sigma}_{11,L}}{\bar{\rho}\bar{c}_1} + \pi. \tag{13.35}$$

Then there exists a unique root $\theta^* \in (\theta_m, \theta_M]$ *of the function* $g : (\theta_m, \theta_M] \to \mathbb{R}$ *and therefore a unique solution of the FSC-RP.*

Proof For $\theta \leq 0$ we have by definition of g on the shock branch (13.32)

$$\lim_{\theta \to \theta_m} g(\theta) = \infty.$$

If $\theta \geq 0$ and either $(\gamma - 1)/2 \in (0, \infty]\backslash\mathbb{N}$ or $(\gamma - 1)/2 \in \mathbb{N}$ even then $\theta_M = \rho_R$ and $g(\theta_M) \leq 0$ provided that

$$\overline{\sigma}_{11,L} + \overline{\rho}\overline{c}_1\left(v_r - \frac{2}{\gamma - 1}c_R - \overline{v}_{1,L}\right) \leq \pi$$

holds. This inequality is satisfied by assumption (13.35). If $\frac{\gamma - 1}{2} \in \mathbb{N}$ is odd then $\theta_M = \infty$ and $\lim_{\theta \to \theta_M} g(\theta) = -\infty$. According to Lemma 1 g is strictly monotonically decreasing in the interval $(\theta_m, \theta_M]$. Thus, by inspecting all cases of limits of g towards the domain θ_m and θ_M the assertion is proven. □

Remark 2 Note that the fluid portion of condition (13.35) corresponds to the right hand side of the vacuum condition for the classical RP, cf. [22].

13.4　A Numerical Example

We perform a simulation of the FSC in one spatial dimension where for the solid and the fluid we choose the following material parameters. The constant density of the steel is $\overline{\rho} = 7800\,\text{kg/m}^3$, the Lamé coefficients are $\overline{\lambda} = 9.3288 \times 10^{10}\,\text{N/m}^2$ and $\overline{\mu} = 9.3288 \times 10^{10}\,\text{N/m}^2$. The dilatation wave velocity is set to $\overline{c}_1 = 5990\,\text{m/s}$, while the shear wave velocity $\overline{c}_2 = 3458\,\text{m/s}$. The ratio of specific heats for water is $\gamma = 2.85$, the formation enthalpy is $Q = 0\,\text{J/kg}$ and the pressure stiffness is $\pi = 10^9\,\text{N/m}^2$. Finally, the specific heat is chosen as $c_v = 1816\,\text{J/(kg} \cdot \text{K)}$ and $Q' = 0\,\text{J/(kg} \cdot \text{K)}$.

We simulate a shock in the fluid moving towards the solid. Thus, the piecewise constant initial data

$$\begin{cases} \overline{\boldsymbol{u}}_L & \text{if } x < 0 \\ \boldsymbol{u}_1 & \text{if } 0 \leq x < x_S, \\ \boldsymbol{u}_2 & \text{if } x \geq x_S \end{cases} \tag{13.36}$$

is chosen such that a shock wave located at $x_S = 0.01$ m in the fluid travels to the left in the direction of the interface Γ located at $x = 0$ m, see Fig. 13.6. The state to the left of the shock wave is steady both in steel and in water with continuous pressure and negative stress, i.e. the initial conditions fulfil the coupling conditions trivially. The values for $\overline{\boldsymbol{u}}_L$, $\boldsymbol{u}_1$ and $\boldsymbol{u}_2$ are listed in Table 13.1. The datum $\boldsymbol{u}_2$ is obtained by fixing $\boldsymbol{u}_1$ and increasing the density by 100 along the shock branch of the forward 1-Lax curve, i.e. $\boldsymbol{u}_2 = L_1^+(100; \boldsymbol{u}_1)$.

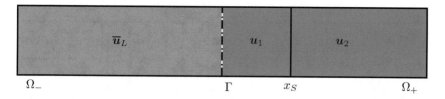

Fig. 13.6 Sketch of the initial conditions for the convergence study

Table 13.1 Initial data for the 1D validation test case

Steel		Water			
	$\overline{\mathbf{u}}_L$			$\boldsymbol{u}_1$	$\boldsymbol{u}_2$
$\overline{v}_1$ [m/s]	0	v_1 [m/s]	0	−128.46	
$\overline{\sigma}_{11}$ [N/m²]	−350000	p [N/m²]	350000	$2.5778 \cdot 10^8$	
		ρ [kg/m³]	1200	1300	

The simulation is performed by means of a third-order RKDG scheme using polynomial elements of order $p = 3$ with a third-order SSP-RK timestepping with three stages, cf. [3, 10, 16]. The final time is set to $T = 30\,\mu s$. The coupling conditions are realised by applying Algorithm 1 with iterative Newton-type solvers from the GSL library for the scalar equation $g(\theta) = 0$, see (13.31a). The choice of the initial value $\theta_{init} = 0$ for the iterative method works well in practice.

The CFL number is set to 0.1 and the timestep in the solid and the fluid is synchronised to the minimum of the respective timesteps. The spatial domains $\Omega_- = [-0.2, 0]\,m$ for the solid part and $\Omega_+ = [0, 0.2]\,m$ for the fluid part are discretised adaptively with 16 cells on the coarsest level $l = 0$ each. Therefore, the cell width on level l is 16×2^{-l}, i.e. the higher the level the smaller the cell. In each domain we use at most $L = 8$ levels of refinement. Thus, the fully refined grid consists of 32×2^8 cells. However, due to grid adaptation the locally refined grid consists of about 200 cells while the accuracy of the solution is comparably to the simulation on the fully refined grid, cf. [10].

Figure 13.7 shows the behaviour of the pressure p in the fluid and the negative stress $-\sigma$ in the solid part of the solution. When the shock in the fluid part hits the interface Γ, two shocks emerge at Γ travelling in opposite directions. The collision causes high pressure and stress states around Γ. We note that the pressure and the negative stress are continuous at the interface Γ, in agreement with the coupling conditions (13.18). The same holds for the velocity that was omitted here for the sake of brevity.

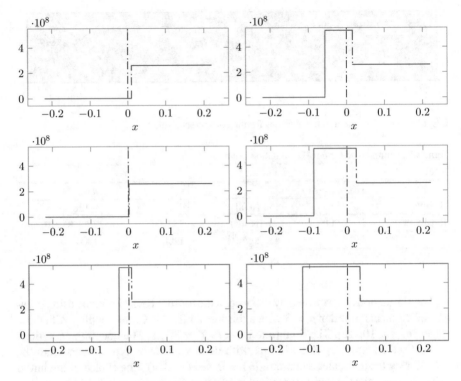

Fig. 13.7 Pressure p (right) and negative stress $-\overline{\sigma}$ (left). From top to bottom, left to right the times are $t = 0\,\mu s, \dots, 25.78\,\mu s$ equidistantly distributed

13.5 Conclusions

Condition (13.35) on the initial data demanded in the previous theorem may be rewritten using the fact that the initial data are subsonic (13.19).

Lemma 2 *Let* $\gamma \in (1, 3)$, $\rho_R, \overline{\rho}, \overline{c}_1 > 0$, $p_R > -\pi$ *and the initial data in the fluid part be subsonic, i.e. Eq. (13.19) holds. Then*

$$\overline{v}_{1,L} \geq \frac{\overline{\sigma}_{11,L}}{\rho c_1} - \pi \tag{13.37}$$

is a sufficient condition for a unique solution of the RP for the FSC.

Proof The subsonic condition (13.19) implies

$$v_R \leq c_R < c_R \frac{2}{\gamma - 1}.$$

Thus, with (13.37) condition (13.35) is satisfied.   □

We have derived a strategy for the FSC problem employing half-RPs and have shown that under reasonable conditions on the subsonic flow there exists a unique solution of the FSC-RP. Moreover, we have validated our implementation of the coupling Algorithm 1 for a one-dimensional setup of a shock colliding with the interface. In addition, our strategy has been applied in the coarse of numerical investigations of a bubble collapse, see [20]. Furthermore, we have shown that the framework of RPs for coupled conservation laws can be extended from the perfect gas EoS to its linear modification, the stiffened gas EoS. This might motivate to extend the results in [13] using this strategy to resolve fine-scale effects in high-pressure gas networks employing various EoS applicable to the material properties of natural gas mixtures.

References

1. An, S., Li, Q., Gedra, T.W.: Natural gas and electricity optimal power flow. In: *2003 IEEE PES Transmission and Distribution Conference and Exposition (IEEE Cat. No.03CH37495)*, vol. 1, pp. 138–143 (2003)
2. Brouwer, J., Gasser, I., Herty, M.: Gas pipeline models revisited: model hierarchies, nonisothermal models, and simulations of networks. Multiscale Model. Simul. **9**(2), 601–623 (2011)
3. Cockburn, B., Shu, C.-W.: The Runge-Kutta discontinuous Galerkin method for conservation laws V: Multidimensional systems. J. Comput. Phys. **141**(2), 199–224 (1998)
4. Colombo, R.M., Garavello, M.: On the cauchy problem for the p-system at a junction. SIAM J. Math. Anal. **39**(5), 1456–1471 (2008)
5. Colombo, R.M., Herty, M., Sachers, V.: On 2×2 conservation laws at a junction. SIAM J. Math. Anal. **40**(2), 605–622 (2008)
6. Courant, R., Friedrichs, K.O.: Supersonic Flow and Shock Waves. Applied Mathematical Sciences. Springer, New York (1976)
7. Dafermos, C.M.: Hyperbolic conservation laws in continuum physics. Grundlehren der Mathematischen Wissenschaften [Fundamental Principles of Mathematical Sciences], vol. 325, 2nd edn. Springer, Berlin (2005)
8. Dickopp, C., Gartz, R., Müller, S.: Coupling of elastic solids with compressible two-phase fluids for the numerical investigation of cavitation damaging. Int. J. Finite Vol. **10**, 1–39 (2013). hal-01121991
9. Flåtten, T., Morin, A., Munkejord, S.T.: On solutions to equilibrium problems for systems of stiffened gases. SIAM J. Appl. Math. **71**(1), 41–67 (2011)
10. Gerhard, N., Iacono, F., May, G., Müller, S., Schäfer, R.: A high-order discontinuous Galerkin discretization with multiwavelet-based grid adaptation for compressible flows. J. Sci. Comput. **62**(1), 25–52 (2015)
11. Godlewski, E., Raviart, P.-A.: Numerical Approximation of Hyperbolic Systems of Conservation Laws. Applied Mathematical Sciences. Springer (1996)
12. Herty, M., Müller, S., Gerhard, N., Xiang, G., Wang, B.: Fluid-structure coupling of linear elastic model with compressible flow models. Int. J. Numer. Methods Fluids **86**(6), 365–391 (2018)
13. Herty, M., Müller, S., Sikstel, A.: Coupling of compressible Euler equations. In: Vietnam Journal of Mathematics, pp. 1–24 (2019)
14. Hou, G., Wang, J., Layton, A.: Numerical methods for fluid-structure interaction - a review. Commun. Comput. Phy. **12**(2), 337–377 (2012)
15. Menikoff, R., Plohr, B.: The Riemann problem for fluid flows of real materials. Rev. Mod. Phys. **61**, 75–130 (1989)

16. Müller, S., Sikstel, A.: Multiwave. Forschung/multiwave, Institut für Geomtrie und Praktische Mathematik, RWTH Aachen (2018). https://www.igpm.rwth-aachen.de/
17. Nasrifar, K., Bolland, O.: Prediction of thermodynamic properties of natural gas mixtures using 10 equations of state including a new cubic two-constant equation of state. J. Pet. Sci. Eng **51**(3–4), 253–266 (2006)
18. Plohr, B.J.: Shockless acceleration of thin plates modeled by a tracked random choice method. AIAA J. **26**(4), 470–478 (1988)
19. Ríos-Mercado, R.Z., Borraz-Sánchez, C.: Optimization problems in natural gas transportation systems: a state-of-the-art review. Appl. Energy **147**, 536–555 (2015)
20. Sikstel, A.: Analysis and Numerical Methods for the Coupling of Hyperbolic Problems. Ph.D. thesis, RWTH Aachen University, to appear in (2020)
21. Thein, F.: Results for Two Phase Flows with Phase Transition. Ph.D. thesis, Universität Magdeburg (2019)
22. Toro, E.F.: Riemann Solvers and Numerical Methods for Fluid Dynamics: A Practical Introduction, 3rd edn. Springer, Berlin (2009)
23. Zlotnik, A., Roald, L., Backhaus, S., Chertkov, M., Andersson, G.: Control policies for operational coordination of electric power and natural gas transmission systems. In: 2016 American Control Conference (ACC), pp. 7478–7483 (2016)

Chapter 14
District Heating Networks – Dynamic Simulation and Optimal Operation

Jan Mohring, Dominik Linn, Matthias Eimer, Markus Rein, and Norbert Siedow

Abstract District heating networks will play a prominent role in sector coupling. On the one hand, they can help compensating for fluctuations in renewable power generation. On the other hand, they allow to use waste heat from industrial processes instead of natural gas. However, this new role of district heating will also require new operating modes, deeper insight into the network and, consequently, more sophisticated simulation and optimization tools. Here, we deal with an optimal control problem which is dominated by time-varying delays between heat source and consumers: optimal preheating. The goal is to satisfy a periodic demand with a constant heat supply. This problem is still simple, as no devices like turbines are involved. But it contains already all challenges of simulating dynamic networks and, therefore, represents an ideal benchmark. We present a suitable mathematical model, some illustrative analytic examples, an efficient numerical scheme, and a solution to optimal preheating for a real municipal network. The model is both, accurate enough to predict pressure drop or cooling, but also simple enough to allow for fast numerical solution by a method of characteristics. Automatic differentiation is used for both, computing exact Jacobians within Newton's method and providing an optimizer with sensitivities.

J. Mohring (✉) · D. Linn · M. Eimer · M. Rein · N. Siedow
Fraunhofer ITWM, Fraunhoferpl. 1, D-67663 Kaiserslautern, Germany
e-mail: jan.mohring@itwm.fraunhofer.de

D. Linn
e-mail: dominik.linn@itwm.fraunhofer.de

M. Eimer
e-mail: matthias.eimer@itwm.fraunhofer.de

M. Rein
e-mail: markus.rein@itwm.fraunhofer.de

N. Siedow
e-mail: norbert.siedow@itwm.fraunhofer.de

© Springer Nature Switzerland AG 2021
S. Göttlich et al. (eds.), *Mathematical Modeling, Simulation and Optimization for Power Engineering and Management*, Mathematics in Industry 34,
https://doi.org/10.1007/978-3-030-62732-4_14

14.1 Introduction

District heating networks (DHN) will play a prominent role in sector coupling. On the one hand, they can help compensating for fluctuations in renewable power generation. On the other hand, they allow to use waste heat from industrial processes and thus save natural gas. However, this new role of district heating requires new operating modes, deeper insight into the network and, consequently, more sophisticated simulation and optimization tools.

Modern district heating plants use combined heat and power generation. Operators can decide which part of the hot steam is used for power generation and which part for district heating. The heating network behaves more like a storage device allowing delays between feed-in and consumption. This enables the operator to react to sudden changes in the demand for electric power by redirecting temporarily the hot steam.

Another task is optimal preheating. In some periods of the year, the power of the main heating source is enough to cover the demand on average, but not during the load peaks. Here, additional gas boilers have to be fired, which is expensive due to long start-up and shut-down phases. To some extent, this can be avoided by intelligent preheating. At some point, however, the contractually guaranteed connection pressures and temperatures can no longer be maintained.

So far, operators have been able to master these tasks thanks to many years of experience. In the future, however, district heating networks will no longer be supplied by a single source, but also by waste heat of industrial plants. At that point at the latest, simulation will have to be used to get an idea of the present state of the network and its future development.

Dynamic run-time effects play a central role for all of these tasks, while current software does usually not take them into account. Products for operational optimization usually treat the entire network as a sink without structure [1]. On the other hand, thermo-hydraulic simulators are made for network design and provide only quasi-stationary solutions [2]. This means that the temperature field is calculated as if the current consumption had always been the same before. In particular, no temperature fields can be represented that are warmer inside the network than at the source.

But also academic codes which are based on classical numerical methods for hyperbolic transport [3] reach their limits, since the shortest pipes and highest flow velocities define the calculation grid in time and place, while smoothness of real consumption profiles is not exploited. The following example may illustrate the challenge. The city net of Technische Werke Ludwigshafen (TWL) comprises about 1000 consumers and 6000 pipes. Here, we will address a periodic optimal control problem for known consumption ignoring issues like reconstructing states from noisy measurements. In the future, however, TWL intends to apply model-predictive control. Periods of two or three days have to be simulated several times within a quarter of an hour. Using a standard first-order upwind method, due to stability, one has to deal with more than 1.9×10^9 degrees of freedom in space and time, which is far too much for solving problems in time.

In this chapter we present a mathematical model of district heating networks, some illustrative analytic examples, an efficient numerical scheme, and a solution to optimal preheating based on real world data. The model is both, accurate enough to predict pressure drop or cooling, but also simple enough to allow for fast numerical solution. The hydraulic subsystem is treated in a quasi stationary manner, while heat transport is described by the method of characteristics. Automatic differentiation (AD) is used for both, computing exact Jacobians within Newton's method and providing an optimization tool with sensitivities.

Other interesting topics such as predicting the consumption or optimizing all resources of a district heating plant are ignored due to the limited space.

Related Work

Before discussing the models and methods used here, we will give a more general overview of related literature. Early approaches to computer aided operational optimization of district heating networks date back to 1995 [4]. The authors identified the significant role of time delays in distribution networks, and the resulting complexity of an accurate numerical solution. The network dynamics are approximated using the so-called *node method* - a *Lagrangian scheme* based on time series of temperatures at nodes (i.e. intersection points between different components) of the the network. A detailed modeling approach for heating networks is presented in [3], featuring complex network geometries with loops, non-constant thermodynamical properties of water, and unsteady friction models. The main focus of this work lies on optimal network design. A common trait, which most modeling approaches for DHN share, is the description of energy transport in the distribution network as an advection process. In [5, 6], the authors analyze this class of models, including an experimental validation. In [7], an embedding into the framework of *Port-Hamiltonian systems* is presented, with the aim to provide a mathematically and physically rigorous basis to model *coupling of sectors*.

Explicit high order numerical schemes for general *hyperbolic conservation laws* on networks are discussed in [8, 9]. The presented numerical methods can easily be applied to the field of district heating. As length scales can vary vastly in district heating networks, a global time step restriction can result in large computational costs. In [10], *local time stepping* (LTS) schemes for networks are used to circumvent this restriction. More common in the field, however, is the use of classical methods, e.g. explicit or implicit Upwind [11] or a modification of the QUICKEST scheme [12].

As already observed, an accurate numerical simulation is essential if sharp estimates of transport time delays are required. On the other hand, such models tend to be computationally intensive and are often too complex for usage in optimization. The *model predictive control* (MPC) approach in [13] tackles this problem by coupling a full nonlinear forward simulation with a simplified, linear approximation, where the former serves as predictor, and the latter is used as an optimization model. In [14], a similar method is used to solve optimal control problems involving multiple, switchable energy sources, resulting in *mixed integer linear programs* (MILP). A nonlinear, instantaneous control approach for district heating networks with power constraints is presented in [15]. An outline of the optimization algorithm, which is a part of the approach we present in this article, is presented in [16]. Based on a combination of a nonlinear, and a linearized model as well, it eliminates the internal state variables from the optimization model, hence reducing its dimension. This makes it possible to optimize across long prediction horizons without the necessity of using subdividing the problem.

An alternative to linearized models for optimization lies in *model order reduction*, as it is depicted in [17, 18]. Moment matching is applied to approximate the full, high resolution model by one with significantly fewer degrees of freedom, while preserving the system's essential dynamic behavior.

In this work we have chosen still another approach to keep simulation times short. Temperatures, or more precisely, energy densities are represented by low order polynomials on both, local and temporal inflow boundaries of a pipeline, and followed along characteristics. As long as information is propagated only from the local inflow boundary, the method is similar to [4]. As far as we know, the mixed approach has not been described in the literature, yet. The resulting differential algebraic equations are integrated by classical collocation [19] and adaptive time stepping is controlled by Richardson extrapolation [20].

While all previously mentioned methods are based on physical models involving ordinary or partial differential equations, the authors of [21] present an approach based on stochastic model predictive control, which has successfully been implemented as an on-line controller.

Besides a numerical simulation of the distribution network, a good load prediction is essential for precise numerical simulation and optimization. Here, standardized load profiles [22], which originally have been developed in the context of gas networks, may serve as a good starting point, if no detailed measurements are available.

14.2 Model

Now, we start to present our approach for modeling and simulating district heating. First, we list a set of transient equations modeling district heating networks based on the equations used in [3]. In contrast to that work, heat transport is formulated in terms of energy density rather than temperature, which allows for an analytic solution of heat transport along characteristics, cf. Eq. (14.23). Moreover, material properties are expressed as polynomial approximations derived from NIST databases [23] and friction is formulated in a modern way using the Lambert W-function. Particular attention is payed to justifying why certain terms may be neglected which appear in the original Navier-Stokes equations (Fig. 14.1).

The pipes are completely filled with hot water under high pressure, which avoids evaporation. The network is composed of two parts, hot flow (F) and colder return (R), linked by sources and consumers. Sources feed energy into the flow and consumers draw energy from it via heat exchangers. The water circuit itself is closed. At the main source, volume variations due to thermal expansion are compensated by a pressure

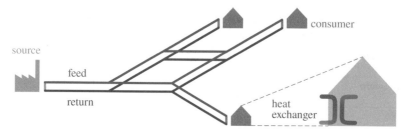

Fig. 14.1 Sketch of a district heating network

Table 14.1 Deviations from standard notation

Quantity	Here	Standard	Unit	Quantity	Here	Standard	Unit
Inner energy per volume	e	u	$\frac{GJ}{m^3}$	Mass flux	q	$\dot{m}$	$\frac{kg}{s}$
Inner energy per mass	ε	e	$\frac{MJ}{kg}$	Heat flux	Q	$\dot{Q}$	$\frac{MJ}{s}$

maintenance device. Note that our notation differs from standard notation for the following quantities (Table 14.1):

This is because heat transport may be described most easily in terms of the volume based energy density, but the symbol u is needed later for inputs. Moreover, the dots in the standard symbols $\dot{m}$ and $\dot{Q}$ could mislead to the interpretation that differentiation is needed, which is not the case.

14.2.1 Water Properties

We consider the following range of pressures and temperatures:

$$p \subset [5,\ 25]\,\text{bar}\,, \quad T \in [50,\ 130]°\text{C}\,. \tag{14.1}$$

The easiest way to describe heat transport is in terms of the inner energy per volume $e = e(t, x)$. The corresponding range is $e \in [0.2067,\ 0.5104]\,\frac{GJ}{m^3}$. Here, material properties of water are almost independent of pressure. All necessary material properties are described as polynomials in e. The coefficients in Table 14.2 are fitted using data from [23]. Polynomial orders are chosen such that relative errors are similar to those ignoring pressure dependence:

Table 14.2 Properties of water as function of the non-dimensional inner energy per volume ξ

Quantity	Approximation	Unit	Rel. error
Inner energy per volume	$e = \xi$	$\frac{GJ}{m^3}$	0
Temperature	$T(e) = 59.2453\,\xi^2 + 220.536\,\xi + 1.93729$	°C	$1.2 \cdot 10^{-3}$
Density	$\rho(e) = -0.208084\,\xi^2 - 0.025576\,\xi + 1.0028$	$\frac{t}{m^3}$	$6.0 \cdot 10^{-4}$
Inner energy per mass	$\varepsilon(e) = 0.306746\,\xi^3 - 0.0482792\,\xi^2 + 1.01734\,\xi - 0.00187485$	$\frac{MJ}{kg}$	$5.7 \cdot 10^{-4}$
Kinematic viscosity	$\nu(e) = 11.9285\,\xi^4 - 22.8079\,\xi^3 + 17.6559\,\xi^2 - 7.00355\,\xi + 1.42624$	$\frac{mm^2}{s}$	$9.9 \cdot 10^{-4}$

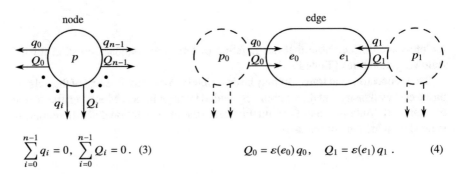

$$\sum_{i=0}^{n-1} q_i = 0, \quad \sum_{i=0}^{n-1} Q_i = 0. \quad (3) \qquad\qquad Q_0 = \varepsilon(e_0)\,q_0, \quad Q_1 = \varepsilon(e_1)\,q_1 . \qquad (4)$$

Fig. 14.2 Scalar variables of nodes and edges and equations independent of edge type

$$\frac{|\Delta T|}{T} \leq 2.8 \cdot 10^{-4}, \quad \frac{|\Delta \rho|}{\rho} \leq 5.3 \cdot 10^{-4}, \quad \frac{|\Delta \varepsilon|}{\varepsilon} \leq 5.4 \cdot 10^{-4}, \quad \frac{|\Delta \nu|}{\nu} \leq 4.8 \cdot 10^{-4} .$$

$$(14.2)$$

Note that density and temperature are no longer independent state variables, but just algebraic functions of the remaining state variable $e = e(t, x)$.

14.2.2 Non-pipe Devices

We describe a district heating network by nodes and edges. A node links two or more edges. An edge links two nodes. Edges represent devices like sources, pumps, consumers, or pipes. Only pipes are modeled by fields depending on time and location, cf. Sect. 14.2.3. Other devices are described by algebraic equations for scalar variables depending only on time. The states are pressure p, heat per volume e, mass flux q, and heat flux Q. Nominally, flows are directed from nodes to edges. Formulating equations we use local indices, i.e. the same quantity may have a different index seen from a node or an edge, cf. Fig. 14.2. Conservation of mass and heat is claimed for nodes. Flows of mass and heat are related by the mass based energy density $\varepsilon(e) = e/\rho(e)$. Equations (14.5) are particular to special devices.

$$
\begin{aligned}
&\text{main source } (R \rightarrow F): p_0 = p_R, && p_1 - p_0 = \Delta p, && T(e_1) = T_F . \\
&\text{other source } (R \rightarrow F): q_0 + q_1 = 0, && p_1 - p_0 = \Delta p, && T(e_1) = T_F . \\
&\text{pump} \qquad\quad (F \rightarrow F): q_0 + q_1 = 0, && p_1 - p_0 = \Delta p, && e_1 - e_0 = 0 . \\
&\text{consumer} \quad\;\; (F \rightarrow R): q_0 + q_1 = 0, && \begin{cases} q_0 = q_C^{\max} \text{ if } T(e_0) < T_C \\ Q_0 + Q_1 = Q_C \text{ else} \end{cases} && T(e_1) = T_R .
\end{aligned}
$$

$$\text{(14.5)}$$

Arrows indicate the orientation. $F \rightarrow R$ means that node 0 is in the flow and node 1 is in the return. The right-hand sides represent time dependent boundary conditions: base pressure p_R, pressure lift Δp, feed temperature T_F, desired consumption Q_C, and return temperature T_R. They may be different for different instances of a device. Note that no conservation of mass is claimed for the main source. A customer is modeled to take instantaneously the desired amount Q_C whenever the inlet temperature is bigger than a critical value $T_C > T_R$. Otherwise, some maximum mass flux $q_C^{\max}$ is taken. Without this differentiation, cooling may lead to unfeasible situations.

14.2.3 Pipes

A pipe is characterized by the following constants: length l [m], diameter d [m], increase in altitude Δh [m], roughness κ [m], heat transfer coefficient k [W/m²K], ground temperature T_∞ [°C], and gravity constant $g = 9.81$ m/s. Typical values may be found in Sect. 14.2.3.2. In particular, we consider pipes where friction dominates inertia effects and convective transport dominates diffusion.

States are represented by velocity v, pressure p, and energy density e. In general, they are functions of time and position $(t, x) \in [0, t_{\text{end}}] \times [0, l] =: \Omega$. Conservation of mass, momentum, and heat are modeled by simplified Euler equations on Ω:

$$\partial_x v = 0 \tag{14.6a}$$

$$-\partial_x p = \frac{v|v|}{2d} \rho(e)\,\lambda(v,e) + g\,\frac{\Delta h}{l}\,\rho(e) \tag{14.6b}$$

$$\partial_t e + v\partial_x e = -\frac{4k}{d}\,(T(e) - T_\infty)\;. \tag{14.6c}$$

The underlying simplifications and the Darcy-Weisbach friction factor λ will be discussed later. The final equations describing a pipe are formulated in non-dimensional variables $u' = \frac{u}{u_r}$, where u is the original variable and u_r is a unit. The units of e and T are chosen such that we can use the polynomial representation of T of Table 14.2:

$$T(e) = T_r T'(e') \text{ with } T'(e') = T_2 e'^2 + T_1 e' + T_0 , \quad T_r = 1^\circ C , \quad e_r = 1\tfrac{\text{GJ}}{\text{m}^3} . \tag{14.7}$$

The units of ρ, ε, and ν are also chosen as in Table 14.2. Let Δt be a typical time step, e.g. 15 min, and $A = \frac{\pi d^2}{4}$ the area of the pipe. Then the remaining units are chosen

$$t_r = \Delta t \quad x_r = l , \quad v_r = l/\Delta t , \quad q_r = A\,v_r\,\rho_r , \quad Q_r = A\,v_r\,e_r . \tag{14.8}$$

From now on we will use only the non-dimensional variables and skip the primes. Equation (14.6a) implies $v = v(t)$. Integrating (14.6b) with respect to x gives that p is only needed at inlet and outlet, i.e. the only remaining field depending on x is $e(t, x)$. Together with (4), the non-dimensional equations describing a pipe read:

$$Q_0 = e_0\,v , \quad Q_0 = \varepsilon(e_0)\,q_0 , \quad Q_1 = -e_1\,v , \quad Q_1 = \varepsilon(e_1)\,q_1 \quad \text{on } [0, t_{\text{end}}] \tag{14.9a}$$

$$p_0 - p_1 = \alpha\,v|v| \int_0^1 \rho(e)\,\lambda(v,e)\,dx + \beta \int_0^1 \rho(e)\,dx \quad \text{on } [0, t_{\text{end}}] \tag{14.9b}$$

$$\partial_t e + v\partial_x e = -\gamma\frac{T(e) - T_\infty}{T_2} \quad \text{on } \Omega$$

$$e(\cdot, 0) = e_0 , \quad e(\cdot, 1) = e_1 \quad \text{on } [0, t_{\text{end}}] , \quad e(0, \cdot) = \tilde{e} \quad \text{on } [0, 1] \tag{14.9c}$$

with initial energy density $\tilde{e}$ and non-dimensional factors

$$\alpha = \frac{l^3 \rho_r}{2 d \Delta t^2 p_r} , \quad \beta = \frac{g\,\Delta h\,\rho_r}{p_r} , \quad \gamma = \frac{4 k \Delta t\,T_r\,T_2}{d\,e_r} . \tag{14.10}$$

14.2.3.1 Darcy-Weisbach Friction Factor

For laminar flow ($Re < 2000$) the friction factor λ is inverse to the Reynolds number:

$$\lambda = \frac{64}{Re}, \quad Re(v,e) = \alpha_R \frac{v}{v(e)}, \quad \alpha_R = \frac{l\,d}{\Delta t\, v_r}, \quad \text{viscosity } v(e) \text{ as in Table 2.}$$

(14.11)

In the turbulent regime ($Re > 4000$) we solve the Colebrook White equation

$$\frac{1}{\sqrt{\lambda}} = -2\log_{10}\left(\frac{2.51}{Re\sqrt{\lambda}} + \frac{\kappa}{3.7\,d}\right)$$

(14.12)

using the Lambert W function defined by $W(x)\,e^{W(x)} = x$ similar to [24]:

$$\lambda = (cx - ab)^{-2}, \quad \partial_{Re}\lambda = \frac{-2\lambda}{Re\,(1+x)}$$

$$x = W\left(\frac{a}{c}\,e^{\frac{ab}{c}}\right), \quad a = \frac{Re}{2.51}, \quad b = \frac{\kappa}{3.7\,d}, \quad c = \frac{2}{\ln 10}.$$

(14.13)

The derivative of λ is needed to construct an Hermite spline gapping intermediate Reynolds numbers. All flow regimes are present in real district heating networks (laminar/turbulent, smooth/rough pipe). The numeric constants in (14.12) are empirical fits to measurements leading to deviations of λ of up to 5% for smooth pipes and up to 10% for rough pipes [25].

14.2.3.2 Benchmark Pipe

The following pipe has been analyzed statically and results have been validated with STANET® [2]. For data $l = 10.5$ m, $d = 27.3$ mm, $\Delta h = -0.02$ m, $\kappa = 0.047$ mm, $k = 2.12\,\frac{W}{Km^2}$, $T_\infty = 8\,°C$, $p_0 = 0.1104$ bar, $p_1 = 0$ bar, $T_0 = 117.9\,°C$ we find $v = 0.5171\,\frac{m}{s}$, $q_0 = 0.2862\,\frac{kg}{s}$, $Q_0 = 1.0303$ MW, $e_1 = 0.4665\,\frac{GJ}{m^3}$, $T_1 = 117.7090\,°C$.

14.2.3.3 Simplifications

Equations (14.6) are based on a number of simplifications. We list which terms have been ignored and why this is justified.

The 3D flow in a pipe is described by 1D equations for the cross sectional mean values of the state variables. This is justified as pipe lengths are much larger than diameters or symmetry disturbing junctions. Boundary layer theory relates the friction factor λ to the viscosity of the fluid [26].

The incompressibility assumption (14.6a) means that a test volume moving with the flow keeps its density ($\partial_t \rho + v\partial_x \rho = 0$). This does *not* mean that density is independent of temperature, but only that heat losses when passing through the pipeline have a negligible influence on density. Passing along the benchmark pipe, the density increases from 945.43 to 945.59 $\frac{kg}{m^3}$, i.e. by a relative factor of 1.7×10^{-4}.

In Eq. (14.6b) we have ignored acceleration ($\rho \, \partial_t v$). In order to check if this is reasonable, we have simulated the city net of TWL with real data for supply and demand over several days in March 2018. In fact, it turns out that the total force on the moving water by inertia is, on average, only 5×10^{-4} the force by friction. The maximal quotient observed over time is 5×10^{-3}.

The heat transport equation (14.6c) models advection and cooling by the environment, but ignores diffusion. According to the database [23], the thermal diffusivity of water at 5 bar pressure in the temperature range $50\,°C - 130\,°C$ takes values within $D_T \in [1.56, 1.72] \times 10^{-7} \, \frac{m^2}{s}$. Assuming typical velocities of $0.1 \, \frac{m}{s}$ and short pipes of $l = 1$ m, the resulting Péclet number measuring the ratio of advective and diffusive heat transport $Pe = \frac{vl}{D_T}$ is in the range $[5.8, 6.4] \times 10^5$. For longer pipes and higher velocities, advection is even more dominant.

14.3 Analytic Solutions

We consider analytic solutions illustrating two important effects to be observed also in real district heating networks: preheating and change of flow direction. Simulation software should be able to treat these phenomena. In both examples, source terms are omitted , i.e. cooling due to imperfect isolation is neglected. While not changing the general behavior of the solution, the source terms would make the analytical solutions much more involved.

14.3.1 Preheating

Many district heating plants in Germany obtain heat from a waste incineration plant. During the summer this is more than sufficient to supply the customers with heat and part of the energy can be converted to electric power via steam turbines. On a cold winter day the situation is opposite: A gas boiler has to be constantly fired to close the heat gap. In spring and autumn, however, there are days when the constant heat flow from the waste incineration plant could cover the demand on daily average, but not during the peaks in the morning and in the evening. In these hours a gas boiler has to be fired, too. This causes high economic and ecological costs and preheating is a strategy to avoid them.

The basic idea is simple and exploits two facts. (1) Heat flow is the product of energy density and volume flow. (2) Heat needs some time to travel from plant to customer, while the corresponding volume flows change simultaneously. In the hours before a peak the operator fills the network with water of maximum temperature. This can be done at moderate power, as the demand is low and so is the volume flow. In the peak period, customers draw the high network temperature at a high volume flow. However, reducing the inlet temperature at the plant to a minimum, the feed power is still moderate. The following analytic example illustrates how preheating allows to satisfy a demand with peak by a constant input power.

Consider a single pipe of length 1 km, say, with a source at one end and a consumer at the other end. In this example, we neglect thermal losses, so the energy transport follows

$$\partial_t e + v \partial_x e = 0, \quad 0 \le t, \ 0 \le x \le 1. \tag{14.14}$$

We assume a periodic consumer behavior with two constant levels during the day

$$Q_1(t) = v(t) \cdot e(t, x = 1) = \begin{cases} Q^* & t_0 \le t \ (\mathrm{mod}\ 24) \le t_1 \\ Q_* & else \end{cases} \tag{14.15}$$

where the time unit is 1h, $Q^* > 1$ is the high consumption in the morning and $Q_* < 1$ is a constant lower consumption during the rest of the day. With $t_1 = t_0 + \frac{24(1-Q_*)}{Q^* - Q_*}$ we get a mean power of $\bar{Q} = 1$. By choosing a supply at the source of the form

$$e_0(t) = a + bt, \quad v(t) = \frac{\bar{Q}}{e_0(t)} = \frac{1}{a + bt} \tag{14.16}$$

the power at the source is constant: $Q_0 = v\, e_0 = 1$. A feed-in at the source of this form produces a piece-wise constant power signal with two different levels at the end of the pipe. We can now determine the parameters a and b such that the PDE solution at the consumer site fulfills exactly the demand given by (14.15). Therefore, we need to solve the following system

$$Q^* = e^{-bl'} \frac{a+b}{b}, \quad Q_* = e^{-bl'}, \quad l' = 1 - \left(\frac{1}{b}(ln(a+b) - ln(a)) \right). \tag{14.17}$$

For $Q^* = 1.25$, $Q_* = 0.95$, and $t_0 = 5.5$ we get the solution $t_1 = 9.5$, $a = 0.3257$, $b = 1.0315$, $l' = 0.1575$, and the signals shown in Fig. 14.3.

14.3.2 Flow Reversal

In a network with loops, the energy density may show discontinuities traveling along a pipe, even if all boundary conditions are continuous. If such a discontinuity reaches a consumer, velocities will jump all over the network as a consequence of the consumer model in (14.5). Therefore, inertia effects cannot be included into the model without making the consumer model more complicated. Moreover, applicable numerical schemes must be able to handle discontinuities.

The following analytic solution illustrates a traveling discontinuity and may be helpful in testing numerical schemes. We consider three equal, horizontal, perfectly isolated pipes forming a triangle and assume constant material and friction coeffi-

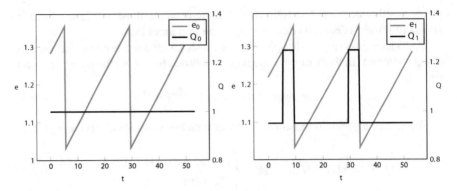

Fig. 14.3 Energy density and heat flux at source (left) and consumer (right)

cients. Pressure is given at node 0 and outflow velocities at the other nodes. Nodes are still described by Eq. (14.3). After suitable scaling, the pipe equations (14.4) and (14.6) read in *local* notation for $t \in [-1, 1]$, $x \in [0, 1]$:

$$q_0(t) = v(t), \quad q_1(t) = -v(t), \quad Q_0(t) = e(t, 0) v(t), \quad Q_1(t) = -e(t, 1) v(t)$$
$$p_0(t) - p_1(t) = v(t) |v(t)|, \quad \partial_t e(t, x) + v(t) \partial_x e(t, x) = 0. \tag{14.18}$$

In the *global* notation of Fig. 14.4, the proposed analytic solution reads:

$$v_1(t) = 1 + \tfrac{t}{6}, \quad v_2(t) = 1 - \tfrac{t}{6}, \quad v_3(t) = -2 \operatorname{sign}(t) \sqrt{\left|\tfrac{t}{6}\right|}$$

$$e_1(t, x) = 3 + 6\sqrt{\left(1 + \tfrac{t}{6}\right)^2 - \tfrac{x}{3}}, \quad e_2(t, x) = 15 - 6\sqrt{\left(1 - \tfrac{t}{6}\right)^2 + \tfrac{x}{3}}$$

$$e_3(t, x) = \begin{cases} 15 - 6\sqrt{\left[1 - \left(\left|\tfrac{t}{6}\right|^{\frac{3}{2}} + \tfrac{x-1}{8}\right)^{\frac{2}{3}}\right]^2} + \tfrac{1}{3} & \text{if } t \geq 0 \text{ and } x \geq 1 - \left(\tfrac{2t}{3}\right)^{\frac{3}{2}} \\[4mm] 3 + 6\sqrt{\left[1 - \left(\left|\tfrac{t}{6}\right|^{\frac{3}{2}} + \tfrac{x}{8}\right)^{\frac{2}{3}}\right]^2} - \tfrac{1}{3} & \text{else} \end{cases}$$

$$p_0 = 2, \qquad v_0 = v_1 + v_2, \quad Q_0 = e_1(\cdot, 0) v_1 + e_2(\cdot, 0) v_2$$
$$p_1 = 2 - v_1^2, \quad v_4 = v_1 - v_3, \quad Q_4 = e_1(\cdot, 1) v_1 - e_3(\cdot, 0) v_3 \tag{14.19}$$
$$p_2 = 2 - v_2^2, \quad v_5 = v_2 + v_3, \quad Q_5 = e_2(\cdot, 1) v_2 + e_3(\cdot, 1) v_3.$$

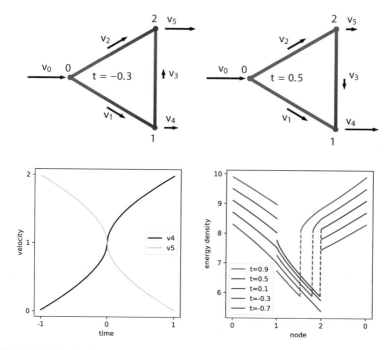

Fig. 14.4 Analytic solution with flow reversal: velocities and energy densities

14.4   Simulation

In the following we describe the numerical approach implemented in *AD-Net District Heating*. AD-Net is a C++ template library for simulating and optimizing general networks like power grids or chemical process chains. The module on district heating has been developed together with *GEF Ingenieur AG* and *TWL* within the publicly funded project *DYNEEF*, cf. Sect. 14.7 The development has been guided by the following requirements:

1. Dynamic run-time effects can be reproduced.
2. Pressure drop and cooling can be predicted with sufficient accuracy.
3. Linked to an optimization tool, the simulation does not only provide values, but also sensitivities with respect to control variables.
4. Time integration allows for large time steps in order to manage optimal control problems with look-ahead times of several days.
5. Pipe models get along without further subdivision to keep the number of unknowns manageable even for large networks with several thousand pipes.
6. STANET®files, which are standard in network design, can be imported.

In *AD-Net District Heating*, we meet these challenges combining a special method of characteristics with automatic differentiation.

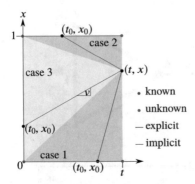

Fig. 14.5 Method of characteristics

Time integration is done by a simple collocation method. Per time step, all states and inputs are represented by polynomials of a scaled local time $t \in [0, 1]$. Except for special events, we assume continuity, i.e. the constant coefficients come from the last time step. The remaining coefficients are the unknowns to be solved for within a Newton iteration. Linear systems are set up by evaluating the non-linear equations of all components at some collocation points including 1. As we employ AD, evaluation gives both, residuals and their Jacobians. The number of collocation points indicates the order of the method. Typical orders are one or two. A constant velocity is assumed per pipe between two collocation points t_{i-1} and t_i. This velocity is linked to pressures and energy densities by conservation of momentum (14.9b) at t_i. The spacial integrals are approximated by Gauss quadrature of appropriate order. Further algebraic equations are contributed by friction, other network components, and nodes. In the next section we concentrate on the heat transport through pipes, which is the only effect described by differential equations, cf. Eq. (14.9c). In the subsequent section we explain in more detail how AD is used.

14.4.1 Method of Characteristics

The method of characteristics computes the energy density $e(t, x)$ by tracing back a sample volume to a previous position (t_0, x_0). This is illustrated in Fig. 14.5. If the velocity is low, the original energy density is found by interpolating known values at $t_0 = 0$. In this case, the method is explicit. In case of high velocities, however, known and unknown energy densities are interpolated at the inflow boundary and the method becomes implicit. As far as we know, this is the first time that such a mixed method of characteristics is applied to district heating networks. In contrast to [3], it allows for long time steps, but it does not require storing long histories of e in the nodes as in [4].

Cooling along the characteristics is computed using Lagrangian coordinates. With $(t, x)(\tau) = (t_0 + \tau, x_0 + \tau v)$ we can formulate Eq. (14.9c) as

Table 14.3 Equation added by method of characteristics of first order with $e(t, x)$ as in Eq. (14.23)

Case			Added equation	t_0	x_0	Interpolated e_*
	vt	$le - 1$	$e_0(t) = e(t, 0)$	$t + \frac{1}{v}$	1	$\left(1 - \frac{t_0}{t}\right) e_1(0) + \frac{t_0}{t} e_1(t)$
$-1 <$	vt	≤ 0	$e_0(t) = e(t, 0)$	0	$-vt$	$(1 - x_0) e_0(0) + x_0 e_1(0)$
$0 \leq$	vt	< 1	$e_1(t) = e(t, 1)$	0	$1 - vt$	$(1 - x_0) e_0(0) + x_0 e_1(0)$
$1 \leq$	vt		$e_1(t) = e(t, 1)$	$t - \frac{1}{v}$	0	$\left(1 - \frac{t_0}{t}\right) e_0(0) + \frac{t_0}{t} e_0(t)$

$$\frac{de}{d\tau} = -\gamma \frac{T(e) - T_\infty}{T_2}.$$ (14.21)

Let e_1 and e_2 be the roots of the polynomial

$$T(e) - T_\infty = T_2 e^2 + T_1 e + T_0 - T_\infty = T_2 (e - e_1)(e - e_2) .$$ (14.22)

Then Eq. (14.21) can be solved analytically by partial fraction decomposition:

$$e(t, x) = e_1 + \frac{\Delta e}{1 - \frac{e_* - e_2}{e_* - e_1} \exp\left[-\gamma \Delta e (t - t_0)\right]} \quad \text{with}$$

$$\Delta e = e_2 - e_1 , \quad e_* = e(t_0, x_0) .$$ (14.23)

Table 14.3 summarizes the method for linear polynomials.

14.4.2 *Automatic Differentiation*

AD is employed in both, computing exact Jacobians within an implicit time integration and propagating control parameters to cost function and constraints. Here, we demonstrate how AD is used to proceed one time step keeping track of how states depend on controls. A general overview of AD is given in [27]. Here, we recapitulate only the principles of *forward mode automatic differentiation*, which is implemented in *AD-Net*.

1. Algebra is done with *linearizations* rather than scalars. A linearization $\bar{x} = (x, \nabla x)$ is composed of a value $x \in \mathbb{R}$ and a gradient $\nabla x \in \mathbb{R}^p$ with respect to p parameters. Let $\mathcal{L}_n \cong \mathbb{R}^{1+p}$ be the vector space of such linearizations.

2. Any equation may be formulated composing basic operators such as $\cdot \times \cdot$ or $\sqrt{\cdot}$. In programming languages like C++, these basic operators can be overloaded such that the gradient of a result is *numerically exact*. For instance, let $f : \mathbb{R}^2 \to \mathbb{R}$ be a differentiable binary operator. Then the lifted mapping implements the chain rule:

$$\bar{f} : \begin{cases} \mathcal{L}_p \times \mathcal{L}_p \to \mathcal{L}_p \\ (\bar{x}, \bar{y}) \mapsto \left(f(x, y), \, \partial_x f(x, y) \nabla x + \partial_y f(x, y) \nabla y \right) \end{cases}. \qquad (14.24)$$

This approach is easily extended to vector-valued linearizations $\bar{\mathbf{x}} = (\mathbf{x}, D\mathbf{x}) \in \mathcal{L}_p^n$. The rows of Jacobian $D\mathbf{x} \in \mathbb{R}^{n \times p}$ are gradients of the scalar linearizations $\bar{x}_i$. A vector-valued function $\mathbf{f} : \mathbb{R}^n \to \mathbb{R}^m$ is lifted to

$$\bar{\mathbf{f}} : \begin{cases} \mathcal{L}_p^n \to \mathcal{L}_p^m \\ \bar{\mathbf{x}} \mapsto (\mathbf{f}(\mathbf{x}), \, D\mathbf{f}(\mathbf{x}) \, D\mathbf{x}) \end{cases} \qquad (14.25)$$

providing exact Jacobians in a single run. In this case, it is recommended to make use of a vectorization library (e.g. Eigen, or any other BLAS interface) to speed up computations. In the case of district heating networks, the interactions between all variables are local, such that the Jacobian of the model equations is sparse. This can be utilized by switching to a sparse representation of tangential vectors, which speeds up the computation of derivatives even further.

Now we come to computing a time step with AD. Let $\bar{\varphi} \in \mathcal{L}_p^m$ comprise initial and boundary conditions with known derivatives for the control parameters $\bar{\mathbf{u}} = (\mathbf{u}, I_p) \in \mathcal{L}_p^p$. Moreover, let $\bar{\mathbf{x}} \in \mathcal{L}_p^n$ be the searched coefficients of states on the time step, and $\bar{\mathbf{f}} : \mathcal{L}_p^n \times \mathcal{L}_p^m \to \mathcal{L}_p^n$ the residuals of all equations relating states with initial and boundary conditions. Then the theorem on implicit functions suggests the following procedure to find $\bar{\mathbf{x}}$ from an initial guess $\mathbf{x}_0$:

> Set $\bar{\mathbf{x}} = (\mathbf{x}_0, I_n)$, $\bar{\psi} = (\varphi, 0)$.
>
> Solve $0 = \mathbf{f}(\mathbf{x}, \psi)$ for $\mathbf{x}$ by repeating Newton iterations
>
> $\quad \bar{\mathbf{y}} = \bar{\mathbf{f}}(\bar{\mathbf{x}}, \bar{\psi})$, $\bar{\mathbf{x}} = (\mathbf{x} - D\mathbf{y}^{-1}\mathbf{y}, I_n)$.
>
> Set $\bar{\mathbf{x}} = (\mathbf{x}, 0)$, $\bar{\mathbf{z}} = \bar{\mathbf{f}}(\bar{\mathbf{x}}, \bar{\varphi})$ and finally $\bar{\mathbf{x}} = (\mathbf{x}, -D\mathbf{y}^{-1}D\mathbf{z})$. $\qquad (14.26)$

14.4.3 Initialization

Special care has to be taken initializing the simulation. We start from a simplified stationary solution with constant temperatures in feed and return, respectively, and heat transfer coefficients set to zero. From this initial state we simulate dynamically some hours before the period of interest driving transfer coefficients and feed temperatures to their true initial values along a ramp (homotopy method). This is necessary to prevent inlet temperatures dropping below outlet temperatures at customers within overshooting Newton steps.

14.5 Optimization

We consider an optimal control problem which dynamic simulation is essential for: preheating of a real network. In contrast to the analytic example in Sect. 14.3.1, the delay times of the customers are all different. Moreover, the operator has to supply consumers with a guaranteed minimal temperature $\underline{T}$, which kills legionella, and a minimal pressures drop $\Delta \underline{p}$ needed by heat exchangers to work. The capacity of pumps limits the possible pressure lift $\Delta \overline{p}$ at sources. For a complete list of the connection conditions at TWL cf. [28]. The main goal of preheating is to avoid additional gas firing, which is assumed to occur, if a critical feed power $\overline{Q}$ is exceeded. Typical values of all limits are given in the next section. Within these constraints, we will look for a common day-periodic feed temperature u with minimal value and variation. The former induces cooling losses. The latter damages the network by making pipes move in the ground. Let time unit Δt be an integer fraction of a day, t_d the length of a day, and $t_{end} = n_d\, t_d$ an integer multiple of a day. Then the discretized optimal control problem reads

$$\min_{\mathbf{u} \in \mathbb{R}^p} \quad \mathcal{J}(\mathbf{u}) = \int_0^{t_{end}} (\partial_t u(t))^2 + \alpha\, u(t)^2 dt \quad \text{subject to}$$

$$T\,(e_0(t, \mathbf{u})) \geq \underline{T} \;, \quad p_0(t, \mathbf{u}) - p_1(t, \mathbf{u}) \geq \underline{p} \quad \text{at consumers}$$
$$Q_1(t, \mathbf{u}) \leq \overline{Q}\;, \qquad \Delta p(t, \mathbf{u}) \qquad \leq \overline{p} \quad \text{at sources}$$

$$\text{for all } t \in \mathbb{N}_0 \text{ with } 0 \leq t \leq t_{end} \text{ and } \alpha > 0 \;. \tag{14.27}$$

$u(t) = \sum_{i=1}^{p} u_i\, \phi_i(t\,(\mathrm{mod}\,t_d))$ is a day-periodic feed temperature made from non-negative smooth basis functions ϕ_i with compact support, e.g. cubic splines, and α a small positive regularization parameter. States like $e_0(t, \mathbf{u})$ are solutions of a forward simulation computed as in Eq. (14.26) for control parameters $\mathbf{u}$. The cost function is quadratic. As α is positive, the Hessian is positive definite:

$$\mathcal{J}(\mathbf{u}) = \tfrac{1}{2}\mathbf{u}^t H\mathbf{u} \quad \text{with} \quad h_{ij} = n_d \int_0^{t_d} \dot{\phi}_i(t)\,\dot{\phi}_j(t) + \alpha\phi_i(t)\,\phi_j(t)\,dt \;. \tag{14.28}$$

Note that pressures do not appear as controls. This is because optimal pressures are computed directly within the forward simulation. First, a time step is performed with zero pressure drop at all sources. This is possible as the pressure levels in feed and return are separated. Let $\Delta p_* < 0$ be the resulting minimal pressure drop at consumers. Then pressures in the feed are corrected adding $\Delta \underline{p} - \Delta p_*$, which guarantees pressure drops $\geq \Delta \underline{p}$ at all consumers. This approach claims that pumping power is cheap enough to be ignored in the cost function and that having the same pressure lift at all sources is no severe restriction. The optimization problem is solved by *sequential linear quadratic programming (SLQP)*: The enclosing optimizer suggests a new vector of controls $\mathbf{u}$ for which a forward simulation is performed. Using AD, not only the values of the constrained variables are computed, but also the sensitivity with respect to the controls. As the controls enter directly the quadratic goal function, we have at hand all ingredients to build the next LQP. In detail, the algorithm reads as follows.

1. Perform forward simulation for given $\mathbf{u}$ evaluating state constraints at each time step. AD arithmetic provides values and gradients to be collected in $\mathbf{c}$ and $D\mathbf{c}$.
2. Set up linear-quadratic program

$$\min_{\Delta\mathbf{u}\in\mathbb{R}^p} \tfrac{1}{2}(\mathbf{u}+\Delta\mathbf{u})^t\, H\,(\mathbf{u}+\Delta\mathbf{u}) \text{ subject to } \mathbf{c}+D\mathbf{c}\,\Delta\mathbf{u} \geq 0 \qquad (14.29)$$

and solve it for $\Delta\mathbf{u}$ using the algorithm proposed in [29].
3. Stop if $\Delta\mathbf{u}$ is sufficiently small. Otherwise, set $\mathbf{u} = \mathbf{u} + \Delta\mathbf{u}$ and go to step 1.

14.6 Results

The subsequent results refer to two networks of TWL. The main network comprises about 1000 consumers and 6000 pipes. The power station has four exits supplying the connected subnetworks *Innenstadt*, *Industriestraße*, and two clinics. One clinic is supplied with a constant temperature of 130 °C, the other parts by a common feed temperature ranging from 80 to 100 °C, which is the control. Although the clinics are simulated, results are shown only for the two main parts due data protection. The network is supplied by a waste incineration plant providing a more or less uniform heat flux. The demand, however, is characterized by a strong peak in the morning and a weaker peak in the evening as illustrated by the red curve of Fig. 14.7. Measurements are taken in the period from 03/14/2018 to 03/16/2018. The daily mean temperatures are 8.4, 7.2, and 8.9 °C, respectively. Here, the supply is still sufficient on average, but not during the peaks. Measurements indicate that the gas boiler is activated if the feed power exceeds $\overline{Q} = 17\,\text{MV}$ at one of the exits. Solving the preheating problem (14.27) provides a feed temperature which keeps feed power below this threshold and avoids gas firing. The other optimization parameters are $\underline{T} = 70\,°\text{C}$, $\Delta\underline{p} = 1\,\text{bar}$, $\Delta\overline{p} = 8\,\text{bar}$. The latter constraint is not active (Fig. 14.6).

In a first step, the demand is reconstructed from measurements, as it cannot be explained by standard load profiles. Measurements with sufficient time resolution are available only for the flows leaving and reaching the four exits of the plant. Assuming that the demand of a customer is the product of the known individual annual mean load and a common time profile, the coefficients of this profile are fitted to measured mass flows and return temperatures using the measured feed temperatures and pressures as input of the simulation. The consumption model is as follows:

$$Q(t) = \tilde{Q}\left[c_0\,(t\,(\text{mod}\,t_d);\,\mathbf{p}_0) + c_1\,(t\,(\text{mod}\,t_d);\,\mathbf{p}_1)\,\tilde{T}\,(t,\,\delta t)\right] \qquad (14.30)$$

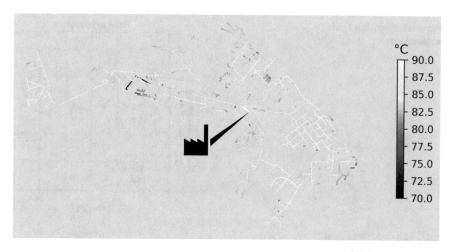

Fig. 14.6 Simulated temperatures in TWL main network at 7:00 a.m. with optimized feed temperature

Fig. 14.7 Heat consumption according to different standard load profiles and reconstructed from measurements

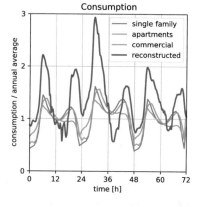

Fig. 14.8 Consumption of gas boiler broken down by daily mean temperatures

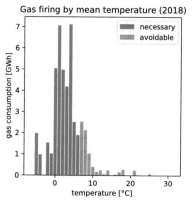

Fig. 14.9 Common feed
temperature

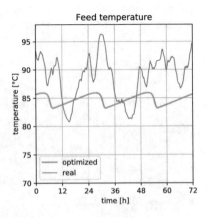

Fig. 14.10 Resulting feed
power, measured and to be
expected from optimized
input

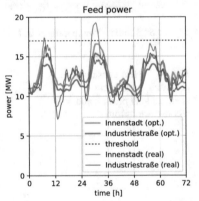

where $Q(t)$ is the consumption of a given customer at time t, $\tilde{Q}$ is his annual mean
consumption, c_i are day-periodic splines with 24 fit parameters p_{ij}, each, t_d is the
length of a day measured in time steps, and $\tilde{T}(t, \delta t)$ is the mean temperature over the
interval $[t - \delta t, t + \delta t]$. A good choice is $\delta t = 2h$. The reconstructed consumption
profile is shown in Fig. 14.7. Usually, the demand is modeled by standard load
profiles [22], which are plotted as thin curves. They depend on the type and the
annual consumption of a customer, and on the daily mean temperature. Note that the
reconstructed consumption turns out to be higher, more volatile, and that the evening
peaks are shifted to the night.

The results of the optimization are illustrated in Figs. 14.6, 14.9, and 14.10.
Figure 14.6 shows a non-stationary state after preheating. The temperatures are still
high inside the network, while they have already been reduced at the inlet. Cooling
effects become visible at customers with low consumption where the water remains
in the pipe for a long while. Actual and optimized feed temperature are depicted in
Fig. 14.9. We recognize that the optimized temperature is smoother and lower than the
actual one and looks like a smoothed version of the ramp shown in Fig. 14.3. In fact,
the optimal feed temperature is just high enough to supply the most distant customer

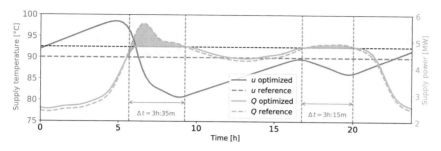

Fig. 14.11 Optimized control (solid lines) versus reference control (dashed lines)

with 70 °C, which reduces losses by cooling, and the special profile redistributes the feed power such that the morning peaks remain subcritical, as shown in Fig. 14.10 (Fig. 14.8).

What is the benefit avoiding additional gas firing? In Fig. 14.8 we have broken down the total consumption of the gas boiler in 2018 by daily mean temperatures. Assuming that additional firing can be avoided whenever the mean temperature is 7 °C or higher, gas consumption could be reduced by 16%, which corresponds to about 1700 t CO_2 and 330 000 € per year – without any additional invest on equipment.

The second network investigated is called *Neubruch* and contains about 300 consumers and 2000 pipes. In reality, it is run at a constant temperature of 90 °C. Figure 14.11 illustrates how preheating is able to cut the feed power (solid orange) at a given threshold (dashed black). The original feed power is essentially identical to the overall consumption (dashed orange) and exceeds the threshold twice a day (green). The optimal feed temperature (solid red) may be interpreted as two ramps – one per peak. Periodic boundary conditions, a daily mean temperature of 8 °C, and standard load profiles are assumed.

14.7 Conclusion

We have presented an environment for solving optimal control problems on district heating networks. It predicts accurately cooling, pressure drop, and run-time effects while allowing for look ahead periods of several days. In particular, an optimal preheating strategy has been found which satisfies a volatile demand by a uniform supply avoiding additional gas firing. Analytical solutions are presented which illustrate typical challenges and may serve as benchmarks. In future projects we will include real time measurements for state correction and integrate the simulator into an optimization framework for sector coupling.

Acknowledgements This chapter contains work founded by the Federal Ministry for Economic Affairs and Energy (*DYNEEF*, FK 03ET1346B) and by the Federal Ministry of Education and Research (*EiFer*, FK 05M18AMB). We thank *TWL*, *GEF Ingenieur AG*, and the academic partners in EiFer for the close cooperation.

References

1. KISTERS Gruppe. ResOpt – Software zur Ressourcenoptimierung in der Energiewirtschaft. https://energie.kisters.de/loesungen-produkte/resopt-optimierung. Accessed: 2020-02-15
2. Fischer-Uhrig, Ingenieurbüro. STANET Netzberechnung (2018). http://stafu.de/de/home. html. Accessed: 2018-12-19
3. Köcher, R.: Beitrag zur Berechnung und Auslegung von Fernwärmenetzen. Ph.D. thesis, TU Berlin (2000)
4. Benonysson, A., Bøhm, B., Ravn, H.F.: Operational optimization in a district heating system. Energy Convers. Manag. **36**(5), 297–314 (1995)
5. van der Heijde, B., Fuchs, M., Ribas Tugores, C., Schweiger, G., Sartor, K., Basciotti, D., Müller, D., Nytsch-Geusen, C., Wetter, M., Helsen, L.: Dynamic equation-based thermo-hydraulic pipe model for district heating and cooling systems. Energy Convers. Manag. **151**, 158–169 (2017)
6. van der Heijde, B., Aertgeerts, A., Helsen, L.: Modelling steady-state thermal behaviour of double thermal network pipes. Int. J. Therm. Sci. **117**, 316–327 (2017)
7. Hauschild, S.A., Marheineke, N., Mehrmann, V., Mohring, J.,Moses Badlyan, A., Rein, M., Schmidt, M.: Port-hamiltonian modeling of district heating networks (2019)
8. Borsche, R., Kall, J.: Ader schemes and high order coupling on networks of hyperbolic conservation laws. J. Comput. Phys. **273**, 658–670 (2014)
9. Borsche, R.: Numerical schemes for networks of hyperbolic conservation laws. Appl. Numer. Math. **108**, 157–170 (2016)
10. Borsche, R., Eimer, M., Siedow, N.: A local time stepping method for thermal energy transport in district heating networks. Appl. Math. Comput. **353**, 215–229 (2019)
11. Vivian, J., Monsalvete Álvarez [de Uribarri], P., Eicker, U., Zarrella, A.: The effect of discretization on the accuracy of two district heating network models based on finite-difference methods. Energy Procedia **149**, 625–634, : 16th International Symposium on District Heating and Cooling, DHC2018, 9–12 September 2018. Hamburg, Germany (2018)
12. Grosswindhager, S., Voigt, A., Kozek, M.: Linear finite-difference schemes for energy transport in district heating networks. In: Proceedings of the 2nd International Conference on Computer Modelling and Simulation, pp. 5 – 7 (2011)
13. Sandou, G., Font, S., Tebbani, S.,Hiret, A., Mondon, C., Tebbani, S., Hiret, A., Mondon, C.: Predictive control of a complex district heating network. In: Proceedings of the 44th IEEE Conference on Decision and Control, , Seville, Spain, IEEE. pp. 7372–7377 (2005)
14. Giraud, L.,Merabet, M., Baviere, R., Vallée, M.: Optimal Control of District Heating Systems using Dynamic Simulation and Mixed Integer Linear Programming, pp. 141–150 (2017)
15. Krug, R., Mehrmann, V., Schmidt, M.: Nonlinear optimization of district heating networks (2019)
16. Linn, D., Mohring, J., Siedow, N.: Optimal control of district heating networks. PAMM **19**(1), e201900491 (2019)
17. Rein, M., Mohring, J., Damm, T., Klar, A.: Model order reduction of hyperbolic systems at the example of district heating networks (2019). arXiv:1903.03342
18. Rein, M., Mohring, J., Damm, T., Klar, A.: Optimal control of district heating networks using a reduced order model (2019). arXiv:1907.05255
19. Iserles, A.: A First Course in the Numerical Analysis of Differential Equations. Number 44. Cambridge university press (2009)

20. Richardson, L.F.: Ix. the approximate arithmetical solution by finite differences of physical problems involving differential equations, with an application to the stresses in a masonry dam. Philos. Trans. R. Soc. Lond. Ser. A, Contain. Pap. Math. Phys. Character **210**(459-470), 307–357 (1911)
21. Nielsen, T., Madsen, H., Holst, J., Søgaard, H.: Predictive control of supply temperature in district heating systems (2002)
22. Bundesverband der deutschen Gas- und Wasserwirtschaft (BGW). Anwendung von Standardlastprofilen zur Belieferung nichtleistungsgemessener Kunden (2006). Last accessed: 2019/07/23
23. National Institute of Standards and Technology. Thermophysical Properties of Fluid Systems (2016)
24. More, A.: Analytical solutions for the colebrook and white equation and for pressure drop in ideal gas flow in pipes. Chem. Eng. Sci. **61**(16), 5515–5519 (2006)
25. Moody, L.F.: Friction factors for pipe flow. Trans. Asme **66**, 671–684 (1944)
26. Schlichting, H., Gersten, K.: Grenzschicht-Theorie, 10th edn. Springer, Berlin (2006)
27. Griewank, A., Walther, A.: Evaluating Derivatives: Principles and Techniques of Algorithmic Differentiation. Society for Industrial and Applied Mathematics, Philadelphia, PA, USA, second edition (2008)
28. Technische Werke Ludwigshafen am Rhein AG. Technische Anschlussbedingungen Heizwasser (2019). https://www.twl.de/dummystorage/user_upload/2019-03-28_Wb_TAB_2019.pdf. Accessed: 2020-02-26
29. Goldfarb, D., Idnani, A.: A numerically stable dual method for solving strictly convex quadratic programs. Math. Program. **27**(1), 1–33 (1983)

Printed in the United States
by Baker & Taylor Publisher Services